Studies in Fuzziness and Soft Computing

Volume 424

The series "Studies in Fuzziness and Soft Computing" contains publications on various topics in the area of soft computing, which include fuzzy sets, rough sets, neural networks, evolutionary computation, probabilistic and evidential reasoning, multi-valued logic, and related fields. The publications within "Studies in Fuzziness and Soft Computing" are primarily monographs and edited volumes. They cover significant recent developments in the field, both of a foundational and applicable character. An important feature of the series is its short publication time and world-wide distribution. This permits a rapid and broad dissemination of research results.

Indexed by SCOPUS, DBLP, WTI Frankfurt eG, zbMATH, SCImago.

All books published in the series are submitted for consideration in Web of Science.

John N. Mordeson · Sunil Mathew · G. Gayathri

Fuzzy Graph Theory

Applications to Global Problems

Springer

John N. Mordeson
Department of Mathematics
Creighton University
Omaha, NE, USA

Sunil Mathew
Department of Mathematics
National Institute of Technology Calicut
Calicut, Kerala, India

G. Gayathri
National Institute of Technology Calicut
Calicut, Kerala, India

ISSN 1434-9922 ISSN 1860-0808 (electronic)
Studies in Fuzziness and Soft Computing
ISBN 978-3-031-23107-0 ISBN 978-3-031-23108-7 (eBook)
https://doi.org/10.1007/978-3-031-23108-7

This Springer imprint is published by the registered company Springer Nature Switzerland AG
The registered company address is: Gewerbestrasse 11, 6330 Cham, Switzerland

John N. Mordeson would like to dedicate the book to his wonderful wife Pat.

Sunil Mathew would like to dedicate the book to Prof. John N. Mordeson, Professor Emeritus, Creighton University, who motivated and inspired many.

G. Gayathri would like to dedicate the book to her father M. Gangadharan, mother Sarasu Gangadharan, and husband Pratheesh K.

Preface I

Human trafficking is a multi-billion dollar criminal industry that denies freedom to nearly 25 million people around the world. The importance of many topics in the book is to help the reader understand the magnitude and complexity of the problem.

Accurate data concerning the flow of trafficking in persons is impossible to obtain due to the very nature of the problem. The goal of the trafficker is to be undetected. The size of the problem also makes it very difficult to obtain accurate data. There are many other reasons for the scarcity of data. Among the most important are the victims' reluctance to report crimes or testify for fear of reprisals, disincentives, both structural and legal, for law enforcement to act against traffickers, a lack of harmony among existing data sources, and an unwillingness of some countries and agencies to share data. Due to the lack of accurate data, the concepts of mathematics of uncertainty provide a valuable way to study the problems of human trafficking and illegal immigration.

The notions of vulnerability of countries and the government response of these countries have been used to determine the susceptibility of the routes to trafficking. The vulnerability and government response data of countries has been made available by various studies. However, none of these studies involved the amount of flow from country to country.

The flow of trafficking from country to country or region to region can be modeled by the use of directed graphs. These graphs can contain source countries, destination countries, and transit countries. They may contain cycles and feedback loops, for example, immigrants being returned to a preceding country. One method of assigning a flow from one country to another has been introduced in a particular study. The flow was reported in linguistic terms, namely (very) low, medium, (very) and high. The terms assigned were determined by the number of times flow between countries was reported by certain sources. It is well-known that the use of fuzzy logic is an ideal way to model situations described linguistically. Because of the complexity of a directed graph used to model trafficking and because of the lack of accurate data to measure the flow, fuzzy graphs provide an ideal method of modeling trafficking of persons and illegal immigration.

Knowing the structure of a fuzzy-directed graph can be used in many ways to deal with the flow. One way is to determine countries that could be targeted for the purpose of reducing the flow. Another way would be to determine the countries that could be targeted to increase their government response or decrease their vulnerability.

In a world experiencing climate change, past assumptions about the weather no longer hold true. Climate data may be available, but it is often hard to find, understand, and apply to decision making. Climate scientists around the world are contributing to simulation models of the future climate. Their aim is to produce critical information to assist decision-makers struggling to effectively plan for the future, but much of their output remains beyond the understanding of end-users and thus cannot be integrated into policies. Thus due to the lack of precise data available, techniques from mathematics of uncertainty may be useful. To overcome the challenges faced by climate change, cooperation among various agencies, companies, and scholars is needed. Techniques from mathematics of uncertainty may be helpful.

In the fuzzy graph theory part of the book, the relatively new concepts of fuzzy soft semigraphs and graph structures are used to study human trafficking, as well as is time intuitionistic fuzzy sets that have been introduced to model forest fires. The notion of legal and illegal incidence strength is used to analyze immigration to the USA. The examination of return refugees to their origin countries is undertaken. The neighborhood connectivity index is determined for trafficking in various regions of the world. The cycle connectivity measure for the directed graph of the flow from South America to the USA is calculated. It is determined that there is a need for improvement in government response by countries.

Outside the area of fuzzy graph theory, a new approach to examine climate change is introduced. Social network theory is used to study feedback processes that affect climate forcing. Tipping points in climate change are considered. The relationship between terrorism and climate change is examined. Ethical issues concerning the obligation of business organizations to reduce carbon emissions are also considered.

Nonstandard analysis is a possible new area that could be used by scholars of mathematics of uncertainty. A foundation is laid to aid the researcher in the understanding of nonstandard analysis. In order to accomplish this, a discussion of some basic concepts from first-order logic is presented as some concepts of mathematics of uncertainty. An application to the theory of relativity is presented.

Omaha, USA — John N. Mordeson
Calicut, India — Sunil Mathew
Calicut, India — G. Gayathri

Preface II

Climate change increases the risk of natural disasters and thus creates poverty and can cause situations of conflict and instability. Displacement can occur giving traffickers an opportunity to exploit affected people. In this book, we examine some issues involving climate change, human trafficking, and other serious world challenges made worse by climate change.

Chapter 1 discusses some of the basic material required for the development of this book, especially for the smooth reading of Chaps. 6–10. Fundamental definitions and results from fuzzy sets, fuzzy relations, fuzzy graphs, and fuzzy incidence graphs are presented.

In Chap. 2, we lay a foundation for a new research area in fuzzy mathematics, namely nonstandard analysis. In order for a scholar to fully understand nonstandard analysis, an understanding of order first logic is necessary. Consequently, we begin this chapter with a discussion of first-order logic and a proof of the transfer principle. We follow this by proving some of the basic results of nonstandard analysis. We then introduce some concepts of mathematics of uncertainty to nonstandard analysis. The chapter is concluded by using concepts of mathematics of uncertainty to the theory of relativity.

In Chap. 3, we introduce a new approach by introducing methods from social network theory to model feedback processes in climate change. Feedback processes amplify or diminish the effect of each climate forcing, i.e., a change which may push the climate system in the direction of warming or cooling. We also consider the opinion that global climate change is an ethical issue. In particular, we consider issues concerning the obligation of business organizations in reducing carbon emissions.

The world faces very serious challenges, namely human trafficking, human slavery, terrorism, and global poverty to name only a few. However, climate change may be the most serious of all. Climate change causes poverty which makes all the other challenges worse. Even more important than this, climate change could make the planet uninhabitable if governments don't meet certain guidelines. In Chap. 4, we determine the similarity of country rankings of countries with respect a country's vulnerability ranking by the ND-Gain Scores and the ranking of countries concerning climate risk of Fragile Planet. We conclude the chapter by finding the similarity of country rankings with respect to global terror, global peace, and climate risk.

In Chap. 5, we use the notion of a time intuitionistic fuzzy set first introduced to model forest fires in order to apply these ideas to study human trafficking. We also use soft set theory to study problems concerning human trafficking by introducing soft set theory to fuzzy semigraphs and graph structures. The social progress index ranks countries with respect to their providing the social and environmental needs of their citizens. The fragile states index ranks countries with respect to their vulnerability to conflict or collapse. Freedom of the world ranks countries with respect to certain categories dealing with issues concerning freedom. We determine the similarity of these rankings.

Chapter 6 focuses on a new development in fuzzy graph theory called directed fuzzy incidence graphs, abbreviated as DFIG. This new model is very effective in dealing with networks influenced by external parameters. Concepts like legal flow and illegal flow are discussed in detail with a hint to the study of human trafficking. Modern networks like Internet and big highway systems can be modeled using this concept. Legal flow enhancing and illegal flow reduction techniques are discussed using different nodes, arcs and pairs of the network. An application related to the migration of people from different parts of the globe to the USA is also provided.

The most important problem of networking theory is the enhancement of effective flow from one node to another. Chapter 7 concentrates on results and discussions to improve flow in directed fuzzy incidence networks (DFIN). Concepts like effective flow and maximum flow are discussed. Flow enhancement and saturation are other major topics considered. A DFIN version of max-flow min-cut theorem also is presented.

Chapter 8 mainly deals with two new parameters associated with fuzzy graphs termed as cycle connectivity and cycle cogency. Reachability is the most desired quality of any network. If two nodes are reachable in two different directions, they are said to be cyclically reachable. Cyclic reachability is the theme of Chap. 8. Several different types of graphs are also investigated. Concepts like cyclically balanced and cyclically fair fuzzy graphs are also discussed. The problem of return of refugees is discussed as the application part.

In Chap. 9, a fuzzy graph parameter named as neighborhood connectivity index (NCI) is discussed. It is effective in dealing with the local imbalance problems of a network. NCI of different types of products of fuzzy graphs is also presented. A human trafficking-related application dealing with illegal flow of humans between different locations of the globe is also studied.

The final chapter deals with cyclic connectivity index and integrity index of fuzzy graphs. These graph parameters reflect the cyclic reachability and average cyclic reachability of the fuzzy graph. Algorithms for the computation of the indices are provided. A new sequence termed as cyclic status sequence connecting graph space to sequence space is studied. Applications in human trafficking and Internet are also discussed.

The authors are grateful to all those who have been directly or indirectly involved in this project. We hope that this work will be beneficial to both students and scientists.

Omaha, USA John N. Mordeson
Calicut, India Sunil Mathew
Calicut, India G. Gayathri

Acknowledgments The authors are grateful to the editorial board and production staffs of Springer International Publishing, especially to Janusz Kacprzyk. The authors are indebted to journals of Fuzzy Sets and Systems, Information Sciences, IEEE Transaction on Fuzzy systems, Iranian Journal of Fuzzy Systems and New Mathematics and Natural Computation.

Contents

About the Authors

Dr. John N. Mordeson is Professor Emeritus of Mathematics at Creighton University. He received his B.S., M.S., and Ph.D. from Iowa State University. He is a member of Phi Kappa Phi. He has published 20 books and over 200 journal articles. He is on the editorial board of numerous journals. He has served as an external examiner of Ph.D. candidates from India, South Africa, Bulgaria, and Pakistan. He has referred for numerous journals and granting agencies. He is particularly interested in applying mathematics of uncertainty to combat the problem of human trafficking.

Dr. Sunil Mathew is a faculty member in the Department of Mathematics, NIT Calicut, India. He has acquired his master's from St. Joseph's College Devagiri, Calicut, and Ph.D. from the National Institute of Technology Calicut in the area of fuzzy graph theory. He has published more than 120 research papers and written 10 books. He is a member of several academic bodies and associations. He is an editor and reviewer of several international journals. He has an experience of 20 years in teaching and research. His current research topics include fuzzy graph theory, bio-computational modeling, graph theory, fractal geometry, and chaos.

Ms. G. Gayathri is currently pursuing research in the Department of Mathematics, National Institute of Technology Calicut, India. She took her master's degree in Mathematics from Government College Kasaragod, Kasaragod. She has got several publications in prestigious journals.

Chapter 1
Preliminaries

1.1 Fuzzy Sets and Relations

This section covers the fundamentals of fuzzy sets and fuzzy relations. In 1965, Lotfy Zadeh [1] introduced the concept of fuzzy sets using fuzzy logic to address the problems of ambiguity and vagueness. Fuzzy set theory facilitates the inclusion of elements in a set with partial memberships ranging from 0 to 1, which is not allowed in classical set theory. Throughout this book, we use A^c or $X \setminus A$ to denote the complement of a subset A of a set X. We denote the cardinality of A by $|A|$. We denote infimum and supremum by $\wedge$ and $\vee$, respectively. Most of the contents of this section are taken from [2].

Definition 1.1.1 Let X be a set. A **fuzzy subset** σ of X is a function $\sigma : X \to [0, 1]$.

In the literature, different notations for a fuzzy set are used. We follow the notation σ given by Zadeh [1]. If there is no confusion about X, the term fuzzy subset can be simply replaced by fuzzy set.

Consider a fuzzy set σ. Let σ^* denote the **support** of σ, defined by $\{x \in X : \sigma(x) > 0\}$. For any $t \in [0, 1]$, a crisp set called the t**-cut** of σ can be defined as $\{x \in X : \sigma(x) \geq t\}$. If $\{x \in X : \sigma(x) > t\}$, then it is a **strong** t**-cut**. Clearly, support of a fuzzy set is a strong 0-cut. A 1−cut is known as the **core** of the fuzzy set. The **height** $h(\sigma)$ and **depth** $d(\sigma)$ of σ can be defined as $h(\sigma) = \vee\{\sigma(x) : x \in X\}$ and $d(\sigma) = \wedge\{\sigma(x) : x \in X\}$, respectively. If $h(\sigma) = 1$, then the fuzzy set σ is **normal** and **subnormal** otherwise.

Example 1.1.2 Consider Fig. 1.1, which shows a trapezoidal fuzzy set σ defined on $\mathbb{R}$. Its membership function is defined by

$$\sigma(x) = \begin{cases} 0 & \text{if } x \leq a \\ \frac{x-a}{b-a} & \text{if } a \leq x \leq b \\ 1 & \text{if } b \leq x \leq c \\ \frac{d-x}{d-c} & \text{if } c \leq x \leq d \\ 0 & \text{otherwise.} \end{cases}$$

J. N. Mordeson et al., *Fuzzy Graph Theory*, Studies in Fuzziness and Soft Computing 424, https://doi.org/10.1007/978-3-031-23108-7_1

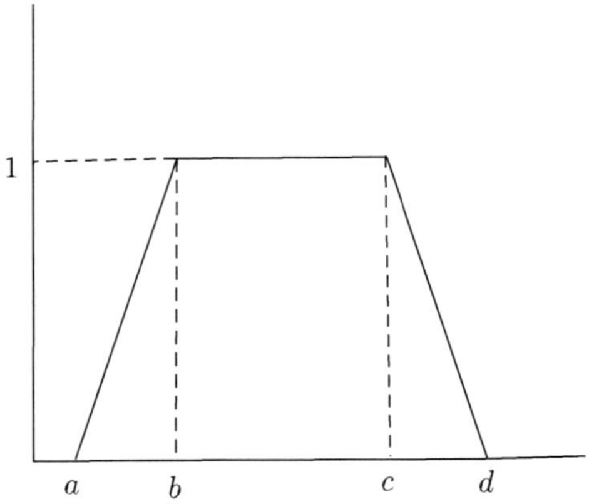

Fig. 1.1 A trapezoidal fuzzy set

Note that, σ is a normal fuzzy set with $\sigma^* = (a, d)$.

Next, we move onto some set theoretical operations on fuzzy sets. If σ and μ are two fuzzy subsets of a set X, then $\mu \subseteq \nu$ if for all $x \in X$, $\mu(x) \leq \nu(x)$. If $\mu \subseteq \nu$ and there exists $x \in X$ such that $\mu(x) < \nu(x)$, we write $\mu \subset \nu$. We define $\mu \cap \nu$ as $(\mu \cap \nu)(x) = \mu(x) \wedge \nu(x)$ for all $x \in X$. We define $\mu \cup \nu$ as $(\mu \cup \nu)(x) = \mu(x) \vee \nu(x)$ for all $x \in X$.

Definition 1.1.3 A function $\eta : [0, 1] \times [0, 1] \to [0, 1]$ is called a ***t*-norm** if it satisfies the following conditions.

(1) $\eta(1, x) = x, \forall x \in [0, 1]$ (Identity element).
(2) $\eta(x, y) = \eta(y, x), \forall x, y \in [0, 1]$ (Commutativity).
(3) $\eta(x, \eta(y, z)) = \eta(\eta(x, y), z), \forall x, y, z \in [0, 1]$ (Associativity).
(4) $w \leq x$ and $y \leq z$ implies $\eta(w, y) \leq \eta(x, z), \forall w, x, y, z \in [0, 1]$ (Monotonicity).

There are several classes of t-norms depending on the nature of the function η. For example, a continuous η is called a **continuous *t*-norm**. A strictly monotonic and continuous η is called a **strict *t*-norm**.

Example 1.1.4 The following are examples of t-norms.

(1) Drastic t-norm: $\eta(x, y) = \begin{cases} x \wedge y \text{ if } x \vee y = 1, \\ 0 \text{ otherwise.} \end{cases}$
(2) Lukasiewicz t-norm: $\eta(x, y) = 0 \vee (x + y - 1)$.
(3) $\eta(x, y) = \frac{xy}{2-(x+y-xy)}$.
(4) Product t-norm: $\eta(x, y) = xy$.
(5) Minimum t-norm: $\eta(x, y) = x \wedge y$.

The t-norm defined in (5) is known as the **standard intersection** for fuzzy sets. Also, among all t-norms, drastic t-norm is the smallest and minimum t-norm is the largest.

Definition 1.1.5 A function $\zeta : [0, 1] \times [0, 1] \to [0, 1]$ is called a ***t*-conorm** if it satisfies the following conditions.

(1) $\zeta(0, x) = x, \forall x \in [0, 1]$ (Identity element).
(2) $\zeta(x, y) = \zeta(y, x), \forall x, y \in [0, 1]$ (Commutativity).
(3) $\zeta(x, \zeta(y, z)) = \zeta(\zeta(x, y), z), \forall x, y, z \in [0, 1]$ (Associativity).
(4) $w \le x$ and $y \le z$ implies $\zeta(w, y) \le \zeta(x, z), \forall w, x, y, z \in [0, 1]$ (Monotonicity).

Example 1.1.6 The following are examples of t-conorms.

(1) Standard union: $\zeta(x, y) = x \vee y$.
(2) Algebraic sum: $\zeta(x, y) = x + y - xy$.
(3) Bounded sum: $\zeta(x, y) = 1 \wedge (x + y)$.
(4) Drastic union: $\zeta(x, y) = \begin{cases} x \text{ if } y = 0, \\ y \text{ if } x = 0, \\ 1 \text{ otherwise.} \end{cases}$

Now, we define the concept of complement as follows.

Definition 1.1.7 A function $c : [0, 1] \to [0, 1]$ is called a **fuzzy complement** if the following conditions hold.

(1) $c(0) = 1$ and $c(1) = 0$ (Boundary conditions).
(2) $\forall x, y \in [0, 1], x \le y$ implies $c(x) \ge c(y)$ (Monotonicity).

Two desirable properties for a fuzzy complement c are continuity and be **involutive**. By involutive nature, we mean $c(c(x)) = x \quad \forall x \in [0, 1]$. Standard complement is an example of an involutive fuzzy complement. That is, $c(x) = 1 - x$ for all $x \in [0, 1]$.

Now recall that a relation on a set S is a subset of $S \times S$. We can extend this concept into fuzzy relation on a set S as a fuzzy subset of $S \times S$. Consider the following definition of fuzzy relation on a fuzzy set.

Definition 1.1.8 Let σ be a fuzzy subset of a set S and μ, a fuzzy relation on S. μ is called a **fuzzy relation** on σ if $\mu(x, y) \le \sigma(x) \wedge \sigma(y)$ for every $x, y \in S$.

Definition 1.1.9 If S and T are two sets and σ and τ are fuzzy subsets of S and T, respectively, then a fuzzy relation μ from the fuzzy subset σ into the fuzzy subset τ is a fuzzy subset μ of $S \times T$ such that $\mu(x, y) \le \sigma(x) \wedge \tau(y)$ for every $x \in S$ and $y \in T$.

Definition 1.1.10 Let $\mu : S \times T \to [0, 1]$ be a fuzzy relation from a fuzzy subset σ of S into a fuzzy subset τ of T and $\nu : T \times U \to [0, 1]$ be a fuzzy relation from the fuzzy subset ρ of T into a fuzzy subset η of U. Define $\mu \circ \nu : S \times U \to [0, 1]$ by $\mu \circ \nu(x, z) = \vee\{\mu(x, y) \wedge \nu(y, z) : y \in T\}$ for every $x \in S, z \in U$. Then $\mu \circ \nu$ is called the **max-min composition** of μ and ν.

Note that, whenever μ and ν are two fuzzy relations on a fuzzy set σ, then $\mu \circ \nu$ is also a fuzzy relation on σ. Clearly, the max-min composition $\mu \circ \mu$ is a fuzzy relation on σ. It is denoted as μ^2. For any two fuzzy relations μ and ν on a finite set S and any $t \in [0, 1]$, $(\mu \circ \nu)^t = \mu^t \circ \nu^t$.

Definition 1.1.11 If μ is a fuzzy relation defined on a fuzzy subset σ of a set S, then the **complement** μ^c of μ is defined as $\mu^c(x, y) = 1 - \mu(x, y)$ for every $x, y \in S$.

Definition 1.1.12 Let $\mu : S \times T \to [0, 1]$ be a fuzzy relation from a fuzzy subset σ of S into a fuzzy subset ν of T. Then $\mu^{-1} : T \times S \to [0, 1]$, the **inverse** of μ from ν into σ is defined as $\mu^{-1}(y, x) = \mu(x, y)$ for all $x, y \in T \times S$.

If μ is a fuzzy relation on a fuzzy set σ, defined over S, then μ is said to be **reflexive** if $\mu(x, x) = \sigma(x)$ for every $x \in S$. μ is said to be **symmetric** if $\mu(x, y) = \mu(y, x)$ for every $x, y \in S$ and **transitive** if $\mu^2 \subseteq \mu$. A fuzzy relation μ on a fuzzy subset σ of a set S is said to be a **fuzzy equivalence relation** if it is reflexive, symmetric and transitive.

1.2 Fuzzy Graphs

Fuzzy graphs are mathematical structures that help to overcome the inadequacy of graphs to portray many real-world problems. Kaufmann [3] proposed the basic definition of a fuzzy graph in 1973 using fuzzy relations on fuzzy sets. Rosenfeld [4] further developed it by defining several fuzzy graph parameters. Several authors made significant contributions to the theoretical development of fuzzy graph theory. Most of the basic results on fuzzy graphs are included from [2]. Because of the wide range of applications in science and technology, fuzzy graph theory has become a dominant area of research in mathematics.

For a set V, consider a subset E of its power set such that every set in E has exactly two elements. Simply we write zw for $\{z, w\} \in E$. Clearly $zw = wz$.

Definition 1.2.1 A **fuzzy graph** $G = (V, \sigma, \mu)$ is a triple consisting of a set V, a fuzzy set σ on V and a fuzzy set μ on E such that $\mu(zw) \leq \sigma(z) \wedge \sigma(w)$ for every $z, w \in V$.

From the above definition, it is clear that μ is a fuzzy relation on V. Unless otherwise mentioned, we assume V is finite, and μ is reflexive and symmetric.

We let $G^* = (\sigma^*, \mu^*)$ to denote the underlying graph of G where $\sigma^* = \{z \in V : \sigma(z) > 0\}$ and $\mu^* = \{zw \in E : \mu(zw) > 0\}$. A fuzzy graph $G = (V, \sigma, \mu)$ is **trivial** when G^* is trivial. The members of σ^* are known as the **vertices** and the members of μ^* are known as the **edges** of the fuzzy graph. If the set V is well defined, we use the abbreviations G or (σ, μ) or $G = (\sigma, \mu)$ to denote a fuzzy graph.

Definition 1.2.2 A fuzzy graph $H = (V, \tau, \nu)$ is called a **partial fuzzy subgraph** of $G = (V, \sigma, \mu)$ if $\tau(u) \leq \sigma(u)$ for every vertex $u \in \sigma^*$ and $\nu(uv) \leq \mu(uv)$ for every $uv \in \mu^*$. In particular, we call $H = (V, \tau, \nu)$, a **fuzzy subgraph** of $G = (V, \sigma, \mu)$ if $\tau(u) = \sigma(u)$ for every $u \in \tau^*$ and $\nu(uv) = \mu(uv)$ for every $uv \in \nu^*$. A fuzzy subgraph $H = (V, \tau, \nu)$ is said to **span** the fuzzy graph $G = (V, \sigma, \mu)$ if $\tau = \sigma$. The fuzzy graph $H = (V, \tau, \nu)$ is called a **fuzzy subgraph of** G **induced by** P if $P \subset \sigma^*$, $\tau(u) = \sigma(u)$ for all u in P and $\nu(uv) = \mu(uv)$ for every $u, v \in P$.

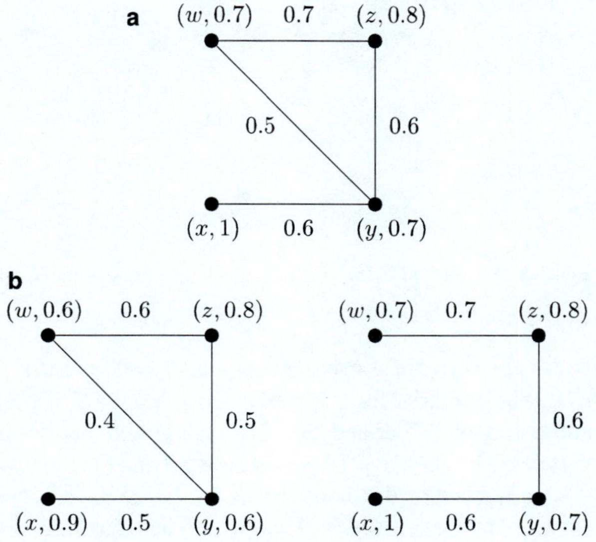

Fig. 1.2 **a** Fuzzy graph G in Example 1.2.5. **b** Partial fuzzy subgraph G_1 and fuzzy subgraph G_2 of G

For a fuzzy graph G, and every $t \in [0, 1]$, we can define an associated graph, called the threshold graph of G corresponding to t.

Definition 1.2.3 Let $G = (\sigma, \mu)$ be a fuzzy graph and let $0 \leq t \leq 1$. Let $\sigma^t = \{x \in \sigma^* : \sigma(x) \geq t\}$ and $\mu^t = \{e \in \mu^* : \mu(e) \geq t\}$. Then $H = (\sigma^t, \mu^t)$ is a graph with vertex set σ^t and edge set μ^t, called the threshold graph of G corresponding to t.

Proposition 1.2.4 Let $G = (\sigma, \mu)$ be a fuzzy graph and $0 \leq s < t \leq 1$. Then the threshold graph (σ^t, μ^t) is a subgraph of (σ^s, μ^s).

Example 1.2.5 Let Fig. 1.2(a) illustrates a fuzzy graph $G = (\sigma, \mu)$ with $\sigma^* = \{x, y, z, w\}$ and $\mu^* = \{xy, yz, zw, yw\}$. A partial fuzzy subgraph G_1 and a fuzzy subgraph G_2 of G are given in Fig. 1.2b. The fuzzy subgraph induced by the subset $P = \{y, z, w\}$ of σ^* and the threshold graph of G corresponding to $t = 0.7$ are given in Fig. 1.3.

Consider G_1 given in Fig. 1.2b. It is a partial fuzzy subgraph of G, because $\tau(z) = \sigma(z)$ and $\tau(a) < \sigma(a)$ for all other vertices $a \in \sigma^*$. Also, $\nu(e) < \mu(e)$ for all edges $e \in \mu^*$. If we consider G_2, then it is a fuzzy subgraph of G. Because, for every vertex $a \in \tau^*$, $\tau(a) = \sigma(a)$ and for every edge in ν^*, $\nu(e) = \mu(e)$. Here, G_2 is also a partial fuzzy subgraph of G. But, G_1 is not a fuzzy subgraph of G. Moreover, both G_1 and G_2 span G as $\tau^* = \sigma^*$.

Let $P = \{y, z, w\}$. Then G_3 of Fig. 1.3 is the fuzzy subgraph induced by P. For G_4 given in Fig. 1.3, $\sigma^{0.7} = \{x, y, z, w\}$ and $\mu^{0.7} = \{wz\}$. Here, $G_4 = (\sigma^{0.7}, \mu^{0.7})$ is the threshold graph of G corresponding to $t = 0.7$.

We let $G - e$ to denote the **edge deleted fuzzy graph** of $G = (\sigma, \mu)$ obtained by deleting an edge $e \in \mu^*$ from G. It is defined by the fuzzy subgraph $H = (\tau, \nu)$

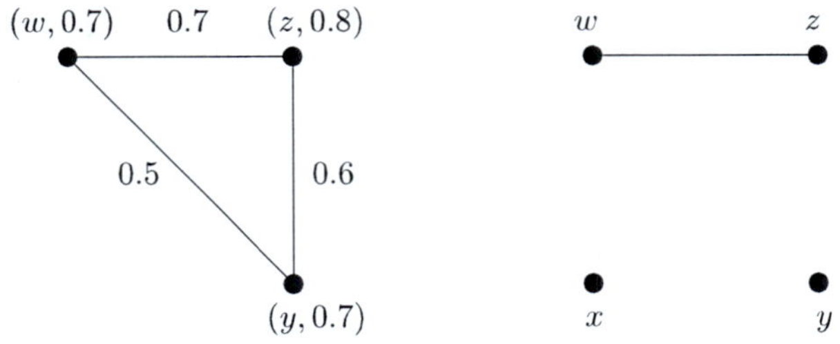

Fig. 1.3 An induced fuzzy graph G_3 and a threshold graph G_4 of G

with $\tau(z) = \sigma(z)$ for every $z \in \sigma^*$, $\nu(e) = 0$, and $\nu(e') = \mu(e')$ for all other edges $e' \in \mu^*$. Similarly, for a vertex $v \in \sigma^*$, we let $G - v$ to denote the **vertex deleted subgraph** of G defined by $H = (\tau, \nu)$ with $\tau(v) = 0, \tau(z) = \sigma(z)$ for all other vertices in σ^*, $\nu(vz) = 0$ for every $z \in \sigma^*$ and $\nu(e) = \mu(e)$ for all other edges $e \in \mu^*$.

A sequence of distinct vertices $P : z_0, z_1, \cdots, z_n$ with $\mu(z_{i-1}z_i) > 0, i = 1, 2, \cdots, n$ is called a **path** P of length n. The degree of membership of a weakest edge in P is defined as its strength. The path P becomes a **cycle** if z_0 coincides with z_n.

Definition 1.2.6 The **strength of connectedness** between two vertices z and w of a fuzzy graph G is defined as the maximum of the strengths of all paths between z and w and is denoted by $CONN_G(z, w)$. It is also denoted as $\mu^\infty(z, w)$. A $z - w$ path P is called a **strongest** $z - w$ path if its strength equals $CONN_G(z, w)$. A fuzzy graph $G = (\sigma, \mu)$ is said to be **connected** if for every $z, w \in \sigma^*$, $CONN_G(z, w) > 0$.

For example, there are two paths connecting y and w in the fuzzy graph given in Example 1.2.5 (Fig. 1.2a). The path yzw and the edge yw are $y - w$ paths where yzw is the unique strongest $y - w$ path. So, $CONN_G(y, w) = 0.6$.

We can refer the strength of connectedness between two vertices z and w of a graph network as the maximum bandwidth between the vertices z and w or the maximum width between the vertices z and w. It has several applications in internet routing problems, QoS problems and several other areas.

Proposition 1.2.7 Let $G = (\sigma, \mu)$ be a connected fuzzy graph and $H = (\tau, \nu)$ be a partial fuzzy subgraph of G. Then $CONN_H(x, y) \leq CONN_G(x, y)$ for every $x, y \in \sigma^*$.

A cycle C is said to be a **fuzzy cycle** if C has more than one weakest edge. A cycle C in a fuzzy graph G is called **locamin** if every vertex of C is adjacent with a weakest edge of the fuzzy graph G. C is called **multimin** if it has more than one weakest edge.

Note that, every locamin cycle is a multimin. But, the converse need not be true.

Example 1.2.8 Consider the fuzzy graphs G_5 and G_6 given in Fig. 1.4. Clearly, G_5 is a fuzzy cycle as it has 2 weakest edges. It is also a multimin. But G_5 is not a locamin. The fuzzy graph G_6 is locamin as every vertex lies on a weakest edge.

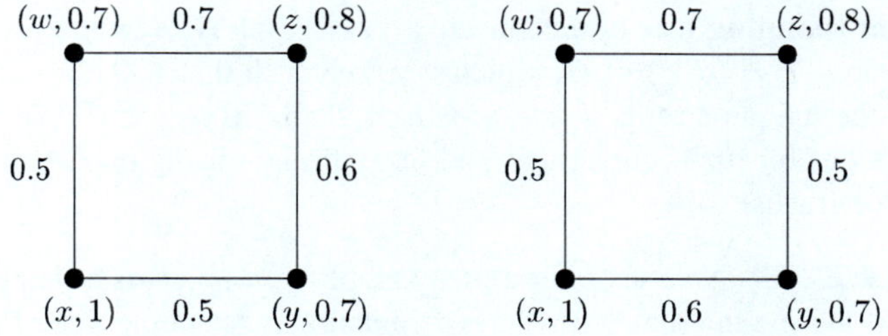

Fig. 1.4 A fuzzy cycle G_5 and a locamin cycle G_6

Fig. 1.5 A fuzzy graph having fuzzy cutvertices and fuzzy bridges

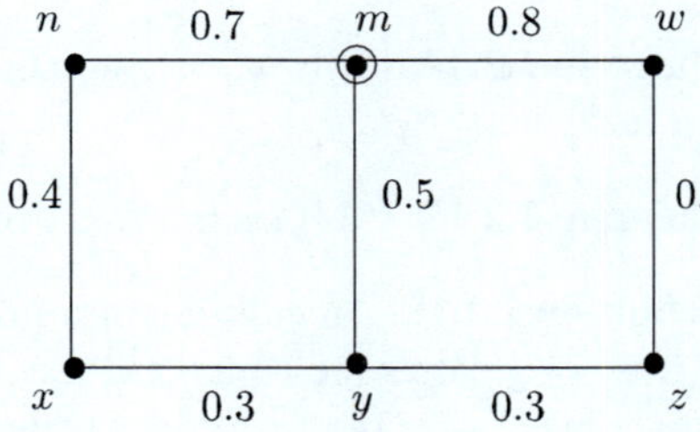

Definition 1.2.9 Let $G = (\sigma, \mu)$ be a fuzzy graph. An edge xy is called a **fuzzy bridge** of G if its removal reduces the strength of connectedness between some pair of vertices in G. That is, $CONN_{G-e}(u, v) < CONN_G(u, v)$ for some $u, v \in \sigma^*$. Similarly a **fuzzy cutvertex** w is a vertex in σ^* whose removal from G reduces the strength of connectedness between some pair of distinct vertices different from w. That is, $CONN_{G-w}(u, v) < CONN_G(u, v)$ where $u, v \in \sigma^*$ such that $u \neq w \neq v$. A fuzzy graph is said to be a **fuzzy block** or simply a **block** if it has no fuzzy cutvertices.

Example 1.2.10 Consider the fuzzy graph $G = (\sigma, \mu)$ in Fig. 1.5. Let $\sigma^* = \{x, y, z, w, m, n\}$ and $\sigma(a) = 1$ for every $a \in \sigma^*$.

In the case of Fig. 1.5, the encircled vertex m is a fuzzy cutvertex. $CONN_G(w, n) = 0.7$, whereas $CONN_{G-m}(w, n) = 0.3$. Moreover, n and w are also fuzzy cutvertices of G. Also, all the edges except xy and yz are fuzzy bridges of G.

In graphs, at least one of the end vertices of a bridge will be a cutvertex. But this is not true for fuzzy bridges and fuzzy cutvertices. Next is a useful characterization for fuzzy bridges of a fuzzy graph.

Theorem 1.2.11 [4] Let $G = (\sigma, \mu)$ be a fuzzy graph. Then the following statements are equivalent.

(1) xy is a fuzzy bridge.
(2) $CONN_{G-xy}(x, y) < \mu(xy)$.
(3) xy is not the weakest edge of any cycle.

A **maximum spanning tree** of a connected fuzzy graph $G = (\sigma, \mu)$ is a spanning fuzzy subgraph $T = (\sigma, \nu)$ of G, which is a tree such that $CONN_G(u, v)$ is the strength of the unique strongest $u - v$ path in T for all $u, v \in G$. Next we have characterizations for fuzzy cutvertices and fuzzy bridges using maximum spanning trees of fuzzy graphs.

Theorem 1.2.12 [2] A vertex w of a fuzzy graph $G = (\sigma, \mu)$ is a fuzzy cutvertex if and only if w is an internal vertex of every maximum spanning tree of G.

Theorem 1.2.13 [2] An edge uv of a fuzzy graph $G = (\sigma, \mu)$ is a fuzzy bridge if and only if uv is in every maximum spanning tree of G.

Theorem 1.2.14 [5] If w is a common vertex of at least two fuzzy bridges, then w is a fuzzy cutvertex.

Theorem 1.2.15 [5] If uv is a fuzzy bridge, then $CONN_G(u, v) = \mu(uv)$.

Definition 1.2.16 An edge xy of a fuzzy graph G is called **strong** if its weight is at least as great as the connectedness of its end vertices in the edge deleted fuzzy subgraph $G - xy$. If xy is a strong edge, then x and y are said to be **strong neighbors**. A vertex z is called a **fuzzy endvertex** if it has exactly one strong neighbor in G. An $x - y$ path P is called a **strong path** if P contains only strong edges.

The **degree** of a vertex v is defined as $d(v) = \sum_{u \neq v} \mu(uv)$. The **minimum degree** of G is $\delta(G) = \wedge\{d(v) : v \in \sigma^*\}$ and the **maximum degree** of G is $\Delta(G) = \vee\{d(v) : v \in \sigma^*\}$. The **strong degree** of a vertex $v \in \sigma^*$ is defined as the sum of membership values of all strong edges incident at v. It is denoted by $d_s(v)$. Also if $N_s(v)$ denotes the set of all strong neighbors of v, then $d_s(v) = \sum_{u \in N_s(v)} \mu(uv)$. The **minimum strong degree** of G is $\delta_s(G) = \wedge\{d_s(v) : v \in \sigma^*\}$ and **maximum strong degree** of G is $\Delta_s(G) = \vee\{d_s(v) : v \in \sigma^*\}$.

Definition 1.2.17 A connected fuzzy graph $G = (\sigma, \mu)$ is a **fuzzy tree** if it has a fuzzy spanning subgraph $F = (\sigma, \nu)$, which is a tree, where for all edges xy not in F there exists a path from x to y in F whose strength is more than $\mu(xy)$.

Note that F is the unique maximum spanning tree of G. Also, if G is not connected, and satisfies the property of Definition 1.2.17, then it is called a **fuzzy forest**.

For example, the fuzzy graph G given in Fig. 1.2a is a fuzzy tree.

Definition 1.2.18 A connected fuzzy graph $G = (\sigma, \mu)$ is a **complete fuzzy graph** if $\mu(xy) = \sigma(x) \wedge \sigma(y)$ for every x and y in σ^*.

Definition 1.2.19 Fig. 1.6 is a complete fuzzy graph $G = (\sigma, \mu)$ with $|\sigma^*| = 4$. Define $\sigma(x) = 0.4$, $\sigma(y) = 0.5$, $\sigma(z) = 0.6$, and $\sigma(w) = 0.9$. For every $a, b \in \sigma^*$, we can see that $\mu(ab) = \sigma(a) \wedge \sigma(b)$.

Theorem 1.2.20 [6] If $G = (\sigma, \mu)$ is a complete fuzzy graph, then for any edge $uv \in \mu^*$, $CONN_G(u, v) = \mu(uv)$.

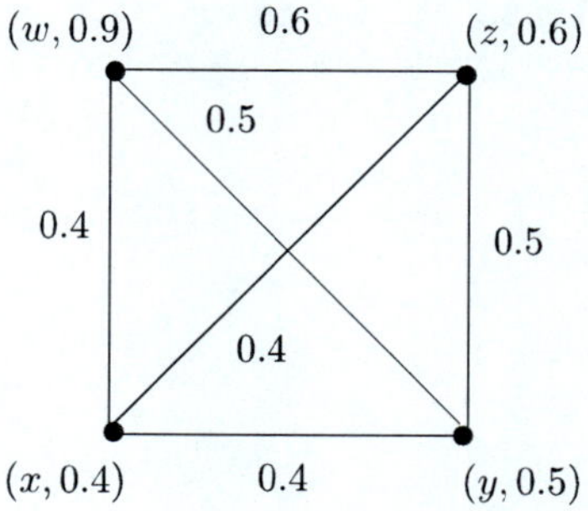

Fig. 1.6 A complete fuzzy graph

Proposition 1.2.21 [4] If $G = (\sigma, \mu)$ is a fuzzy tree, then the edges of its maximum spanning tree $F = (\tau, \nu)$ are just the fuzzy bridges of G.

Theorem 1.2.22 [7] Let $G = (\sigma, \mu)$ be a fuzzy graph such that G^* is a cycle. Then G is a fuzzy cycle if and only if G is not a fuzzy tree.

Theorem 1.2.23 [5] Let $G = (\sigma, \mu)$ be a connected fuzzy graph with no fuzzy cycles. Then G is a fuzzy tree.

Theorem 1.2.24 [5] If G is a fuzzy tree, then the internal vertices of F are fuzzy cutvertices of G.

Theorem 1.2.25 [5] Let $G = (\sigma, \mu)$ be a fuzzy graph. Then G is a fuzzy tree if and only if the following conditions are equivalent for all $u, v \in V$.

(1) uv is a fuzzy bridge.
(2) $CONN_G(u, v) = \mu(uv)$.

Theorem 1.2.26 [5] A fuzzy graph is a fuzzy tree if and only if it has a unique maximum spanning tree.

Based on the strength of connectedness between the end vertices of an edge, edges of fuzzy graphs can be divided into three categories as given below.

Definition 1.2.27 [8] An edge xy in a fuzzy graph $G = (\sigma, \mu)$ is called **α-strong** if $\mu(xy) > CONN_{G-xy}(x, y)$, **$\beta$-strong** if $\mu(xy) = CONN_{G-xy}(x, y)$ and **δ-edge** if $\mu(xy) < CONN_{G-xy}(x, y)$.

The δ-edges of a fuzzy graph G, whose μ values are more than that of the weakest edge of G are called δ^*-edges. A path in a fuzzy graph is called an **α-strong path** if all its edges are α-strong and is said to be a **β-strong path** if all its edges are β-strong.

G is said to be **α-saturated**, if at least one α-strong edge is incident at every vertex $v \in \sigma^*$. G is called **β-saturated**, if at least one β-strong edge is incident at every vertex. G is called **saturated**, if it is both α-saturated and β-saturated. That is, at least one α-strong edge and one β-strong edge is incident on every vertex $v \in \sigma^*$. Also a fuzzy graph which is not saturated is called **unsaturated**.

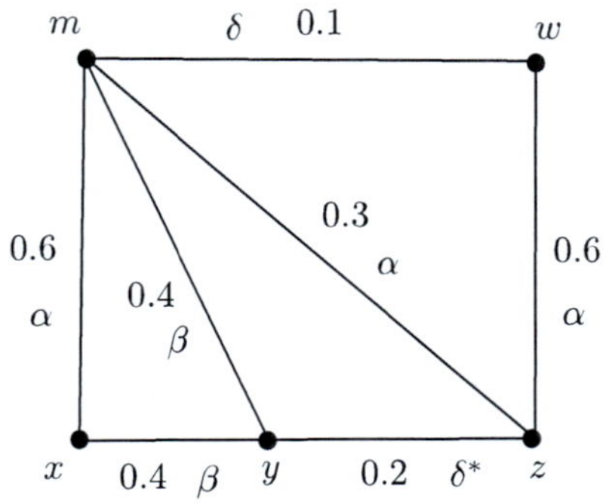

Fig. 1.7 Fuzzy graph with edge classification

Example 1.2.28 The fuzzy graph given in Fig. 1.7 contains all three type of edges including δ^*-edges.

Theorem 1.2.29 [8] Let $G = (\sigma, \mu)$ be a fuzzy graph. Then an edge xy of G is a fuzzy bridge if and only if it is α-strong.

Theorem 1.2.30 [8] A connected fuzzy graph G is a fuzzy tree if and only if it has no β-strong edges.

Theorem 1.2.31 [8] An edge xy of a fuzzy tree $G = (\sigma, \mu)$ is α-strong if and only if xy is an edge of the maximum spanning tree $F = (\sigma, \nu)$ of G.

Theorem 1.2.32 [8] A complete fuzzy graph has no δ-edges.

Definition 1.2.33 Let $G = (\sigma, \mu)$ be a connected fuzzy graph. A set of vertices $X = \{v_1, v_2, \cdots, v_m\} \subset \sigma^*$ is said to be a **fuzzy vertex cut** or **fuzzy node cut (FNC)** of G if either, $CONN_{G-X}(x, y) < CONN_G(x, y)$ for some pair of vertices $x, y \in \sigma^*$ such that both $x, y \neq v_i$ for $i = 1, 2, \cdots, m$ or $G - X$ is trivial.

Let X be a fuzzy vertex cut of G. The **strong weight** of X, denoted by $s(X)$ is defined as $s(X) = \sum_{x \in X} \mu(xy)$, where $\mu(xy)$ is the minimum of the weights of strong edges incident at x.

Definition 1.2.34 The **fuzzy vertex connectivity** of a connected fuzzy graph G is defined as the minimum strong weight of fuzzy vertex cuts of G. It is denoted by $\kappa(G)$.

Definition 1.2.35 Let $G = (\sigma, \mu)$ be a fuzzy graph. A set of strong edges $E = \{e_1, e_2, \cdots, e_n\}$ where $e_i = u_i v_i, i = 1, 2, \cdots, n$ is said to be a **fuzzy edge cut** or **fuzzy arc cut (FAC)** of G if either $CONN_{G-E}(x, y) < CONN_G(x, y)$ for some pair of vertices $x, y \in \sigma^*$ with at least one of x or y different from both u_i and $v_i, i = 1, 2, \cdots, n$, or $G - E$ is disconnected.

If there are n edges in E, then it is called an n-FAC. Among all fuzzy edge cuts, an edge cut with one edge (1-FAC) is a special type of fuzzy bridge, called a **fuzzy bond**. At least one of the end vertices of a fuzzy bond is always a fuzzy cutvertex. The strong weight of a fuzzy edge cut E is defined as $s'(E) = \sum_{e_i \in E} \mu(e_i)$.

Definition 1.2.36 The **fuzzy edge connectivity** $\kappa'(G)$ of a connected fuzzy graph G is defined to be the minimum strong weight of fuzzy edge cuts of G.

Theorem 1.2.37 [9] In a fuzzy tree $G = (\sigma, \mu)$, $\kappa(G) = \kappa'(G) = \wedge\{\mu(xy) : xy$ is a strong edge in $G\}$.

Theorem 1.2.38 [9] In a connected fuzzy graph $G = (\sigma, \mu)$, $\kappa(G) \leq \kappa'(G) \leq \delta_s(G)$.

In a CFG, $G = (\sigma, \mu)$, $\kappa(G) = \kappa'(G) = \delta_s(G)$.

Definition 1.2.39 Let u and v be any two vertices of a fuzzy graph $G = (\sigma, \mu)$ such that the edge uv is not strong. A subset $S \subseteq \sigma^*$ of vertices is said to be a $u - v$ **strength reducing set of vertices** if $CONN_{G-S}(u, v) < CONN_G(u, v)$, where $G - S$ is the fuzzy subgraph of G obtained by removing all vertices in S.

Similarly, a set of edges $E \subseteq \mu^*$ is said to be a $u - v$ **strength reducing set of edges** if $CONN_{G-E}(u, v) < CONN_G(u, v)$ where $G - E$ is the fuzzy subgraph of G obtained by removing all edges in E.

Theorem 1.2.40 [10] Let $G = (\sigma, \mu)$ be a fuzzy cycle with $|\sigma^*| \geq 3$. If no two α-strong edges of G are adjacent, then $\kappa(G) = 2\mu(xy)$, where xy is a weakest edge of G, otherwise $\kappa(G) = \mu(uv)$, if $\mu(uv) < 2\mu(xy)$, else $\kappa(G) = 2\mu(xy)$, where xy is a weakest edge of G and uv is an edge with minimum μ value among all α-strong edges adjacent with one another.

Theorem 1.2.41 [10] Let $G = (\sigma, \mu)$ be a fuzzy graph with $|\sigma^*| = n$. If $H = (\sigma, \nu)$ is a partial fuzzy subgraph of G, then $\kappa(H) \leq \kappa(G)$.

Theorem 1.2.42 [11] Let $G = (\sigma, \mu)$ be a fuzzy graph with $|\sigma^*| = n$. If $H = (\sigma, \nu)$ is a partial fuzzy subgraph of G having the same vertex set of G, then $\kappa'(H) \leq \kappa'(G)$.

Theorem 1.2.43 [12] **(Generalization of the vertex version of Menger's Theorem)** Let $G = (\sigma, \mu)$ be a fuzzy graph. For any two vertices $u, v \in \sigma^*$ such that uv is not strong, the maximum number of internally disjoint strongest $u - v$ paths in G is equal to the number of vertices in a minimal $u - v$ strength reducing set.

Definition 1.2.44 Let G be a connected fuzzy graph and $t \in (0, \infty)$. G is called t**-connected** if $\kappa(G) \geq t$ and G is called t-edge connected if $\kappa'(G) \geq t$.

Theorem 1.2.45 [12] Let G be a connected fuzzy graph. Then G is t-connected if and only if $mCONN_G(u, v) \geq t$ for every pair of vertices u and v in G, where m is the number of internally disjoint strongest $u - v$ paths in G.

Theorem 1.2.46 [12] Let G be a connected fuzzy graph. Then G is t-edge connected if and only if $mCONN_G(u, v) \geq t$ for every pair of vertices u and v in G, where m is the number of edge disjoint strongest $u - v$ paths in G.

In 2013, Mathew and Sunitha introduced a new connectivity parameter called cycle connectivity, which is given in Definition 1.2.48. Up to Definitions 1.2.53 are from [13].

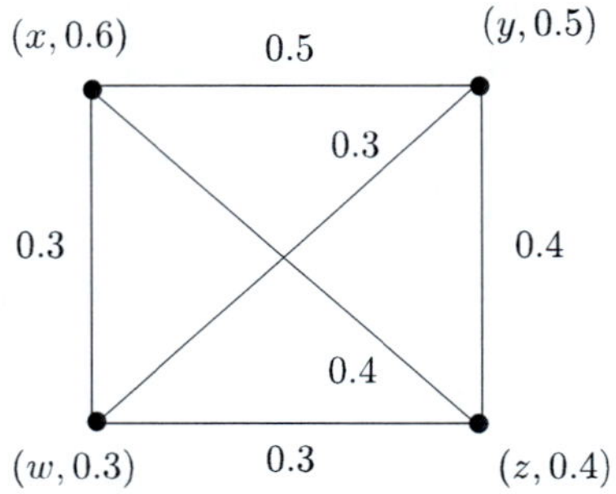

Fig. 1.8 Illustration of cycle connectivity

Definition 1.2.47 Let $G = (\sigma, \mu)$ be a fuzzy graph. Then for any two vertices u and v of G, there associated a set $\theta(u, v)$ called the **θ-evaluation** of u and v defined as $\theta(u, v) = \{\alpha : \alpha \in (0, 1]$, where α is the strength of a strong cycle passing through both u and $v\}$.

Note that if there are no strong cycles passing both u and v, then $\theta(u, v) = \phi$.

Definition 1.2.48 Let $G = (\sigma, \mu)$ be a fuzzy graph. Then, $\vee\{\alpha | \alpha \in \theta(u, v); u, v \in \sigma^*\}$, is defined as the **cycle connectivity** between u and v in G and denoted by $C^G_{u,v}$. If $\theta(u, v) = \phi$ for some pair of vertices u and v, define the cycle connectivity between u and v to be 0.

Example 1.2.49 Consider the fuzzy graph $G = (\sigma, \mu)$ given in Fig. 1.8 with weights defined as in the figure. Here G is a complete fuzzy graph. $\theta\{x, z\} = \{0.4, 0.3\}$ and hence $C^G_{x,z} = 0.4$.

Definition 1.2.50 Let $G = (\sigma, \mu)$ be a fuzzy graph. **Cycle connectivity** of G is defined as $CC(G) = Max\{C^G_{u,v} : u, v \in \sigma^*\}$.

Definition 1.2.51 A node w in a fuzzy graph is called a **cyclic cutvertex** if $CC(G - w) < CC(G)$.

Definition 1.2.52 An edge uv of a fuzzy graph is called a **cyclic bridge** if $CC(G - (u, v)) < CC(G)$.

Proposition 1.2.53 In a fuzzy graph, if edge uv is a cyclic bridge, then u and v are cyclic cutvertices.

Definition 1.2.54 [14] A **cyclic vertex cut** (CVC) of a fuzzy graph $G = (\sigma, \mu)$ is a set of vertices $X \subseteq \sigma^*$ such that $CC(G - X) < CC(G)$, provided $CC(G) > 0$.

Definition 1.2.55 [14] Let X be a cyclic vertex cut of G. The strong weight of X is defined as $S_c(X) = \sum_{x \in X} \mu(xy)$, where $\mu(xy)$ is the minimum weights of strong edges incident on x.

Definition 1.2.56 [14] **Cyclic vertex connectivity** of a fuzzy graph G, denoted by $\kappa_c(G)$, is the minimum of the cyclic strong weights of cyclic vertex cuts in G.

Theorem 1.2.57 [2] A cycle C in a fuzzy graph G is called a **strongest strong cycle (SSC)** if C is the union of two strongest strong $u - v$ paths for every pair of vertices u and v in C except when uv is a fuzzy bridge of G in C.

Theorem 1.2.58 [2] Let $G = (\sigma, \mu)$ be a fuzzy cycle. Then the following are equivalent.

(1) G is either saturated or β-saturated.
(2) G is a block.
(3) G is a strongest strong cycle (SSC).
(4) G is a locamin cycle.

Theorem 1.2.59 [15] The following statements are equivalent for a fuzzy graph $G = (\sigma, \mu)$.

(1) G is a block.
(2) Any two vertices u and v such that uv is not a fuzzy bridge are joined by two internally disjoint strongest paths.
(3) For every three distinct vertices of G, there is a strongest path joining any two of them not containing the third.

Theorem 1.2.60 [15] If $G = (\sigma, \mu)$ is a block, then the following conditions hold and are equivalent.

(1) Every two vertices of G lie on a common strong cycle.
(2) Each vertex and a strong edge of G lie on a common strong cycle.
(3) Any two strong edges of G lie on a common strong cycle.
(4) For any two given vertices and a strong edge in G, there exists a strong path joining the vertices containing the edge.
(5) For every three distinct vertices of G, there exists a strong path joining any two of them containing the third.
(6) For every three vertices of G, there exist strong paths joining any two of them which does not contain the third.

Theorem 1.2.61 [2] Let C_n be a fuzzy cycle. Then it is saturated if and only if the following two conditions are satisfied.

(1) $n = 2k$, where k is an integer.
(2) α-strong and β-strong edges appear alternatively on C_n.

Definition 1.2.62 [2] An **isomorphism** $h : G \to G'$ is a map $h : S \to S'$ which is bijective that satisfies $\psi(m) = \psi'(h(m))$ for all $m \in S$, $\gamma(m, p) = \gamma'(h(m), h(p))$ for all $m, p \in S$.

Definition 1.2.63 [2] The **complement** of a fuzzy graph $G = (\psi, \gamma)$ is the fuzzy graph $G^c = (\psi^c, \gamma^c)$ where $\psi^c = \psi$ and $\gamma^c(m, p) = \psi(m) \wedge \psi(p) - \gamma(m, p)$ for all $m, p \in V$.

Definition 1.2.64 [16] Let $G_1 = (\sigma_1, \mu_1)$ and $G_2 = (\sigma_2, \mu_2)$ be two fuzzy graphs with $G_1^* = (V_1, E_1)$ and $G_2^* = (V_2, E_2)$ with $V_1 \cap V_2 = \phi$ and let $G^* = G_1^* \cup G_2^* = (V_1 \cup V_2, E_1 \cup E_2)$ be the union of G_1^* and G_2^*. Then the **union** of two fuzzy graphs G_1 and G_2 is a fuzzy graph $G = G_1 \cup G_2 = (\sigma_1 \cup \sigma_2, \mu_1 \cup \mu_2)$ defined by

$$(\sigma_1 \cup \sigma_2)(u) = \begin{cases} \sigma_1(u) & \text{if } u \in V_1 - V_2 \\ \sigma_2(u) & \text{if } u \in V_2 - V_1 \end{cases}$$

and

$$(\mu_1 \cup \mu_2)(uv) = \begin{cases} \mu_1(uv) & \text{if } uv \in E_1 - E_2 \\ \mu_2(uv) & \text{if } uv \in E_2 - E_1. \end{cases}$$

Definition 1.2.65 [16] Consider the join $G^* = G_1^* + G_2^* = (V_1 \cup V_2, E_1 \cup E_2 \cup E')$ of graphs where E' is the set of al edges joining the vertices of V_1 and V_2 where we assume that $V_1 \cap V_2 = \phi$. Then the **join** of two fuzzy graphs G_1 and G_2 is a fuzzy graph $G = G_1 + G_2 = (\sigma_1 + \sigma_2, \mu_1 + \mu_2)$ defined by

$$(\sigma_1 + \sigma_2)(u) = (\sigma_1 \cup \sigma_2)(u), \ \ u \in V_1 \cup V_2$$

and

$$(\mu_1 + \mu_2)(uv) = \begin{cases} (\mu_1 \cup \mu_2)(uv) & \text{if } uv \in E_1 \cup E_2 \\ \sigma_1(u) \wedge \sigma_2(v) & \text{if} uv \in E'. \end{cases}$$

Definition 1.2.66 [16] Let $G^* = G_1^* \times G_2^* = (V, E'')$ be the Cartesian product of two graphs where $V = V_1 \times V_2$ and $E'' = \{(u, u_2)(u, v_2) : u \in V_1, u_2v_2 \in E_2\} \cup \{(u_1, w)(v_1, w) : w \in V_2, u_1v_1 \in E_1\}$. Then the **Cartesian product** $G = G_1 \times G_2 = (\sigma_1 \times \sigma_2, \mu_1 \times \mu_2)$ is a fuzzy graph defined by

$$(\sigma_1 \times \sigma_2)(u_1, u_2) = \sigma_1(u_1) \wedge \sigma_2(u_2) \ \forall \, (u_1, u_2) \in V$$

$$\mu_1 \times \mu_2((u, u_2)(u, v_2)) = \sigma_1(u) \wedge \mu_2(u_2v_2) \, \forall \, u \in V_1, \ \forall \, u_2v_2 \in E_2,$$

$$\mu_1 \times \mu_2((u_1, w)(v_1, w)) = \sigma_2(w) \wedge \mu_1(u_1v_1) \ \forall \, w \in V_2, \ \forall \, u_1v_1 \in E_1.$$

Definition 1.2.67 [16] Let $G^* = G_1^* \circ G_2^* = (V_1 \times V_2, E)$ be the composition of two graphs, where $E = \{(u, u_2)(u, v_2) : u \in V_1, u_2v_2 \in E_2\} \cup \{(u_1, w)(v_1, w) : w \in V_2, u_1v_1 \in E_1\} \cup \{(u_1, u_2)(v_1, v_2) : u_1v_1 \in E_1, u_2 \neq v_2\}$. Then the **composition** of fuzzy graphs $G = G_1 \circ G_2 = (\sigma_1 \circ \sigma_2, \mu_1 \circ \mu_2)$ is a fuzzy graph defined by

$$(\sigma_1 \circ \sigma_2)(u_1, u_2) = \sigma_1(u_1) \wedge \sigma_2(u_2) \ \forall \, (u_1, u_2) \in V_1 \times V_2$$

$$(\mu_1 \times \mu_2)((u, u_2)(u, v_2)) = \sigma_1(u) \wedge \mu_2(u_2v_2) \ \forall \, u \in V_1, \ \forall \, u_2v_2 \in E_2;$$

$$(\mu_1 \times \mu_2)((u_1, w)(v_1, w)) = \sigma_2(w) \wedge \mu_1(u_1v_1) \ \forall \, w \in V_2, \ \forall \, u_1v_1 \in E_1;$$

$$(\mu_1 \times \mu_2)((u_1, u_2)(v_1, v_2)) = \sigma_2(u_2) \wedge \sigma_2(v_2) \wedge \mu_1(u_1v_1)$$
$$\forall (u_1, u_2)(v_1, v_2) \in E - E'',$$

where

$$E'' = \{(u, u_2)(u, v_2) : u \in V_1, \forall u_2v_2 \in E_2\} \cup$$
$$\{(u_1, w)(v_1, w) : w \in V_2, \forall u_1v_1 \in E_1\}.$$

Definition 1.2.68 [17] For a fuzzy graph $G = (\psi, \gamma)$, the **connectivity Index** (CI) is defined as $CI(G) = \sum^{*}_{m,p\in\psi} \psi(m)\psi(p)CONN_G(m, p)$, where $CONN_G(m, p)$ is the strength of connectedness between m and p.

Definition 1.2.69 [23] For a fuzzy graph $G = (\psi, \gamma)$, the **Wiener Index** (WI) is defined as $WI(G) = \sum_{m,p\in\psi^*} \psi(m)\psi(p)d_S(m, p)$, where $d_S(m, p)$ is the minimum sum of weights of geodesics from m to p.

Theorem 1.2.70 [23] For a complete fuzzy graph $CI(G) = WI(G)$.

1.3 Fuzzy Incidence Graphs

Dinesh [18] introduced the notion of fuzzy incidence graphs (FIG), which were later developed by Mathew and Mordeson [19–21]. These graph structures discussed the relationships between vertices and edges by including the degree of incidence of a vertex on an edge. FIGs are extremely useful while dealing with networks that have extraneous support and flows. In particular, they can be used to model the ramping system in highways in order to control the unpredictable flow between cities and highways. The following preliminaries are taken from [22].

Definition 1.3.1 Let (V, E) be a graph. Then $G = (V, E, I)$ is called an **incidence graph**, where $I \subseteq V \times E$.

It is important to note that if $V = \{u, v\}$, $E = \{uv\}$ and $I = \{(v, uv)\}$, then (V, E, I) is an incidence graph even though $(u, uv) \notin I$.

Definition 1.3.2 Let $G = (V, E, I)$ be an incidence graph. If $(u, vw) \in I$, then (u, vw) is called an **incidence pair** or simply a **pair**. If (u, uv), (v, uv), (v, vw), $(w, vw) \in I$, then uv and vw are called **adjacent edges**.

Definition 1.3.3 An **incidence subgraph** H of an incidence graph G is an incidence graph having its vertices, edges, and pairs in G. If H is an incidence subgraph of G, then G is called an incidence **supergraph** of H.

Definition 1.3.4 Let $G = (V, E, I)$ be an incidence graph. Let $V' \subseteq V$, $E' \subseteq E$, and $I' \subseteq I$. Then $G' = (V', E', I')$ is called a **near incidence subgraph** of G if

(1) $u'v' \in E' \Rightarrow u' \in V'$ or $v' \in V'$ and
(2) $(v', u'v') \in I' \Rightarrow u'v' \in E'$.

Definition 1.3.5 Let $G = (V, E, I)$ be an incidence graph. A sequence

$$v_0, (v_0, v_0v_1), v_0v_1, (v_1, v_0v_1), v_1, ..., v_{n-1}, (v_{n-1}, v_{n-1}v_n), v_{n-1}v_n, (v_n, v_{n-1}v_n), v_n$$

is called a **walk**. It is **closed** if $v_0 = v_n$. If the pairs are distinct, then it is called an **incidence trail**. If the edges are distinct, then it is called a **trail**. If the vertices are distinct, then it is called a **incidence path**.

If a walk in an incidence graph is closed, then it is called a **cycle** if the vertices are distinct.

Let (V, E) be a graph and (V, E, I) an incidence graph. Then $I \subseteq V \times E$. We assume in the following that $I \subseteq \{(u, uv)|uv \in E\}$. Let $E^{(i)} = \{(u, uv)|uv \in E\}$. (Note that since $uv = vu$, $(v, uv) \in E^{(i)}$). Although not allowed here, incidence pairs of the form (u, vw), where $v \neq u \neq w$ also have potential applications in network theory. For example, (u, vw) might represent u's influence on vw with respect to flow from v to w. The flow might be human trafficking between countries or illicit flow of drugs, arms, or money between countries.

An incidence graph in which all pairs of vertices are joined by a path is said to be **connected**.

Definition 1.3.6 An incidence graph without cycles is called a **forest**. If it is connected, then it is called a **tree**.

Since a tree is connected, all pairs of vertices are connected by a path.

Definition 1.3.7 If the removal of an edge in an incidence graph increases the number of connected components, then the edge is called a **bridge**.

Definition 1.3.8 If the removal of a vertex in an incidence graph increases the number of connected components, then the vertex is called a **cutvertex**.

Definition 1.3.9 If the removal of an incidence pair in an incidence graph increases the number of connected components, then the incidence pair is called a **cutpair**.

Definition 1.3.10 Let $G = (V, E)$ be a graph and σ be a fuzzy subset of V and μ a fuzzy subset of $V \times E$. Let Ψ be a fuzzy subset of $V \times E$. If $\Psi(v, e) \leq \sigma(v) \wedge \mu(e)$ for all $v \in V$ and $e \in E$, then Ψ called a **fuzzy incidence** of G.

Definition 1.3.11 Let $G = (V, E)$ be a graph and (σ, μ) be a fuzzy subgraph of G. If Ψ is a fuzzy incidence of G, then $\widetilde{G} = (\sigma, \mu, \Psi)$ is called a **fuzzy incidence graph** of G. $xy \in \text{Supp}(\mu)$ is an edge of the fuzzy incidence graph $\widetilde{G} = (\sigma, \mu, \Psi)$ and if $(x, xy), (y, xy) \in \text{Supp}(\Psi)$, then (x, xy) and (y, xy) are called **pairs**. Two vertices v_i and v_j joined by a path in a fuzzy incidence graph are said to be **connected**.

Example 1.3.12 Consider the example of a fuzzy incidence graph $\widetilde{G} = (\sigma, \mu, \Psi)$ given in Fig. 1.9. $\widetilde{G}$ has $V = \{x, y, z, w\}$ with values of σ, μ and Ψ as defined in the figure.

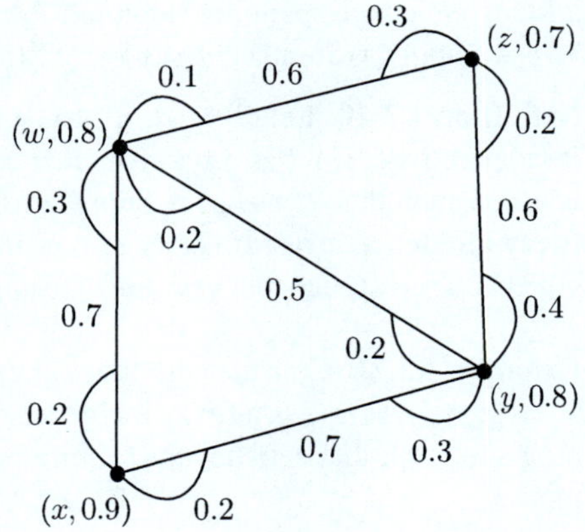

Fig. 1.9 Example of a fuzzy incidence graph

Definition 1.3.13 The **incidence strength** of a path in $\widetilde{G} = (\sigma, \mu, \Psi)$ is the least value of non zero Ψ values of all incidence pairs in $\widetilde{G}$. For any $u, v \in \sigma^* \cup \mu^*$, define $\Psi^{\infty}(u, v)$ to be the incidence strength of the path from u to v of greatest incidence strength.

In Fig. 1.9 of Example 1.3.12, there are three $x - y$ paths in $\widetilde{G}$ given by $P_1 : x, (x, xy), xy, (y, xy), y$ and $P_2 : x, (x, xw), xw, (w, xw), w, (w, wz), wz, (z, wz), z, (z, zy), zy, (y, zy), y$ and $P_3 : x, (x, xw), xw, (w, xw), w, (w, wy), wy, (y, wy), y$. Hence $\Psi^{\infty}(x, y) = \wedge\{0.2, 0.1\} = 0.2$.

Definition 1.3.14 Let $\widetilde{G} = (\sigma, \mu, \Psi)$ be a fuzzy incidence graph. Then $\widetilde{H} = (\tau, \nu, \Omega)$ is a **fuzzy incidence subgraph** of $\widetilde{G}$ if $\tau \subseteq \sigma, \nu \subseteq \mu$, and $\Omega \subseteq \Psi$. A fuzzy incidence subgraph $\widetilde{H}$ of $\widetilde{G}$ is a **fuzzy incidence spanning subgraph** of $\widetilde{G}$ if $\tau = \sigma$.

Definition 1.3.15 The fuzzy incidence graph $\widetilde{G} = (\sigma, \mu, \Psi)$ is a **cycle** if (Supp(σ), Supp(μ), Supp(Ψ)) is a cycle.

Definition 1.3.16 The fuzzy incidence graph $\widetilde{G} = (\sigma, \mu, \Psi)$ is a **fuzzy cycle** if (Supp(σ), Supp(μ), Supp(Ψ)) is a cycle and there exists no unique $xy \in$ Supp(μ) such that $\mu(xy) = \wedge\{\mu(uv)|uv \in \text{Supp}(\mu)\}$.

Definition 1.3.17 The fuzzy incidence graph $\widetilde{G} = (\sigma, \mu, \Psi)$ is a **fuzzy incidence cycle** if it is a fuzzy cycle and there exists no unique $(x, yz) \in$ Supp(Ψ) such that $\Psi(x, yz) = \wedge\{\Psi(u, vw)|(u, vw) \in \text{Supp}(\Psi)\}$.

Definition 1.3.18 The fuzzy incidence graph $\widetilde{G} = (\sigma, \mu, \Psi)$ is a **tree** if (Supp(σ), Supp(μ), Supp(Ψ)) is a tree and is a **forest** if (Supp(σ), Supp(μ), Supp(Ψ)) is a forest.

Definition 1.3.19 Let $\widetilde{G} = (\sigma, \mu, \Psi)$ be a fuzzy incidence graph. Then $\widetilde{G}$ is a **fuzzy tree** if it has a fuzzy incidence spanning subgraph $\widetilde{F} = (\sigma, \nu, \Omega)$ which is also a tree such that $\forall uv \in \text{Supp}(\mu)\backslash\text{Supp}(\nu)$, $\mu(uv) < \nu^{\infty}(uv)$. $\widetilde{G}$ is a **fuzzy forest** if it has a fuzzy incidence spanning subgraph $\widetilde{F} = (\tau, \nu, \Omega)$ which is also a forest such that $\forall uv \in \text{Supp}(\mu)\backslash\text{Supp}(\nu)$, $\mu(uv) < \nu^{\infty}(uv)$.

Definition 1.3.20 Let $\widetilde{G} = (\sigma, \mu, \Psi)$ be a fuzzy incidence graph. Then $\widetilde{G}$ is a **fuzzy incidence tree** if it has a fuzzy incidence spanning subgraph $\widetilde{F} = (\sigma, \nu, \Omega)$ which is a tree such that $\forall (u, vw) \in \text{Supp}(\Psi)\backslash\text{Supp}(\Omega)$, $\Psi(u, vw) < \Omega^{\infty}(u, vw)$. $\widetilde{G}$ is a **fuzzy incidence forest** if $\widetilde{G}$ has a fuzzy incidence spanning subgraph $\widetilde{F} = (\sigma, \nu, \Omega)$ which is a forest such that $\forall (u, vw) \in \text{Supp}(\Psi)\backslash\text{Supp}(\Omega)$, $\Psi(u, vw) < \Omega^{\infty}(u, vw)$.

Example 1.3.21 Consider the fuzzy graph $\widetilde{G} = (\sigma, \mu, \Psi)$ given in Fig. 1.10 with $V = \{x, y, z, w\}$. $\widetilde{G}$ is a fuzzy incidence tree as we can find a fuzzy incidence spanning subgraph that satisfies the requirements in the definition of a fuzzy incidence tree.

Theorem 1.3.22 If there is at most one path with the most incidence strength between any vertex and edge of the fuzzy incidence graph $\widetilde{G} = (\sigma, \mu, \Psi)$, then $\widetilde{G}$ is a fuzzy incidence forest.

Theorem 1.3.23 Let $\widetilde{G} = (\sigma, \mu, \Psi)$ be a cycle. Then $\widetilde{G}$ is a fuzzy incidence cycle if and only if $\widetilde{G}$ is not a fuzzy incidence tree.

Definition 1.3.24 Let $\widetilde{G} = (\sigma, \mu, \Psi)$ be a fuzzy incidence graph and let $xy \in E$. Then xy is called a **bridge** if there exists $uv \in E\backslash\{xy\}$ such that $\mu'^{\infty}(uv) < \mu^{\infty}(uv)$, where μ' is μ restricted to $E\backslash\{xy\}$.

Definition 1.3.25 Let $\widetilde{G} = (\sigma, \mu, \Psi)$ be a fuzzy incidence graph. Let $w \in V$ and E' be the set difference of E and the set of edges with w as an end vertex. Then w is called a **cutvertex** if $\mu'^{\infty}(uv) < \mu^{\infty}(uv)$ for some $uv \in E'$ such that $u \neq w \neq v$, where μ' is μ restricted to E'.

Definition 1.3.26 Let $\widetilde{G} = (\sigma, \mu, \Psi)$ be a fuzzy incidence graph. Let $w \in V$ and E' be the set difference of E and the set of edges with w as an end vertex. Then w is called an **incidence cutvertex** if $\Psi'^{\infty}(u, uv) < \Psi^{\infty}(u, uv)$ for some $(u, uv) \in V \times E'$ such that $u \neq w \neq v$, where Ψ' is Ψ restricted to $V \times E'$.

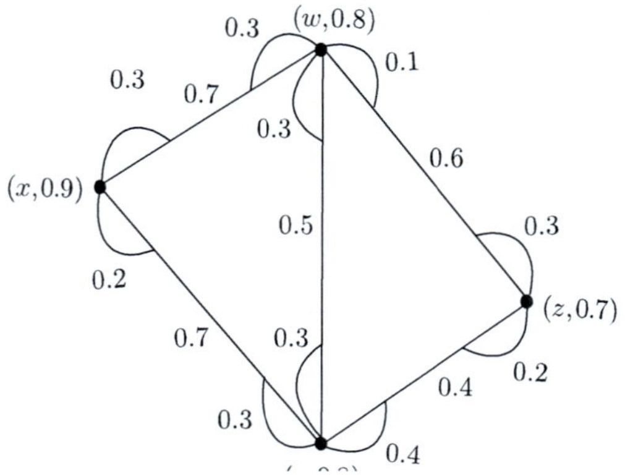

Fig. 1.10 Example of a fuzzy incidence tree

Definition 1.3.27 Let $\widetilde{G} = (\sigma, \mu, \Psi)$ be a fuzzy incidence graph. Then (x, xy) is called an **incidence cutpair** if $\Psi'^{\infty}(u, uv) < \Psi^{\infty}(u, uv)$ for some pair (u, uv) in $\widetilde{G}$, where Ψ' is Ψ restricted to $(V \times E)\backslash\{(x, xy)\}$.

Definition 1.3.28 Let $\widetilde{G} = (\sigma, \mu, \Psi)$ be the fuzzy incidence graph. Then $\widetilde{G}$ is said to **fuzzy incidence complete** if for all $(u, vw) \in V \times E$, $\Psi(u, vw) = \sigma(u) \wedge \mu(vw)$.

Theorem 1.3.29 If $\widetilde{G}$ is a fuzzy incidence forest, then the vertex edge pairs of $\widetilde{F}$ (as in the definition of fuzzy incidence forest) are exactly the incidence cutpairs of $\widetilde{G}$.

Theorem 1.3.30 Let $\widetilde{G}=(\sigma, \mu, \Psi)$ be a fuzzy incidence graph and $\widetilde{G}^* = (\text{Supp}(\sigma), \text{Supp}(\mu), \text{Supp}(\Psi))$ a cycle. Then an edge is a bridge if and only if it is an edge common to two incidence cutpairs.

Theorem 1.3.31 Let $\widetilde{G} = (\sigma, \mu, \Psi)$ be a fuzzy incidence graph and $(u, uv) \in V \times E$. If (u, uv) is an incidence cutpair, then $\Psi(u, uv) = \Psi^{\infty}(u, uv)$.

References

1. Zadeh, L.A.: Fuzzy sets. Inf. Control **8**, 338–353 (1965)
2. Mathew, S., Mordeson, J. N., Malik, D S.: *Fuzzy Graph Theory*, vol. 363, Springer International Publishing (2018)
3. Kaufmann, A.: Introduction to the Theory of Fuzzy Sets. Academic Press Inc., Orlando, Florida (1973)
4. Rosenfeld, A.: Fuzzy graphs. In: Zadeh, L.A., Fu, K.S., Shimura, M. (eds.) Fuzzy Sets and Their Applications, pp. 77–95. Academic Press, New York (1975)
5. Sunitha, M.S., Vijayakumar, A.: A characterization of fuzzy trees. Inf. Sci. **113**, 293–300 (1999)
6. Bhutani, K.R.: On automorphisms of fuzzy graphs. Pattern Recogn. Lett. **9**, 159–162 (1989)
7. Mordeson, J.N., Nair, P.S.: Cycles and cocycles of fuzzy graphs. Inf. Sci. **90**, 39–49 (1996)
8. Mathew, S., Sunitha, M.S.: Types of arcs in a fuzzy graph. Inf. Sci. **179**, 1760–1768 (2009)
9. Mathew, S., Sunitha, M.S.: Node connectivity and arc connectivity in fuzzy graphs. Inf. Sci. **180**, 519–531 (2010)
10. Ali, S., Mathew, S., Mordeson, J.N., Rashmanlou, H.: Vertex connectivity of fuzzy graphs with applications to human trafficking. New Math. Nat. Comput. **14**, 457–485 (2018)
11. Sebastian, A., Mathew, S., Mordeson, J.N.: Generalized fuzzy graph connectivity parameters with application to human trafficking. Mathematics **8**(3), 424–444 (2020)
12. Mathew, S., Sunitha, M.S.: Menger's theorem for fuzzy graphs. Inf. Sci. **222**, 717–726 (2013)
13. Mathew, S., Sunitha, M.S.: Cycle connectivity in fuzzy graphs. J. Intell. Fuzzy Syst. **24**, 549–554 (2013)
14. Jicy, N., Mathew, S.: Connectivity analysis of cyclically balanced fuzzy graphs. Fuzzy Inf. Eng. **7**, 245–255 (2015)
15. Mathew, S., Sunitha, M.S.: Strongest strong cycles and $\theta-$ fuzzy graphs. IEEE Trans. Fuzzy Syst. **21**, 1096–1103 (2013)
16. Mordeson, J.N., Peng, C.S.: Operations on fuzzy graphs. Inf. Sci. **79**, 159–170 (1994)
17. Binu, M., Mathew, S., Mordeson, J.N.: Connectivity index of a fuzzy graph and its application to human trafficking. Fuzzy Sets Syst., 117–136 (2018)
18. Dinesh, T.: A study on graph structures, Incidence Algebras and their Fuzzy Analogues, Ph.D. Thesis, Kannur University, Kerala, India (2012)

19. Mathew, S., Mordeson, J.N.: Connectivity concepts in fuzzy incidence graphs. Inf. Sci. **382**, 326–333 (2017)
20. Mathew, S., Mordeson, J.N.: Fuzzy endnodes in fuzzy incidence graphs. New Math. Natural Comput. **13**(1), 13–20 (2017)
21. Mathew, S., Mordeson, J.N.: Fuzzy incidence blocks and their application in illegal migration problems. New Math. Natural Comput. **13**(3), 245–260 (2017)
22. Malik, D.S., Mathew, S., Mordeson, J.N.: Fuzzy incidence graphs: applications to human trafficking. Inf. Sci. **447**, 244–259 (2018)
23. Binu, M., Mathew, S., Mordeson, J.N.: Wiener index of a fuzzy graph and application to illegal immigration networks. Fuzzy Sets Syst., 132–147 (2019)

Chapter 2
Nonstandard Analysis

The purpose of this chapter is to lay the foundation for a new area of research in fuzzy mathematics. This new area is based on nonstandard analysis. We begin the chapter with a discussion of first order logic and a proof of the transfer principle. In 1960 Abraham Robinson developed a nonstandard analysis by rigorously extending the real numbers $\mathbb{R}$ to a field $\mathbb{R}^*$ which includes infinitesimal numbers and finite numbers, [1]. Our approach replaces the interval $[0, 1]$ with an extension of it to $\mathbb{R}^*$. There are two possible extensions. One is replacing $[0, 1]$ with its natural extension $[0, 1]^*$ or with $]^-0, 1^+[$. We apply our extension to an application of nonstandard analysis to the theory of relativity by extending Herrmann's application, [2], to fuzzy nonstandard analysis. We first provide some basic concepts, definitions, and results from nonstandard analysis. We extend the notion of a fuzzy number to that of a nonstandard fuzzy number. We review some results concerning continuity and differentiability of functions that are pertinent to nonstandard analysis. We apply these results to an application of the theory of relativity. Many results involving $\mathbb{R}^*$ follow immediately from the transfer principle. However, we provide many proofs since we feel this will help the reader understand our extension.

We let $\mathbb{N}$ denote the positive integers and $\mathbb{R}$ the set of all real numbers. We let $\vee$ denote maximum or supremum and $\wedge$ denote minimum or infimum. If X is a set, $\mathcal{P}(X)$ denotes the power set of X. If X an Y are sets, we let $X \backslash Y$ denote set difference. If X is a universal set and $Y \subseteq X$, we sometimes write Y^c for $X \backslash Y$.

2.1 First Order Logic

The following is from [3]. The point of this section is to give the reader a feel for first order logic. Our goal is to aid the reader in understanding the Transfer Principle.

J. N. Mordeson et al., *Fuzzy Graph Theory*, Studies in Fuzziness and Soft Computing 424, https://doi.org/10.1007/978-3-031-23108-7_2

Definition 2.1.1 An **alphabet** of first order logic is a set containing the following elements:

(1) An infinite set of constants
(2) An infinite set of variables
(3) An infinite set of functions
(4) An infinite set of relations
(5) Logical connectives: $\wedge$, $\vee$, $-$, $\rightarrow$, $\leftrightarrow$
(6) Logical quantifiers: $\forall$, $\exists$
(7) The equality symbol: $=$
(8) The two parentheses; (and)

Every function and relation symbol is an n-placed function or relation symbol. The number is commonly referred as the arity of the function or symbol. For example, a function with arity 1 is called unitary, with arity 2 is called binary, etc. We will denote the arity of a function f as $ar(f)$.

Definition 2.1.2 A **term** is a string of symbols from the alphabet that is defined recursively as follows:

(1) Every constant is a term.
(2) Every variable is a term.
(3) If f is a function with arity n and $t_1, ..., t_n$ are terms, then $f(t_1, ..., t_n)$ is a term.
(4) A string of symbols is a term if it can be constructed applying the previous steps finitely many times.

Recall that all elements in (5) and (6) of Definition 2.1.1 can be derived from $\wedge$, $-$, and $\exists$.

Definition 2.1.3 A **formula** is a string of symbols from the alphabet that is defined recursively as follows:

(1) If t_1 and t_2 are terms, then $(t_1 = t_2)$ is a formula.
(2) If R is a relation with arity n and $t_1, ..., t_n$ are terms, then $R(t_1, ..., t_n)$ is a formula.
(3) If φ is a formula, the so is $-\varphi$.
(4) If φ and ψ are formulas, then so is $(\varphi \wedge \psi)$.
(5) If φ is a formula and x is a variable, then $(\exists x)\varphi$ is a formula.
(6) A string of symbols is a formula if it can be constructed by finitely many applications of the previous steps.

For example, $(\psi \vee \varphi)$ is a formula if and only if ψ and φ are formulas by using the fact that $\psi \vee \varphi = -(-(\psi) \wedge -(\varphi))$.

The formulas in (1) and (2) of Definition 2.1.3 are called **atomic** formulas. Note also that (1) can be understood as a binary relation $R = (t_1, t_2)$ is true if and only if $t_1 = t_2$.

One limitation of first order logic is that it does not allow quantification over relations, only variables. This is the key difference between first and higher order logic. While the definitions in this section will focus on developing first order logic, this distinction between first and higher logic statements will be the key in the development of nonstandard analysis.

We say that a variable is **free** if it does not appear next to a quantifier and **bound** otherwise.

Definition 2.1.4 Let φ be a formula. We define the set of free variables of φ, denoted $FV(\varphi)$, inductively as follows:

(1) If $\varphi = (t_1 = t_2)$, then $FV(\varphi) = \{x|x \text{ appears in } t_1 \text{ or } t_2\}$.

(2) If $\varphi = (R(t_1, ..., t_n))$ for some arity n, then $FV(\varphi) = \{x|x \text{ appears in } t_i \text{ for some } 1 \leq i \leq n\}$.

(3) If $\varphi = (-\psi)$, where ψ is a formula, then $FV(\varphi) = FV(\psi)$.

(4) If $\varphi = (\mu \wedge \nu)$, where μ and ν are formulas, then $FV(\varphi) = FV(\mu) \cup FV(\nu)$.

(5) If $\varphi = (\exists x)\psi$, where ψ is a formula, $FV(\varphi) = FV(\psi)\backslash\{x\}$.

Definition 2.1.5 A formula φ is called a **sentence** if it has no free variables, i.e., $FV(\varphi) = \emptyset$.

Definition 2.1.6 A **language** $\mathcal{L}$ is a set containing all logical symbols and quantifiers (including the equality sign and parenthesis) and some arbitrary number of constants, variables, function symbols and relation symbols.

All formulas made from any language $\mathcal{L}$ follow the previous rules.

Definition 2.1.7 Let A be a nonempty set and $V \subseteq \mathcal{L}$ be the set of all variables in $\mathcal{L}$. A **variable assignment** is a function $\beta : V \to A$, which assigns elements of A to all variables in V. Particularly, for some element $k \in A$, some variable $x \in V$ and some assignment β, there is a function $\beta[x, v]$ defined as follows:

$$\beta[x, k](y) = \begin{cases} k \text{ if } x = y, \\ \beta(y) \text{ if } x \neq y. \end{cases}$$

Definition 2.1.8 A **model** or **structure** M for some language $\mathcal{L}$ is an ordered triple $M = (A, I, \beta)$, where A is a nonempty set. β is a variable assignment and I is an interpretation function with domain the set of all constants, relations and function symbols in L such that the following conditions hold:

(1) For every constant symbol $c \in \mathcal{L}$, we have that $I(c) \in A$.

(2) For every function symbol $f \in \mathcal{L}$ with arity n, we have that $I(f) \in A^n \times A$; meaning $I(f)$ is a function of arity n defined on A.

(3) For every relation $R \in \mathcal{L}$ with arity n, we have that $I(R) \subseteq A^n$; meaning $I(R)$ is the set of all n-tuples that satisfy R under I.

A is frequently called the **universe** of M.

Definition 2.1.9 Let $\mathcal{L}$ be a language and $M = (A, I, \beta)$ a model for $\mathcal{L}$. Then the **interpretation** of any term t, denoted as $(t)^{t,\beta}$, of symbols in $\mathcal{L}$ is defined as follows:

(1) If $t = c$ for some constant c, then $(t)^{t,\beta} = I(c)$.

(2) If $t = x$ for some variable x, then $(t)^{t,\beta} = \beta(x)$.

(3) If $t = f(t_1, ..., t_n)$ for some function f of arity n, then

$$(t)^{t,\beta} = I(f)((t_1)^{I,\beta}, ..., (t_n)^{I,\beta}.$$

Definition 2.1.10 Let $\mathcal{L}$ be a language, $M = (A, I, \beta)$ a model for $\mathcal{L}$ and φ some formula in $\mathcal{L}$. Then we say that M **satisfies** φ and write $M \models \varphi$ or $(A, I, \beta) \models \varphi$ if the following conditions hold:

(1) If $\varphi = R(t_1, ..., t_n)$ for some relation $R \in L$ of arity n, meaning φ is atomic, then $(A, I, \beta) \models \varphi$ if $((t_1)^{I,\beta}, ..., (t_n)^{I,\beta}) \in I(R)$.

(2) If $\varphi = -\psi$ for some atomic formula ψ, then $(A, I, \beta) \models \varphi$ if (A, I, β) does not satisfy ψ.

(3) If $\varphi = (\mu \wedge \nu)$ for some atomic formulas μ and ν, then $(A, I, \beta) \models \varphi$ if $(A, I, \beta) \models \mu$ and $(A, I, \beta) \models \nu$.

(4) If $\varphi = (\exists x)\psi$ for some atomic formula ψ, then $(A, I, \beta) \models \varphi$, if there exists some $k \in A$ such that $(A, I, \beta[x, k]) \models \psi$,where x is a free variable on ψ.

2.2 Ultrafilters

Definition 2.2.1 A **filter** F on a set I is a set $F \subseteq \mathcal{P}(I)$ such that

(1) $I \in F$.

(2) If $X \in F$ and $X \subseteq Y$, then $Y \in F$ for all $X, Y \in \mathcal{P}(I)$.

(3) If $X, Y \in F$, then $X \cap Y \in F$ for all $X, Y \in \mathcal{P}(I)$.

It follows easily that $\{I\}$ and $\mathcal{P}(I)$ are filters on I.

Definition 2.2.2 Let $x \in A$. Then $F_x = \{Y \in \mathcal{P}(I) | x \in Y\}$ is called the **principal filter** of x over I.

Lemma 2.2.3 Let $x \in A$. Then the principal filter of x over I is a filter of I.

Proof Since $x \in I$, $I \in F_x$. Let $X \in F_x$ and $X \subseteq Y$. Then $x \in Y$ and so $Y \in F_x$. Let $X, Y \in F_x$. Then $x \in X$ and $x \in Y$. Thus $x \in X \cap Y$ and so $X \cap Y \in F_x$. □

Definition 2.2.4 A nontrivial and nonprincipal filter is called a **free filter**.

Theorem 2.2.5 A filter F over some I is free if and only if $\cap_{A \in F} A = \emptyset$.

Proof Assume that $\cap_{A \in F} A \neq \emptyset$. Then there exists $x \in I$ such that $x \in Y$ for all $Y \in F$. Since F is a filter, it is upwardly closed and so must contain every set containing x and hence $F = \{Y \in \mathcal{P}(I) | x \in Y\}$ Thus F is principal.

Conversely, suppose $\cap_{A \in F} A = \emptyset$. Suppose F is principal. Then there is some $x \in I$ such that $F = \{Y \in \mathcal{P}(I) | x \in Y\}$. Hence $\cap_{A \in F} A = \{x\}$, a contradiction. Thus F is nonprincipal. □

Since filters are closed under finite intersection, the previous result implies there are no free filters on finite sets.

Definition 2.2.6 The **Frechet filter** on a set I is defined as $F_I = \{X \subseteq I | X$ is cofinite$\}$.

Lemma 2.2.7 Let I be an infinite set. Then the Frechet filter over I is a free filter.

Proof Assume that the Frechet Filter over I is principal. Then by Theorem 2.5, $\cap_{A \in F} A \neq \emptyset$. Let $k \in \cap_{A \in F} A$. Let $X \in F$. Then $x \in X$. Since X is cofinite, $X \backslash \{k\}$ is cofinite. Thus $X \backslash \{k\} \in F$. Thus $k \notin \cap_{A \in F} A$, a contradiction. □

Lemma 2.2.8 Every free filter contains the Frechet Filter.

Proof Let F be a free filter over some infinite set I. Let F_I denote the Frechet Filter over I. Fix some $Y \in F_I$. Then it follows that $I \backslash Y$ is a finite set. Since F is a free filter, for every $x \in I \backslash Y$ there exists some set $K_x \in F$ such that $x \notin K_x$. Since F is closed under finite intersection, it follows that $\cap_{x \in I \backslash Y} K_x \in F$. and $\cap_{x \in I \backslash Y} K_x \subseteq Y$. Thus $Y \in F$ since F is upwardly closed. Thus $F_I \subseteq F$. □

Definition 2.2.9 A filter $\mathcal{U}$ on a set I is an **ultrafilter** of I if for all $X \subseteq I$ either $X \in \mathcal{U}$ or $I \backslash X \in \mathcal{U}$, but not both.

Theorem 2.2.10 Let $x \in A$. Then the principal filter of x over I is an ultrafilter of I.

Proof Let F_x be the principal filter of x over I , Then F_x is a filter. Let Y be a subset of I. If $x \in Y$, then $x \notin I \backslash Y$ so $Y \in F_x$ and $Y \notin F_x$. If $x \notin Y$, then $x \in I \backslash Y$ and so $Y \notin F_x$ and $I \backslash Y \in Fx$. □

Definition 2.2.11 A set $G \subseteq \mathcal{P}(I)$ has the **finite intersection property** (FIP) if the intersection if any finite number of elements of G is nonempty.

Note that every filter has the finite intersection property.

Theorem 2.2.12 Every $S \subseteq \mathcal{P}(I)$ with the FIP has a proper filter containing it.

Proof Let $\mathcal{S} = \{F \subseteq \mathcal{P}(I) | S \subseteq F$ and F is a proper filter on $I\}$ and $\mathcal{F} = \cap_{F \in \mathcal{S}} F$. Note that $\emptyset \notin \mathcal{F}$ and $I \in \mathcal{F}$ since $\mathcal{F}$ is the intersection of proper filters and S has the FIP. Let $X \in \mathcal{F}$ and $X \subseteq Y$. It follows that X is an element of every filter and therefore so is Y. Hence $Y \in \mathcal{F}$. Let $X, Y \in \mathcal{F}$. Then $X, Y \in F$ for every $F \in \mathcal{S}$. Hence $X \cap Y \in F$ for every $F \in \mathcal{S}$ and so $X \cap Y \in \mathcal{F}$. Thus $\mathcal{F}$ is a filter on I and $S \subseteq \mathcal{F}$. □

The filter $\mathcal{F}$ in the previous result is called the **filter generated** by S.

Theorem 2.2.13 Every proper filter $E \subseteq P(I)$ is contained in some ultrafilter.

Proof Let $\mathcal{F} = \{F \subseteq \mathcal{P}(I) | F \supseteq E$ and F is a proper filter on $I\}$. Note that $\mathcal{F}$ is a partially ordered set with $\subseteq$ as a partial order relation. Let $\mathcal{C} \subseteq \mathcal{F}$ be an arbitrary chain (totally ordered subset) of $\mathcal{F}$. To show $\mathcal{C}$ is bounded we consider $\cup_{C \in \mathcal{C}} C$. It suffices to show $\cup_{C \in \mathcal{C}} C \in \mathcal{F}$. Then $\mathcal{F}$ has a maximum element by Zorn's Lemma.

Since every element in $\mathcal{C}$ is a subset of $\mathcal{P}(I)$, it follows that $\cup_{C \in \mathcal{C}} C \subseteq \mathcal{P}(I)$. Let $X \in \cup_{C \in \mathcal{C}} C$ and let $X \subseteq Y \subseteq \mathcal{P}(I)$. Since $X \in \cup_{C \in \mathcal{C}} C$, there exists some filter F such that $X \in F$. Therefore, $Y \in F$. Thus $Y \in \cup_{C \in \mathcal{C}} C$. Let $X, Y \in \cup_{C \in \mathcal{C}} C$. Then there are filters F, F' such that $X \in F$ and $Y \in F'$. Since C is a chain, we can assume without loss of generality that $F \subseteq F'$. Thus $X, Y \in F'$ and so $X \cap Y \in F'$. Hence $X \cap Y \in \cup_{C \in \mathcal{C}} C$. Thus $\cup_{C \in \mathcal{C}} C$ is a filter and so $\cup_{C \in \mathcal{C}} C \in \mathcal{F}$. Thus by Zorn's Lemma, $\mathcal{F}$ has a maximal element, say $\mathcal{U}$.

We next show that U is an ultrafilter. Clearly, $\emptyset \notin \mathcal{U}$ and so $\mathcal{U}$ cannot contain both a subset of $\mathcal{P}(I)$ and its complement. Assume that U is not an ultrafilter. Then there exists some $A \subseteq I$ such that $A \notin \mathcal{U}$ and $I \backslash A \notin \mathcal{U}$. It follows that $\mathcal{U} \cup \{A\}$ has the finite intersection property since $\mathcal{U}$ has the finite intersection property and $I \backslash A \notin \mathcal{U}$. Hence no subset of $I \backslash A$ is an element of $\mathcal{U}$. Let S be the filtered generated by $\mathcal{U} \cup \{A\}$. Clearly, $\mathcal{U} \subseteq S$ which implies that $\mathcal{U}$ is not maximal in $\mathcal{F}$, a contradiction. □

Corollary 2.2.14 There exists a free filter on $\mathbb{N}$.

Proof Let $\mathcal{F}_{\mathbb{N}}$ be the Frechet filter on $\mathbb{N}$. Since $\mathbb{N}$ is infinite such a filter exists. By Theorem 2.2.13, there exists some ultrafilter of $\mathbb{N}$, say $\mathcal{U}$, containing $\mathcal{F}_{\mathbb{N}}$. Thus $\cap_{A \in \mathcal{U}} A \subseteq \cap_{A \in \mathcal{F}_{\mathbb{N}}} A$. Therefore, $\cap_{A \in \mathcal{U}} A = \emptyset$. Hence $\mathcal{U}$ is free. □

In the remainder of this presentation, we assume I is an infinite set with some free ultrafilter U and $\{A_i\}_{i\in I}$ is a collection of nonempty sets.

Let $\mathcal{U}$ be a free ultrafilter on some indexing set I and let $\{A_i\}_{i\in I}$ be a collection of nonempty subsets. Then the **arbitrary product** of the collection is defined as

$$\prod_{i\in I} A_i = \{f \mid f \text{ has domain } I \text{ and } f(i) \in A_i \text{ for all } i \in I\}.$$

Definition 2.2.15 Two functions $f, g \in \prod_{i\in I} A_i$ are said to be **equivalent modulo** $\mathcal{U}$ if $\{i \in I \mid f(i) = g(i)\} \in \mathcal{U}$. We write $f =_{\mathcal{U}} g$ to indicate this relationship.

Lemma 2.2.16 $=_{\mathcal{U}}$ is an equivalence relation on $\prod_{i\in I} A_i$.

Proof Let $f \in \prod_{i\in I} A_i$. Then $\{i \in I \mid f(i) = f(i)\} = I \in \mathcal{U}$. Thus $=_{\mathcal{U}}$ is reflexive. Since $f(i) = g(i)$ if and only if $g(i) = f(i)$ it follows that $=_{\mathcal{U}}$ is symmetric. Suppose $f =_{\mathcal{U}} g$ and $g =_{\mathcal{U}} h$. Then $K = \{i \in I \mid f(i) = g(i)\} \in \mathcal{U}$ and $H = \{i \in I \mid g(i) = h(i)\} \in \mathcal{U}$. Since U is closed under finite intersection, $K \cap H \in U$. Thus $f =_U h$. □

Definition 2.2.17 We let $[f]_{\mathcal{U}}$ denote the equivalence class of $f \in \prod_{i\in I} A_i$ for the equivalence relation $=_{\mathcal{U}}$.

Definition 2.2.18 The ultraproduct of $\{A_i\}_{i\in I}$ modulo $\mathcal{U}$ is

$$(\prod_{i\in I} A_i)/\mathcal{U} = \{[f]_U \mid f \in \prod_{i\in I} A_i\}.$$

Definition 2.2.19 Let $\mathcal{U}$ be a free filter of $\mathbb{N}$. Then the set of hyperreal numbers $\mathbb{R}^*$ is the ultrapower of $\mathbb{R}$ modulo $\mathcal{U}$. That is,

$$\mathbb{R}^* = (\prod_{n\in\mathbb{N}} \mathbb{R})/\mathcal{U}.$$

Note that in Definition 2.2.19, $I = \mathbb{N}$ and $A_n = \mathbb{R}$ for all $n \in \mathbb{N}$.

2.3 Structure of Ultraproducts

In this section, we prove Los's Theorem which leads to the Transfer Principle.

Definition 2.3.1 Let $\mathcal{L}$ be a language. Then a **theory** of $\mathcal{L}$ is a set of sentences of $\mathcal{L}$.

Definition 2.3.2 Let $\mathcal{L}$ be a language, M a model for $\mathcal{L}$, and T a theory. We say that M satisfies T, written $M \models T$ if $M \models \varphi$ for all $\varphi \in T$.

Definition 2.3.3 Let $\mathcal{L}$ be a language and M a model for $\mathcal{L}$. The theory of M, written $Th(M)$, is the set of all sentences of $\mathcal{L}$ such that $M \models \varphi$.

Definition 2.3.4 Let I be an index set with some ultrafilter $\mathcal{U}$ on I. Let $M_i = (A_i, I_i, \beta_i)$ be a model for some language $\mathcal{L}$ for all $i \in I$. Then the ultraproduct $M^* = ((\prod_{i \in I} A_i)/\mathcal{U}, I^*, \beta^*)$ is a model of $\mathcal{L}$ with an interpretation function I^* and variable assignment function β^* defined as follows:

(1) If x is a variable in $\mathcal{L}$, then $\beta^*(x) = [(\beta_i(x))]_{\mathcal{U}}$.

(2) If c is a constant in $\mathcal{L}$, then $I^*(c) = [I_i(c)]_{\mathcal{U}}$.

(3) If f is a function symbol of arity n, then

$$I^*(f)([g_1]_{\mathcal{U}}, ..., [g_n]_{\mathcal{U}}) = [(I_i(f)(g_1(i), ..., g_n(i))]_{\mathcal{U}}.$$

(4) If R is a relation symbol of arity n, then $([g_1]_{\mathcal{U}}, ..., [g_n]_{\mathcal{U}}) \in I^*(R)$ if and only if $\{i \in I | (g_1(i), ..., g_n(i))\} \in I_i(R)\} \in \mathcal{U}$.

Proposition 2.3.5 The definitions in Definition 2.3.4 do not depend on the choices of $[g_i]_U$.

Proof Let $g_1, ..., g_n, g_1', ..., g_n' \in \prod_{i \in I} A_i$ be such that $g_1 =_{\mathcal{U}} g_1', ..., g_n =_{\mathcal{U}} g_n'$ and $([g_1]_{\mathcal{U}}, ..., [g_n]_{\mathcal{U}}) \in I_i(R)$ for some relation symbol R of arity n. Now $S = \{i \in I | g_i(i) = g_1'(i), ..., g_n(i) = g_n'(i)\} \in \mathcal{U}$ since $\mathcal{U}$ is closed under finite intersection. Thus

$$(\forall i \in S)((g_1(i), ..., g_n(i)) \in I_i(R) \Leftrightarrow (g_1'(i), ..., g_n'(i)) \in I_i(R).$$

Hence $\{i \in I | (g_1'(i), ..., g_n'(i)) \in I_i(R)\} \in \mathcal{U}$ since it is a superset of S. Thus $([g_1']_{\mathcal{U}}, ..., [g_n']_{\mathcal{U}}) \in I^*(R)$. Hence R is well-defined.

In the case of functions, we get a similar result. Let $g_1, ..., g_n, g_1', ..., g_n' \in \prod_{i \in I} A_i$ be such that $g_1 =_{\mathcal{U}} g_1', ..., g_n =_{\mathcal{U}} g_n'$ and $([g_1]_{\mathcal{U}}, ..., [g_n]_{\mathcal{U}}) \in I_i(R)$ for some function symbol f of arity n. Define S as above and so $S \in \mathcal{U}$. Then

$$(\forall i \in S)(I_i(f)(g_1(i), ...g_n(i)) = I_i(f)(g_1'(i), ...g_n'(i))$$

since f is a function and all the inputs are the same. Therefore, the sequence $(I_i(f)(g_1(i), ...g_n(i))$ modulo $\mathcal{U}$ is equivalent to the sequence $(I_i(f)(g_1(i), ...g_n(i))$ and so both belong to the same equivalence class. Thus f is well-defined. □

Theorem 2.3.6 (Los's Theorem) Let $\mathcal{L}$ be a language, I be a set with some ultrafilter $\mathcal{U}$ on I, and $M_i = (A_i, I_i, \beta_i)$ be a model for $\mathcal{L}$ for all $i \in I$. Then for all $\varphi \in \mathcal{L}$, $M^* = ((\prod_{i \in I} A_i)/\mathcal{U}, I^*, \beta^*) \models \varphi$ if and only if $\{i \in I | M_i \models \varphi\} \in \mathcal{U}$.

Proof (1) If φ is an atomic formula, the result holds by the previous definition.

(2) Let $\varphi = (\mu \wedge \nu)$, where μ and ν are atomic formulas. Suppose $M^* \models \mu \wedge \nu$. Then $M^* \models \mu$ and $M^* \models \nu$ by definition. Therefore, since μ and ν are atomic formulas, it follows that $\{i \in I | M_i \models \mu\} \in \mathcal{U}$ and $\{i \in I | M_i \models \nu\} \in \mathcal{U}$. Since $\mathcal{U}$ is closed under finite intersection, we have that $\{i \in I | M_i \models (\mu \wedge \nu)\} = \{i \in I | M_i \models \mu\} \cap \{i \in I | M_i \models \nu\} \in \mathcal{U}$.

Conversely, suppose that $\{i \in I | M_i \models (\mu \wedge \nu)\} \in \mathcal{U}$. Now $\{i \in I | M_i \models (\mu \wedge \nu)\} \subseteq \{i \in I | M_i \models \mu\}$ and $\{i \in I | M_i \models (\mu \wedge \nu)\} \subseteq \{i \in I | M_i \models \nu)\}$. Since $\mathcal{U}$ is upwardly closed, it follows that $\{i \in I | M_i \models \mu\} \in \mathcal{U}$ and $\{i \in I | M_i \models \nu\} \in \mathcal{U}$. Thus $M^* \models \mu$ and $M^* \models \nu$. Hence $M^* \models \mu \wedge \nu$.

(3) Let $\varphi = (-\psi)$, where ψ is an atomic formula. Suppose $M^* \models \varphi$. Then M^* does not model ψ. Thus $\{i \in I | M_i \models \psi\} \notin \mathcal{U}$ since ψ is atomic. Hence $\{i \in I | M_i$ does not model $\psi\} \in U$ since U is an ultrafilter. However, $\{i \in I | M_i$ does not model $\psi\} = \{i \in I | M_i \models \varphi\} \in \mathcal{U}$. Note that all steps in the proof are reversible. Hence the biconditional holds.

(4) Suppose that $\varphi = (\exists x)\psi$, where ψ is an atomic formula and x is a free variable. Suppose $M^* \models (\exists x)\psi$. Then there is a $[g]_U \in (\prod_{i \in I} A_i)/\mathcal{U}$ such that $(\prod_{i \in I} A_i)/\mathcal{U}, I^*, \beta^*[x, [g]_\mathcal{U}]) \models \psi$. Since ψ is atomic, $\{i \in I | A_i, I_i, \beta_i[x, g(i)]) \models \psi\} \in \mathcal{U}$. By definition, it follows that $\{i \in I | A_i, I_i, \beta_i[x, g(i)]) \models \psi\} = \{i \in I | M_i \models (\exists x)\psi\} \in \mathcal{U}$.

Conversely, suppose $\{i \in I | M_i \models (\exists x)\psi\} \in \mathcal{U}$. Define a function $g : I \to \cup_{i \in I} A_i$ such that for all $i \in \{i \in I | M_i \models (\exists x)\psi\}$, $g(i)$ is such that $(A_i, I_i, \beta_i[x, g(i)]) \models \psi$ and $g(i) \in A_i$ otherwise. Since such $g(i)$ exist by assumption, this step also requires the axiom of choice. Furthermore, it is clear that $[g]_U \in (\prod_{i \in I} A_i)/\mathcal{U}$ by the definition of ultra product. Therefore, $(\prod_{i \in I} A_i)/\mathcal{U}, I^*, \beta^*[x, [g]_U]) \models \psi$ by the definition of g. Hence $M^* \models (\exists x)\psi$. □

All formulas of $\mathcal{L}$ are obtained by finite application of the steps above.

Corollary 2.3.7 (Transfer Principle) Let $\mathcal{L}$ be a language, I be a set with some ultrafilter $\mathcal{U}$ on I, and $M = (A, I, \beta)$ be a model for $\mathcal{L}$. Then for all φ of $\mathcal{L}$, $M^* = ((\prod_{i \in I} A)/\mathcal{U}, I^*, \beta^*) \models \varphi$ if and only if $M = (A, I, \beta) \models \varphi$. In other words, $M^* \models Th(M)$.

Proof The result follows immediately from Los's theorem. If $M^* \models \varphi$, then $\{i \in I | M_i \models \varphi\} \in \mathcal{U}$. Since $A_i = A$ it follows that $\{i \in I | M_i \models \varphi\} = I$ since otherwise it would equal the empty set. Therefore, $M \models \varphi$. Conversely, if $M \models \varphi$, then $\{i \in I | M_i \models \varphi\} = I \in \mathcal{U}$ and so $M^* \models \varphi$ by Los's theorem. □

Recall from Definition 2.2.19 that the set of hyperreal numbers is $\mathbb{R}^*$ is $(\prod_{n \in \mathbb{N}} \mathbb{R})/\mathcal{U}$, i.e., the set of all equivalence classes under modulo U equivalence. From Corollary 2.3.7 that $\mathbb{R}^*$ satisfies the same first order theory that $\mathbb{R}$ does. Hence if we define addition and multiplication of $\mathbb{R}^*$ as in Definition 2.3.4, $\mathbb{R}^*$ is a field. All field axioms can be expressed as first order logic statements.

Let $[(x_n)]_{\mathcal{U}}$ and $[(y_n)]_{\mathcal{U}}$ be elements of $\mathbb{R}^*$. Then $[(x_n)]_{\mathcal{U}} + [(y_n)]_{\mathcal{U}} = [(x_n + y_n)]_{\mathcal{U}}$ and $[(x_n)]_{\mathcal{U}} \cdot [(y_n)]_{\mathcal{U}} = [(x_n \cdot y_n)]_{\mathcal{U}}$ We define $\leq_U$ on $\mathbb{R}^*$ by $[(x_n)]_{\mathcal{U}} \leq_U [(y_n)]_{\mathcal{U}}$ if and only if $\{n \in N | x_n \leq y_n\} \in \mathcal{U}$. Since $\mathbb{R}$ is totally ordered and $\mathcal{U}$ is an ultrafilter, it follows that $\mathbb{R}^*$ is totally ordered by $\leq_{\mathcal{U}}$.

Theorem 2.3.8 Define $i : \mathbb{R} \to \mathbb{R}^*$ by for all $r \in \mathbb{R}^*$, $i(r) = [(r, r, r, ...)]_{\mathcal{U}}$. Then i is a one-to-one function of $\mathbb{R}$ into $\mathbb{R}^*$ which preserves addition, multiplication, and ordering.

Proof Clearly, i is single valued. Suppose that $i(r) = i(r')$. Then $[(r, r, r, ...)]_{\mathcal{U}} = [(r', r', r', ...)]_{\mathcal{U}}$. Then $[(0, 0, 0, , ,)]_{\mathcal{U}} = [(r, r, r, ...)]_{\mathcal{U}} - [(r', r', r', ...)]_{\mathcal{U}} = [(r, r, r, ...) - (r', r', r', ...)]_{\mathcal{U}} = [(r - r', r - r', r - r', ...)]_{\mathcal{U}}$. Hence $(0, 0, 0, ...) =_{\mathcal{U}} (r - r', r - r', r - r', ...)$. Thus $\{n \in \mathbb{N} | r - r' = 0\} \in \mathcal{U}$. Thus $r = r'$. Hence i is one-to-one. Now $i(r + s) = [(r + s, r + s, ...)]_{\mathcal{U}} = [(r, r, ...)]_{\mathcal{U}} + [(s, s, ...)]_{\mathcal{U}} = i(r) + i(s)$. Thus i preserves addition. Similarly i preserves multiplication. Suppose that $r \leq s$. Then $[(r, r, ...)]_{\mathcal{U}} \leq_{\mathcal{U}} [(s, s, ...)]_{\mathcal{U}}$ since $\{n \in \mathbb{N} | r \leq s\} = \mathbb{N} \in \mathcal{U}$. □

We define the absolute value of an element of $\mathbb{R}^*$ as follows: Let $[(r_n)]_{\mathcal{U}} \in \mathbb{R}^*$. Define $|[(r_n)]_{\mathcal{U}}|$ to be $[(|r_n|)_{\mathcal{U}}$. We have that

$$\begin{aligned}
[(a_n)]_{\mathcal{U}} &= {}_{\mathcal{U}}[(b_n)]_{\mathcal{U}} \Leftrightarrow \{n \in \mathbb{N} | a_n = b_n\} \in \mathcal{U}, \\
&\Rightarrow \{n \in \mathbb{N} |\ |a_n| = |b_n|\} \in \mathcal{U} \\
&\Leftrightarrow [(|a_n|)]_{\mathcal{U}} = [(|b_n|)]_{\mathcal{U}} \\
&\Leftrightarrow |[(a_n)]_{\mathcal{U}}| =_{\mathcal{U}} |[(b_n)]_{\mathcal{U}}|
\end{aligned}$$

where the implication hold since $\{n \in \mathbb{N} | a_n = b_n\} \subseteq \{n \in \mathbb{N} |\ |a_n| = |b_n|\}$. Thus the absolute value on $\mathbb{R}^*$ is single-valued.

Now

$$\begin{aligned}
|[(a_n)]_U + [(b_n)]_U| &= |[(a_n + b_n)]_U| \\
&= [(|a_n + b_n|)]_U \leq [(|a_n| + |b_n|)]_U \\
&= [(|a_n|)]_U + [(|b_n|)]_U \\
&= |[(a_n)]_U| + |[(b_n)]_U|.
\end{aligned}$$

Thus the triangle property holds.

Theorem 2.3.9 (1) There exists a hyperreal number ω such that $|\omega| > i(r)$ for all $r \in \mathbb{R}$.

(2) There exists a hyperreal number ε such that $0 < |\varepsilon| < i(r)$ for all $r \in \mathbb{R}$, $r > 0$.

Proof (1) Let $\omega = [(n)]_{\mathcal{U}}$. Let $r \in \mathbb{R}$. Then $r \leq m$ for some $m \in \mathbb{N}$. Let $i(m) = [(m)]_{\mathcal{U}}$. Then $\{n \in \mathbb{N} | n \leq m\}$ is finite. Thus $\{n \in \mathbb{N} | n > m\} \in \mathcal{U}$ since $\mathcal{U}$ contains all cofinite subsets of $\mathbb{N}$. Thus $\omega > i(m) \geq i(r)$.

(2) Let $r \in \mathbb{R}$, $r > 0$. Let $\varepsilon = [(1, \frac{1}{2}, \frac{1}{3}, ...]_U$. Then $\varepsilon = \omega^{-1} < i(r)^{-1}$. However, this holds for all $r \in \mathbb{R}$ such that $r > 0$. Thus $\varepsilon < i(r)$ for all positive real numbers r. □

Definition 2.3.10 A hyperreal number ω is said to be **unlimited** if $|\omega| > i(r)$ for all $r \in \mathbb{R}^+$., where $\mathbb{R}^+ = \{r \in \mathbb{R} | r > 0\}$.

Definition 2.3.11 A hyperreal number ε is said to be **infinitesimal** if $|\varepsilon| < i(r)$ for all $r \in \mathbb{R}^+$.

Definition 2.3.12 A hypereal number η is said to be **finite** if there exist $r, s \in \mathbb{R}^+$ such that $i(r) \leq |\eta| \leq i(s)$.

Definition 2.3.13 A hypereal number x is said to be **limited** if it is not limited.

Theorem 2.3.14 $i(\mathbb{R})$ is a nonempty bounded set in $\mathbb{R}^*$ without supremum.

Proof Clearly, $i(\mathbb{R})$ is bounded by above by any unlimited hyperreal. Assume on the contrary that $i(\mathbb{R})$ has a supremum, say ω. The ω must be unlimited since otherwise it wouldn't be an upper bound. However $\omega - i(1)$ is unlimited and hence an upper bound of $i(\mathbb{R})$. However, $\omega - i(1) < \omega$. Thus ω is not a supremum, a contradiction. □

2.4 Hyperreals

The following three sections are from [4]. See also [5, 6]. We recall the following definition from sect. 2.2.

Definition 2.4.1 (Free Ultrafilter) A **filter** $\mathcal{U}$ on a set J is a subset of $\mathcal{P}(J)$ satisfying properties (1)–(3). A filter $\mathcal{U}$ is called an **ultrafilter** if it satisfies (4) and an ultrafilter is called **free** if it satisfies (5).

(1) Proper: $\emptyset \notin \mathcal{U}$,

(2) Finite intefsection property: If $A, B \in \mathcal{U}$, then $A \cap B \in \mathcal{U}$.

(3) Superset property: If $A \in \mathcal{U}$ and $A \subseteq B \subseteq J$, then $B \in \mathcal{U}$,

(4) Maximality: For all $A \subseteq J$, either $A \in \mathcal{U}$ or $J \backslash A \in \mathcal{U}$,

(5) Freeness: $\mathcal{U}$ contains no finite subsets.

It is important to note that by (1), (2) and (4), if $A \subseteq J$, then either $A \in \mathcal{U}$ or $J \backslash A \in \mathcal{U}$, but not both.

Lemma 2.4.2 Let U be an ultrafilter on $\mathbb{N}$ and let $\{A_1, ..., A_n\}$ be a finite collection of disjoint subsets of $\mathbb{N}$ such that $\cup_{j=1}^{n} A_i = \mathbb{N}$. Then $A_i \in \mathcal{U}$ for exactly one $i \in \{1, ..., n\}$.

Proof Suppose that $\mathcal{U}$ contains no A_i. Then by (4), $\mathcal{U}$ contains A_i^c for each i. Thus by (2), contains $\cap_{i=1}^{n} A_i^c = (\cup_{i=1}^{n} A_i)^c = \mathbb{N}^c = \emptyset$, contrary to (1). Thus $\mathcal{U}$ contains some A_i. Suppose that $\mathcal{U}$ contains A_i and A_j, $i \neq j$. Then by (2), $\mathcal{U}$ contains $A_i \cap A_j = \emptyset$, contrary to (1). □

Lemma 2.4.3 (Ultrafilter Lemma) Let A be a set and $F_0 \subseteq \mathcal{P}(A)$ be a filter on A. Then F_0 can be extended to an ultrafilter $\mathcal{F}$ on A.

Proof Consider the set Φ of all filters on A which contain F_0. Consider any chain $\mathcal{C}$ of filters in Φ. Let $G = \cup_{F_n \in \mathcal{C}} F_n$. We show G is a filter in Φ. Clearly, $\emptyset \notin F_n$ for $F_n \in \mathcal{C}$ and so $\emptyset \notin G$. Let $C \in G$. Then $C \in F_n$ for some n. Hence for any $D \in G$ such that $C \subseteq D$, we have $D \in F_m$ for some $F_m \supseteq F_n$. Thus $D \in F_m \subseteq G$. Now suppose $C, D \in G$. Then $C \in F_n$ and $D \in F_m$ for some n and m. Either $F_n \subseteq F_m$ or $F_m \subseteq F_n$, say $F_m \subseteq F_n$. Then $C \cap D \in F_n \subseteq G$. Thus G is a filter in Φ. Hence by Zorn's Lemma, Φ has a maximal element, say $\mathcal{F}$.

Let $X \subseteq A$. Suppose that $\mathcal{F}$ contains neither X nor $A \backslash X$. Then $\mathcal{F}$ contains some C such that $C \cap X = \emptyset$ for if not $\mathcal{F} \cup \{X\}$ would be in a filter contradicting the maximality of $\mathcal{F}$. Similarly, $\mathcal{F}$ must contain some D such that $D \cap (A \backslash X) = \emptyset$. However, by the finite intersection property, we have $C \cap D \neq \emptyset$. However, this is impossible since $C \cap X = \emptyset$ and $D \cap (A \backslash X) = \emptyset$. Thus F contains either X or $A \backslash X$. □

Proposition 2.4.4 Free Ultrafilters exist.

Proof Let F be a filter consisting of all cofinite sets (a Frechet filter) and extend it to an ultrafilter $\mathcal{U}$. Since F contains no finite sets and $\mathcal{U}$ contains every set or its complement, clearly $\mathcal{U}$ contains no finite sets and so is free. □

Definition 2.4.5 Let $\mathcal{U}$ be a free ultrafilter on $\mathbb{N}$. Let $\mathbb{R}^{\mathbb{N}}$ denote the set of all real-valued sequences. Define the relation $=_{\mathcal{U}}$on $\mathbb{R}^{\mathbb{N}}$ by $\forall (a_n), (b_n) \in \mathbb{R}^{\mathbb{N}}$, $(a_n) =_{\mathcal{U}} (b_n)$ if and only if $\{n \in \mathbb{N} | a_n = b_n\} \in \mathcal{U}$.

Proposition 2.4.6 $=_{\mathcal{U}}$ is an equivalence relation on $\mathbb{R}^{\mathbb{N}}$.

Proof Clearly, $=_{\mathcal{U}}$ is reflexive and symmetric. Suppose $(a_n) =_{\mathcal{U}} (b_n)$ and $(b_n) =_{\mathcal{U}} (c_n)$. Then $\{n \in \mathbb{N} | a_n = b_n\} \in \mathcal{U}$ and $\{n \in \mathbb{N} | b_n = c_n\} \in \mathcal{U}$. Clearly,$\{n \in \mathbb{N} | a_n = c_n\} = \{n \in \mathbb{N} | a_n = b_n\} \cap \{n \in \mathbb{N} | b_n = c_n\} \in \mathcal{U}$ by the finite intersection property. Thus $+_{\mathcal{U}}$ is transitive. □

Let $[(a_n)]_{\mathcal{U}}$ denote the equivalence class of $=_{\mathcal{U}}$ determined by (a_n). Let $\mathbb{R}^* = \{[(a_n)]_U | a_n \in \mathbb{R}, n = 1, 2, ...\}$. Define addition $+$ and multiplication $\cdot$ on $\mathbb{R}^*$ as follows: $\forall [(a_n)]_{\mathcal{U}}, [(b_n)]_{\mathcal{U}} \in R^*$.

$$[(a_n)]_{\mathcal{U}} +_U [(b_n)]_{\mathcal{U}} = [(a_n + b_n)]_{\mathcal{U}},$$
$$[(a_n)]_{\mathcal{U}} \cdot_U [(b_n)]_{\mathcal{U}} = [(a_n \cdot b_n)]_{\mathcal{U}}.$$

We next show $\mathbb{R}^*$ is a field under these operations. The result actually holds from the Transfer Principle, but the beginner may find it useful to prove these results in order to get a better feel of the hyperreals.

Theorem 2.4.7 $(\mathbb{R}^*, +_{\mathcal{U}}, \cdot_{\mathcal{U}})$ is a field.

Proof We first show $+_{\mathcal{U}}$ and $\cdot_{\mathcal{U}}$ are well defined. Suppose $[(a_n)]_{\mathcal{U}} = [(c_n)]_{\mathcal{U}}$ and $[(b_n)]_{\mathcal{U}} = [(d_n)]_{\mathcal{U}}$. Then $(a_n) =_U (c_n)$ and $(b_n) =_U (d_n)$. Thus $\{n \in \mathbb{N} | a_n = c_n\} \in \mathcal{U}$ and $\{n \in \mathbb{N} | b_n = d_n\} \in \mathcal{U}$. By the finite intersection property,

$$\{n \in \mathbb{N} | a_n = c_n \text{ and } b_n = d_n\} = \{n \in \mathbb{N} | a_n = c_n\} \cap \{n \in \mathbb{N} | b_n = d_n\} \in \mathcal{U}.$$

Now $\{n \in \mathbb{N} | a_n = c_n \text{ and } b_n = d_n\} \subseteq \{n \in \mathbb{N} | a_n + b_n = c_n + d_n\}$. By the superset property, $\{n \in \mathbb{N} | a_n + b_n = c_n + d_n\} \in \mathcal{U}$. Hence $[(a_n + b_n)]_U = [(c_n + d_n)]_U$. Thus $+_U$ is well-defined. Similarly, it can be shown that $\cdot_U$ is well-defined.

The associative and commutative laws follow routinely. We prove the distributive laws. Let $[(a_n)]_U, [(b_n)]_U, [(c_n)]_{\mathcal{U}} \in \mathbb{R}^*$ Then

$$\begin{aligned} &[(a_n)]_{\mathcal{U}} \cdot_{\mathcal{U}} ([(b_n)]_{\mathcal{U}} +_{\mathcal{U}} [(c_n)]_{\mathcal{U}}) \\ &= [(a_n(b_n + c_n))]_{\mathcal{U}} = [(a_n b_n + a_n c_n)]_{\mathcal{U}} \\ &= [(a_n b_n)]_{\mathcal{U}} +_{\mathcal{U}} [(a_n c_n)]_{\mathcal{U}} \\ &= [(a_n)]_{\mathcal{U}} \cdot_{\mathcal{U}} [(b_n)]_{\mathcal{U}} +_{\mathcal{U}} [(a_n)] \cdot_{\mathcal{U}} [(c_n)]_{\mathcal{U}}. \end{aligned}$$

Let 1 be the multiplicative identity for $\mathbb{R}$. Then $[(1)]_{\mathcal{U}}$ is a multiplicative identity for $\mathbb{R}^*$. We show it is unique. Suppose $[(e_n)]_U$ is an multiplicative identity for $\mathbb{R}^*$. Then $[(1)]_{\mathcal{U}} = [(1)]_{\mathcal{U}} \cdot_{\mathcal{U}} [(e_n)]_{\mathcal{U}} = [(1e_n)]_{\mathcal{U}} = [(e_n)]_{\mathcal{U}}$. Thus $1 = 1e_n = e_n$ for $n \in$

$\{n \in N | 1 = e_n\} \in U$. Thus $[(e_n)]_{\mathcal{U}} = [(1)]_{\mathcal{U}}$. Thus $[(1)]_{\mathcal{U}}$ is unique. (Two-sidedness follows since $\cdot_U$ is commutative.) It follows similarly that $\mathbb{R}^*$ has a unique two-sided additive identity.

Let $[a_n)]_{\mathcal{U}} \neq [(0)]_{\mathcal{U}}$. Then $\{n \in \mathbb{N} | a_n = 0\} \notin \mathcal{U}$. Thus $\{n \in \mathbb{N} | a_n \neq 0\} \in \mathcal{U}$. For all $n \in \{n \in \mathbb{N} | a_n \neq 0\}$, there exists $a_n^{-1} \in \mathbb{R}$ such that $a_n a_n^{-1} = 1$. It follows that $\{n \in \mathbb{N} | a_n^{-1} \neq 0\} = \{n \in \mathbb{N} | a_n \neq 0\}$ and so $\{n \in \mathbb{N} | a_n^{-1} \neq 0\} \in \mathcal{U}$. Now $[(a_n)]_{\mathcal{U}} \cdot_{\mathcal{U}} [(a_n^{-1})]_{\mathcal{U}} = [(a_n a^{-1})]_{\mathcal{U}} = [(1)]_{\mathcal{U}}$. That is, every nonzero element of $\mathbb{R}^*$ has a multiplicative inverse. A similar argument shows that every element of $\mathbb{R}^*$ has an additive inverse. □

Definition 2.4.8 Define $\leq_{\mathcal{U}}$ on $\mathbb{R}^*$ as follows: $\forall [(a_n)]_{\mathcal{U}}, [(b_n)]_{\mathcal{U}} \in \mathbb{R}^*$, $[(a_n)]_{\mathcal{U}} \leq_{\mathcal{U}} [(b_n)]_{\mathcal{U}}$ if and only if $\{n \in \mathbb{N} | a_n \leq b_n\} \in \mathcal{U}$.

Let $[(a_n)]_{\mathcal{U}}, [(b_n)]_{\mathcal{U}}, [(b_n]_{\mathcal{U}} \in \mathbb{R}^*$. Suppose that $[(a_n)]_{\mathcal{U}} \leq [(b_n)]_{\mathcal{U}}$ and $[(a_n)]_{\mathcal{U}} \leq [(b_n)]_{\mathcal{U}}$. Then $\{j \in \mathbb{N} | a_j \leq b_j\} \in \mathcal{U}$ and $\{j \in \mathbb{N} | a_j \leq b_j\} \in \mathcal{U}$ By the finite intersection property, it follows that $\{j \in \mathbb{N} | a_j \leq c_j\} \in \mathcal{U}$ and so $[(a_n)]_{\mathcal{U}} \leq [(c_n)]_{\mathcal{U}}$.

Let $[(a_n)]_{\mathcal{U}}, [(b_n)]_{\mathcal{U}} \in \mathbb{R}^*$. Let $X = \{j \in \mathbb{N} | a_j \leq b_j\}$. Then either $X \in \mathcal{U}$ or $\mathbb{N} \backslash X \in \mathcal{U}$. If $X \in \mathcal{U}$, then $[(a_n)]_{\mathcal{U}} \leq_{\mathcal{U}} [(b_n)]_{\mathcal{U}}$. If $X \notin U$, then $\mathbb{N} \backslash X \in \mathcal{U}$, but $\mathbb{N} \backslash X = \{j \in \mathbb{N} | a_n > b_n\}n$ and so $[(a_n)]_{\mathcal{U}} >_U [(b_n)]_{\mathcal{U}}$.
Thus $\leq_U$ is a total ordering on $\mathbb{R}^*$.

We can restate Definitions 2.3.11 and 2.3.13 as follows.

Definition 2.4.9 A hyperreal number $[(a_n)]_{\mathcal{U}}$ in $\mathbb{R}^*$ is said to be **infinitesimal** if $[(a_n)]_{\mathcal{U}} \leq_{\mathcal{U}} [(j)]_{\mathcal{U}}$ for every $j \in \mathbb{N}$ and **infinite** if $[(j)]_{\mathcal{U}} \leq_{\mathcal{U}} [(a_n)]_{\mathcal{U}}$ for every $j \in \mathbb{N}$.

Consider $[(1, 2, 3, ...)]_{\mathcal{U}}$. Let $j \in \mathbb{N}$. Since $\mathcal{U}$ is free, it contains all cofinite subsets. Thus $\mathcal{U}$ contains $\{m \in \mathbb{N} | m \geq j\}$. Hence $[(1, 2, 3, ...)]_{\mathcal{U}} \geq_{\mathcal{U}} [(j, j, j, ...)]_{\mathcal{U}}$ for all $j \in \mathbb{N}$. Thus $\mathbb{R}^*$ contains infinite elements. Similarly, it can be shown that $[(1, \frac{1}{2}, \frac{1}{3}, ...)]_{\mathcal{U}} \leq_{\mathcal{U}} [(\frac{1}{j}, \frac{1}{j}, \frac{1}{j}, ...)]_{\mathcal{U}}$ for fixed $j \in \mathbb{N}$. Hence $\mathbb{R}^*$ contains infinitesimal elements.

Define the function $f : \mathbb{R} \to \mathbb{R}^*$ by for all $a \in \mathbb{R}$, $f(a) = [(a, a, a, ...)]_{\mathcal{U}}$. It is easily shown that f is a one-to-one function of $\mathbb{R}$ into $\mathbb{R}^*$ that preserves addition and multiplication. It also follows easily for all $a, b \in \mathbb{R}$ that $a \leq b$ if and only if $f(a) \leq_{\mathcal{U}} f(b)$.

In [2], Herrmann applies nonstandard analysis to explain issues from special and general relativity and the theory of light-clocks. We extend some of the results in [2] to nonstandard fuzzy analysis. We do this in terms of nonstandard fuzzy functions and nonstandard fuzzy numbers.

We rely heavily on [7] and [3] in the development presented here. We first recall some properties $\mathbb{R}^*$ possesses.

Let $\mathbb{R}^*$ denote a nonstandard universe with the following properties:

$(\mathbb{R}, +, \cdot, 0, 1, <)$ is an ordered subfield of $(\mathbb{R}^*, +, \cdot, 0, 1, <)$.

$\mathbb{R}^*$ has a positive infinitesimal element, that is $\varepsilon \in \mathbb{R}^*$ such that $\varepsilon > 0$, but $\varepsilon < r$ for all positive real numbers r.

For all $n \in \mathbb{N}$ and every function $f : \mathbb{R}^n \to \mathbb{R}$, there is a natural extension $f : (\mathbb{R}^*)^n \to \mathbb{R}^*$. The natural extensions of the field operations $+, \cdot : \mathbb{R}^2 \to \mathbb{R}$ coincide with the field operations in $\mathbb{R}^*$. Similarly, for every $A \subseteq R^n$, there is a subset $A^* \subseteq (\mathbb{R}^*)^n$ such that $A^* \cap \mathbb{R}^n = A$.

$\mathbb{R}^*$, equipped with the above assignments of extensions of functions and subsets, behaves logically like $\mathbb{R}$. $\mathbb{R}^*$ is called the ordered field of hyperreals.

Now ε has an additive inverse $-\varepsilon$. It is easily seen that $-\varepsilon$ is a negative infinitesimal. Since $\varepsilon \neq 0$, it has a multiplicative inverse ε^{-1}. For any positive real number r, $\epsilon^{-1} > r$ since $\varepsilon < r$. Thus ε^{-1} is a positive infinite element and $-\varepsilon^{-1}$ is a negative infinite element.

Recall that the definition of absolute value on $\mathbb{R}^*$ appears just above Theorem 2.3.9.

Definition 2.4.10 (1) Let $\mathbb{R}_{fin} = \{x \in \mathbb{R}^* |\, |x| \leq n \text{ for some } n \in \mathbb{N}\}$. $\mathbb{R}_{fin}$ is called the set of **finite hyperreals**.

(2) Let $\mathbb{R}_{\text{inf}} = \mathbb{R}^* \backslash \mathbb{R}_{fin}$. $\mathbb{R}_{\text{inf}}$ is called the set of **infinite hyperreals**.

(3) Let $\mu(0) = \{x \in \mathbb{R}^* |\, |x| \leq \frac{1}{n} \text{ for all } n \in \mathbb{N}\}$. $\mu(0)$ is called the set of **infinitesimal hyperreals**.

We see that $\mu(0) \subseteq \mathbb{R}_{fin}$, $\mathbb{R} \subseteq \mathbb{R}_{fin}$, and $\mu(0) \cap \mathbb{R} = \{0\}$. If $\delta \in \mu(0)\backslash\{0\}$, then $\delta^{-1} \notin \mathbb{R}_{fin}$.

Proposition 2.4.11 (1) $\mathbb{R}_{fin}$ is a subring of $\mathbb{R}^*$.

(2) $\mu(0)$ is and ideal of $\mathbb{R}_{fin}$.

Proof (1) Let $x, y \in \mathbb{R}_{fin}$. Then there exists $n, m \in \mathbb{N}$ such that $|x| \leq n$ and $|y| \leq m$. Then $|x + y| \leq |x| + |y| \leq n + m$. Thus $x, y \in \mathbb{R}_{fin}$. Also, $|xy| = |x||y| \leq nm$. Thus $xy \in \mathbb{R}_{fim}$. Hence $\mathbb{R}_{fin}$ is closed under addition and multiplication. Now $\mathbb{R}_{fin}$ inherits the remaining necessary properties from $\mathbb{R}$.

(2) Let $n \in \mathbb{N}$. Let $x, y \in \mu(0)$. Then $|x| \leq \frac{1}{2n}$ and $|y| \leq \frac{1}{2n}$. Thus $|x \pm y| \leq |x| + |y| \leq \frac{1}{2n} + \frac{1}{2n} = \frac{1}{n}$. Hence $x \pm y \in \mu(0)$. Let $x \in \mu(0)$ and let $r \in \mathbb{R}_{fin}$. Then $|r| \leq q$ for some $q \in \mathbb{N}$ and $|x| \leq \frac{1}{qn}$. Thus $|rx| = |r||x| \leq q\frac{1}{qn} = \frac{1}{n}$. Hence $rx \in \mu(0)$.

□

Definition 2.4.12 Define the relation $\approx$ on $\mathbb{R}^*$ by for all $x, y \in \mathbb{R}^*$, $x \approx y$ if and only if $x - y \in \mu(0)$. If $x \approx y$, we say that x and y are **infinitely close**.

It follows immediately that $\approx$ is an equivalence relation on $\mathbb{R}^*$. It also follows that $\approx$ is a congruence relation on $\mathbb{R}_{fin}$. This follows since $\mu(0)$ is an ideal of $\mathbb{R}_{fin}$.

Theorem 2.4.13 [7] (Existence of Standard Parts) Let $r \in \mathbb{R}_{fin}$. Then there exists a unique $s \in \mathbb{R}$ such that $r \approx s$. We call s the **standard part** of r and write $st(r) = s$.

Proof Suppose $r > 0$. Let $A = \{x \in \mathbb{R} | x < r\}$. Since $r \in \mathbb{R}_{fin}$, A is bounded above ($r \leq n$ for some $n \in \mathbb{N}$). Since $0 \in A$, $A \neq \emptyset$. By the Completeness Property, sup(A) (in $\mathbb{R}$) exists. Let $s = \sup(A)$. Let δ be a positive real number. Then $s + \delta \notin A$ and so $r \leq s + \delta$. Now $r \geq s - \delta$ since s is a least upper bound of A. Hence $|r - s| \leq \delta$. Since δ was arbitrary, it follows that $r - s \in \mu(0)$. If $r < 0$, the above argument hold for $-r$.

Suppose $s_1, s_2 \in \mathbb{R}$ are such that $r \approx s_1$ and $r \approx s_2$. Then $s_1 \approx s_2$. Hence $s_1 - s_2 \in \mu(0) \cap \mathbb{R} = \{0\}$. That is s is unique. □

Corollary 2.4.14 $\mathbb{R}_{fin} = \mathbb{R} + \mu(0)$ and $\mathbb{R} \cap \mu(0) = \{0\}$.

Corollary 2.4.15 Define $st : \mathbb{R}_{fin} \to \mathbb{R}$ by for all $r \in \mathbb{R}$, $st(r) = s$, where s is the standard part of r. Then st is a homomorphism of $\mathbb{R}_{fin}$ onto $\mathbb{R}$ such that Ker$(st) = \mu(0)$.

Corollary 2.4.16 The quotient ring $\mathbb{R}_{fin}/\mu(0)$ is isomorphic to $\mathbb{R}$, $\mu(0)$ is a maximal ideal of $\mathbb{R}_{fin}$, and is in fact the unique maximal ideal of $\mathbb{R}_{fin}$.

Proof Let $a \in \mathbb{R}_{fin} \backslash \mu(0)$. Then $a^{-1} \in \mathbb{R}^*$. However, $a^{-1} \notin \mathbb{R}^* \backslash \mathbb{R}_{fin}$ since $a \notin \mu(0)$. Thus $a^{-1} \in \mathbb{R}_{fin}$. That is, every element in $\mathbb{R}_{fin}$, but not in $\mu(0)$ has an inverse. □

Let F_0 be the filter consisting of all cofinite subsets of $\mathbb{N}$. Let $\mathcal{U}$ be a free ultrafilter. Let $A \in F_0$. Then A or A^c is in U. However, A^c is not in $\mathcal{U}$ since A^c is finite. Thus $A \in \mathcal{U}$. Hence $F_0 \subseteq \mathcal{U}$.

Let (x_i) and (y_i) be sequences of real numbers. Define the relation $\simeq$ by $(x_i) \simeq (y_i)$ if and only if $\{i \in \mathbb{N} | x_i = y_i\} \in \mathcal{U}$. Then $\simeq$ is an equivalence relation. Let $[(x_i)]_{\mathcal{U}}$ denote the equivalence class of (x_i) with respect to $\simeq$.

Hence $[(x_i)]_{\mathcal{U}} = [(y_i)]_{\mathcal{U}}$ if and only if $\{i \in \mathbb{N} | x_i = y_i\} \in \mathcal{U}$.

$\mathbb{R}^* = \{[(x_i)]_{\mathcal{U}} | x_i \in \mathbb{R}, i = 1, 2, ...\}$.

We restate Definition 2.4.8 as follows.

$[(x_1, x_2, ...)]_{\mathcal{U}} \leq [(y_1, y_2, ...)]_{\mathcal{U}}$ if and only if $\{i \in \mathbb{N} | x_i \leq y_i\} \in \mathcal{U}$.

Define $\geq, <, >$ on $\mathbb{R}^*$ similarly.

We have $\mu(0) = \{[(x_i)]_{\mathcal{U}} | [(x_i)]_{\mathcal{U}} < [(r, r, , ...)]_{\mathcal{U}}$ for all $r \in \mathbb{R}, r > 0\}$ and $[(x_i)]_{\mathcal{U}} \approx [(y_i)]_{\mathcal{U}}$ if and only if $[(x_i)]_{\mathcal{U}} - [(y_i)]_{\mathcal{U}} \in \mu(0)$, i.e., $[(x_i - y_i)]_{\mathcal{U}} < [(r, r, ...)]_{\mathcal{U}}$ for all $r \in \mathbb{R}, r > 0$.

Definition 2.4.17 [[3], p. 10] Let $A \subseteq \mathbb{R}$. The **natural extension** of A to $\mathbb{R}^*$ is the set A^* defined to be the set of all $[(r_n)]_{\mathcal{U}}$ such that $\{n \in \mathbb{N} | r_n \in A\} \in \mathcal{U}$.

Definition 2.4.18 [[3], p. 10] Let $f : X \to \mathbb{R}$, where X is a subset of $\mathbb{R}$. The natural extension of f to $\mathbb{R}^*$ is the function $f^* : X^* \to \mathbb{R}^*$ defined as follows:

$$f^*([(r_n)]_{\mathcal{U}}) = [(f(r_n))]_{\mathcal{U}}.$$

Consequently, the natural extension of $[0, 1]$ to $\mathbb{R}^*$ is $[0, 1]^* = \{x \in \mathbb{R}^* | 0 \leq x \leq 1\}$.

Proposition 2.4.19 Let $[a, b]$ be a closed interval in $\mathbb{R}$. Then $[a, b]^* = \{x \in \mathbb{R}^* | a \leq x \leq b\}$.

Proof We have that

$$\begin{aligned} [(r_n)]_{\mathcal{U}} \in [a, b]^* &\Leftrightarrow \{n \in \mathbb{N} || r_n \in [a, b]\} \in \mathcal{U} \\ &\Leftrightarrow \{n \in \mathbb{N} | a \leq r_n \leq b\} \in \mathcal{U} \\ &\Leftrightarrow a = [(a, a, ...)]_{\mathcal{U}} \leq [(r_n)]_{\mathcal{U}} \leq [(b, b, ...)]_{\mathcal{U}} = b. \end{aligned}$$

Consequently, the natural extension of $[0, 1]$ to $\mathbb{R}^*$ is $[0, 1]^* = \{x \in \mathbb{R}^* | 0 \leq x \leq 1\}$. □

Let $a = [(a, a, a, ...,)]_{\mathcal{U}}$ and $m = [(1, 1/2, 1/3, ...)]_{\mathcal{U}}$. Then $a + m > a$. Define $A(a) = a$ for all $a \in \mathbb{R}$. Let A^* denote the natural extension of A to $\mathbb{R}^*$. Then $A^*([(x_1, x_2, ...)]_{\mathcal{U}}) = [(A(x_1), A(x_2), ...]_{\mathcal{U}} = [(x_1, x_2, ...)]_{\mathcal{U}}$. Thus $A^*(a + m) > A(a)$.

2.5 Fuzzy Numbers

We review some basic results of fuzzy numbers.

Definition 2.5.1 [[8], p. 97] Let A be a fuzzy subset of $\mathbb{R}$. Then A is a fuzzy number if the following conditions hold.

(1) There exists $x \in \mathbb{R}$ such that $A(x) = 1$.

(3) A^{α} is a closed bounded interval for all $\alpha \in (0, 1]$.

(3) The support of A is bounded.

Theorem 2.5.2 [[8], p. 98] Let A be a fuzzy subset of $\mathbb{R}$. Then A is a fuzzy number if and only if there is a closed interval $[c, d]$ and functions $l : (-\infty, c) \to [0, 1]$, $r : (d, \infty) \to [0, 1]$, and $a, b \in \mathbb{R}$, $a \leq c \leq d \leq b$ such that

$$A(x) = \begin{cases} 1 \text{ if } x \in [c, d], \\ l(x) \text{ if } x \in (-\infty, c), \\ r(x) \text{ if } x \in (d, \infty), \end{cases}$$

where l is monotonic increasing, continuous from the right and such that $l(x) = 0$ for $x \in (-\infty, a)$; r is monotonic decreasing, continuous from the left and such that $r(x) = 0$ for $x \in (b, \infty)$.

Theorem 2.5.3 [[8], p. 41] Let A be a fuzzy subset of $\mathbb{R}$. Then $A = \cup_{\alpha \in [0,1]\alpha} A$, where ${}_{\alpha}A(x) = \alpha A^{a}(x)$ and $(\cup_{\alpha \in [0,1]\alpha} A)(x) = \vee\{_{\alpha}(A)(x) | x \in [0, 1]\}$ for all $x \in \mathbb{R}$.

We next present the second method for developing fuzzy arithmetic, which is the extension principle. Employing this principle, standard arithmetic operations on real numbers are extended to fuzzy numbers.

Let $*$ denote any of the four basic arithmetic operations and let A, B denote fuzzy numbers. Then define $A * B$ by for all $z \in \mathbb{R}$.

$$(A * B)(z) = \vee\{A(x) \wedge B(y) | z = x * y, x, y \in \mathbb{R}\}$$

Theorem 2.5.4 [8] Let $* \in \{+, -, \cdot, /\}$ and let A, B denote continuous fuzzy numbers. Then the fuzzy subset $A * B$ is a continuous fuzzy number.

Let A be a fuzzy subset of $\mathbb{R}$. Let A^* be the natural extension of A to $\mathbb{R}^*$. Let $B = 1 - A$, i.e., for all $x \in \mathbb{R}$, $B(x) = 1 - A(x)$. Let B^* be the natural extension of B to $\mathbb{R}^*$. Let $[(x_1, x_2, ..., x_n, ...)]_U$. Then

$$\begin{aligned} B^*([(x_1, x_2, ..., x_n, ...)]_U) &= [(B(x_1), B(x_2), ..., B(x_n), ...]_U \\ &= [(1 - A(x_1), 1 - A(x_2), ..., 1 - A(x_n), ...]_U \\ &= [(1, 1, ..., 1, ...)]_U - [A(x_1), A(x_2), ..., A(x_n), ...]_U \\ &= 1 - A^*([(x_1, x_2, ..., x_n, ...)]_U. \end{aligned}$$

In the following, let A and B be continuous fuzzy numbers. Let A^*, B^*, and $(A + B)^*$ denote the natural extensions of A, B, and $A + B$ to $\mathbb{R}^*$, respectively.

Definition 2.5.5 Define $A^* + B^*$ as follows:

$$(A^* + B^*)(a + m) = (A + B)(a) + m, \text{ if } a \in \mathbb{R}, m \in \mu(0),$$
$$(A^* + B^*)(x) = 0 \text{ if } x \in \mathbb{R}\backslash\mathbb{R}_{fin}.$$

Let $a \in \mathbb{R}$. Then $(A^* + B^*)(a) = (A + B)(a) = (A + B)^*(a)$. Now $(A + B)^*(a + m) \approx (A + B)(a) \approx (A^* + B^*)(a + m)$ since $(A^* + B^*)(a + m) = (A^* + B^*)(a) + m$.

Definition 2.5.6 A^* is a nonstandard fuzzy number if the following properties hold:

(1) There exist $x \in \mathbb{R}^*$ such that $A^*(x) \approx 1$,
(2) $\forall \alpha \in [0, 1]^*$, there exists $c_\alpha, d_\alpha \in [0, 1]^*$ such that $c_\alpha \le d_\alpha$ and $A^{*\alpha} = \{x \in \mathbb{R}^* | c_\alpha \lessapprox x \lessapprox d_\alpha\}$.
(3) There exists $c, d \in \mathbb{R}_{fin}$ such that $c \le d$ and $\text{NSupp}(A^*) \subseteq \{x \in R^* | c \lessapprox x \lessapprox d\}$.

Theorem 2.5.7 Suppose A is continuous. Then A is a fuzzy number if and only if A^* is a nonstandard fuzzy number.

Proof Suppose A is a fuzzy number.

(1) Then there exists $x \in \mathbb{R}$ such that $A(x) = 1$. Hence $A^*(x) = 1$.
(2) Let $\alpha \in [0, 1]^*$ and $y \in \mathbb{R}^*$. Suppose $A^*(y) \gtrapprox \alpha$. Since A^* is microcontinuous, $A^*(y) \approx A(st(y))$ and so $A(st(y)) \gtrapprox \alpha$. Thus $A(st(y)) \ge st(\alpha)$. Hence there exist $c_{st(\alpha)}, d_{st(\alpha)} \in [0, 1]$ such that $c_{st(\alpha)} \le d_{st(\alpha)}$ and $c_{st(\alpha)} \le st(y) \le d_{st(\alpha)}$. Thus $c_{st(\alpha)} \lessapprox y \lessapprox d_{st(\alpha)}$.
(3) Suppose $A^*(y) \notin \mu(0)$, where $y \in \mathbb{R}^*$. Now $A^*(y) \approx A(st(y))$. Thus $A(st(y)) > 0$. Hence there exists $c, d \in R$ such that $c \le st(y) \le d$. Thus $c \lessapprox y \lessapprox d$.

Conversely, suppose A^* is a nonstandard fuzzy number.

(1) Then there exists $y \in \mathbb{R}^*$ such that $A^*(y) \approx 1$. Hence $A(st(y)) = A^*(st(y)) = 1$.
(2) Let $\alpha \in [0, 1]$ and $x \in \mathbb{R}$. Suppose $A(x) \ge \alpha$. Then $A^*(x) = A(x) \ge \alpha$. Thus $A^*(x) \gtrapprox \alpha$. Hence there exists $c_\alpha, d_\alpha \in [0, 1]$ with $c \le d$ such that $c_\alpha \lessapprox x \lessapprox d_\alpha$. Since $x \in R$, $c_\alpha \le x \le d_\alpha$.
(3) Suppose $A(x) > 0$, where $x \in \mathbb{R}$. Then $A^*(x) > 0$ and so $x \notin \mu(0)$. Thus there exists $c, d \in [0, 1]^*$ such that $c \lessapprox x \lessapprox d$. Since $x \in \mathbb{R}$, $st(c) \le x \le st(d)$. Thus $\text{Supp}(A) \subseteq [st(c), st(d)]$.

□

Proposition 2.5.8 Let C and D be nonstandard fuzzy subsets of $\mathbb{R}^*$. If C is a nonstandard fuzzy number and $C(y) \approx D(y)$ for all $y \in \mathbb{R}^*$, then D is a nonstandard fuzzy number.

Proof There exists $y \in R$ such that $A(y) \approx 1$. Thus $B(y) \approx 1$. Let $y \in R^*$. Let $\alpha \in [0, 1]^*$. Then $B(y) \gtrapprox \alpha$ if and only if $A(y) \gtrapprox \alpha$. Thus B^α is bounded.

Now $A(y) \notin \mu(0)$ if and only if $B(y) \notin \mu(0)$ since $A(y) \approx B(y)$. Hence $\text{NSupp}(B) = \text{NSupp}(A)$. Thus $\text{NSupp}(B)$ is bounded. □

Corollary 2.5.9 Let A and B be fuzzy subsets of $\mathbb{R}$. Then $A + B$ is a fuzzy number if and only if $A^* + B^*$ is a nonstandard fuzzy number.

Proof $A + B$ is a fuzzy number if and only if $(A + B)^*$ is a fuzzy number. Now $(A + B)^*(y) \approx (A^* + B^*)(y)$ for all $y \in \mathbb{R}^*$. □

Proposition 2.5.10 Let $a \in \mathbb{R}$ and $m \in \mu(0)$. Let $m = m' + m''$, where $m', m'' \in \mu(0)$. Then $(A^* + B^*)(a + m) = \vee\{(A^* \circ st)(b + m') \wedge (B^* \circ st(c + m'')|a = b + c\}$.

Proof For all such m', m'' (held fixed),

$$\begin{aligned}\vee\{(A^* \circ st)(b + m') \wedge (B^* \circ st(c + m'')|a = b + c\} \\ &= \vee\{A^*(b) \wedge B^*(c)|a = b + c\} \\ &= \vee\{A(b) \wedge B(c)|a = b + c\} \\ &= (A + B)(a).\end{aligned}$$

□

2.6 Continuity and Differentiability

Let A be a fuzzy subset of $\mathbb{R}$. Assume there exist real numbers a, b with $a \leq b$ such that $A(y) = 0$ for all $y \notin [a, b]$.

Proposition 2.6.1 If A^* is the natural extension of A to $\mathbb{R}^*$, then $A^*(y) = 0$ for all $y \in \mathbb{R}^* \backslash [a, b]^*$.

Proof Suppose $[(y_n)]_U \in \mathbb{R}^* \backslash [a, b]^*$. Then $\{n \in \mathbb{N}|a \leq y_n \leq b\} \notin U$ else $[(y_n)]_U \in [a, b]^*$. Hence $\{n \in \mathbb{N}|y_n \notin [a, b]\} \in U$ since either $\{n \in \mathbb{N}|a \leq y_n \leq b\} \in U$ or $\{n \in \mathbb{N}|a \leq y_n \leq b\}^c \in U$, but not both. Thus $\{n \in \mathbb{N}|A(y_n) = 0\} \in U$. Hence $A^*([(y_n)]_U) = [A(y_n)]_U = [(0, 0, \ldots)]_U$. □

It follows that A^* maps every element of $\mathbb{R}^* \backslash \mathbb{R}_{fin}$ to 0.

Definition 2.6.2 [3] (Nonstandard Definition of Continuity) Let $f : \mathbb{R} \to \mathbb{R}$ and $a \in \mathbb{R}$. Then f is continuous at a if and only if $\forall \delta \approx 0$, $f^*(a+\delta) - f(a) \approx 0$, where f^* is the natural extension of f to $\mathbb{R}^*$.

Let $A : \mathbb{R} \to \mathbb{R}$ (or $[0, 1]$) and let A^* be the natural extension of A to $\mathbb{R}^* \to \mathbb{R}^*$(or $[0, 1]^*$). Let $a \in \mathbb{R}$. Then in $\mathbb{R}^*$, $a = [(a, a, ..., a, ..)]_U$ and $A^*(a) = [(A(a), A(a), ..., A(a), ...)]_U = A(a)$. That is, $A^*|_{\mathbf{R}} = A$.

Definition 2.6.3 [[3], p. 11] Let $A \subseteq \mathbb{R}^*$. Then a function $f : A \to \mathbb{R}^*$ is said to be **microcontinuous** at $x_0 \in A$ if $x \approx x_0$ implies $f(x) \approx f(x_0)$ for all $x \in X$.

Let $f : \mathbb{R} \to \mathbb{R}$ and $a \in \mathbb{R}$. If f is continuous at a, then f^* is microcontinuous at $a + m$ for all $m \in \mu(0)$. Note that if f is continuous at a, then $f^*(a+m) \approx f(a)$ for all $m \in \mu(0)$ and so $f(a+m) \approx f(a+m')$ for all $m, m' \in \mu(0)$.

Theorem 2.6.4 [[3], p. 11] A function $f : A \to \mathbb{R}$ is continuous at $c \in \mathbb{R}$ if and only if f^* is microcontinuous at c.

Proof Suppose f is continuous at $c \in \mathbb{R}$. Suppose $x_0 \approx c$. Then

$$(\forall \epsilon \in R^+)(\exists \delta R^+)(\forall x \in A)(|x - c| < \delta \to |f(x) - f(c)| < \epsilon).$$

$\forall$Fix such an ϵ and δ. Then by the transfer principle

$$(\forall x \in A^*)(|x - c| <_{\mathcal{U}} \delta \to |f^*(x) - f^*(c)| <_{\mathcal{U}} \epsilon).$$

Since $x_0 \approx c$, we have that $|x_0 - c| <_{\mathcal{U}} \delta$. Thus $|f^*(x_0) - f^*(c)| <_{\mathcal{U}} \epsilon$. Since ϵ was arbitrary, this holds for all ϵ in $\mathbb{R}^+$. Hence $f^*(x) \approx f^*(c)$.

Conversely, suppose f^* is micrcontinuous at c. Let ϵ belong to $\mathbb{R}^+$. Let δ be an element in $\mathbb{R}^+$ such that $\delta \approx 0$. Then for all x in A^*, $|x - c| <_{\mathcal{U}} \delta$ implies $x \approx c$. Hence $f^*(x) \approx f^*(c)$ and so $|f^*(x) - f^*(c)| <_{\mathcal{U}} \epsilon$. Consequently,

$$(\exists \delta \in \mathbb{R}^{*+})(\forall x \epsilon A^*)(|x - c| <_{\mathcal{U}} \delta \Rightarrow |f^*(x) - f^*(c) <_{\mathcal{U}} \epsilon).$$

Thus by the transfer principle,

$$(\exists \delta \in \mathbb{R}^+)(\forall x \epsilon A)(|x - c| < \delta \Rightarrow |f(x) - f(c) < \epsilon).$$

Hence f is continuous at c. □

Theorem 2.6.5 [[9]. p. 21] The nonstandard definition of continuity given above is equivalent to the classic definition of continuity: f is continuous at a if and only if $\lim_{x \to a} f(x) = f(a)$.

Let A be a continuous fuzzy number. Then $\forall a \in \mathbb{R}$ and $\forall m \in \mu(0)$, $A^*(a+m) \approx A(a)$. Since $A^*(a) = A(a)$ and $A^* : \mathbb{R} \backslash \mathbb{R}_{fin} \to \{0\}$, we have a "picture" of A^*.

Let A^* be the natural extension of A to $\mathbb{R}^*$. We could define A^* to be a nonstandard fuzzy number if A is a fuzzy number. Note $A^*|_{\mathbb{R}} = A$.

We next review some results dealing with the differentiation and integration of fuzzy functions in order to make their connection to the work of [2] concerning relativity.

We denote the space of all fuzzy-valued functions on $[a, b]$ by $\mathcal{F}[a, b]$, or simply $\mathcal{F}$.

Definition 2.6.6 Define the fuzzy subset $\widetilde{F}$ of $\mathbb{R}$ by $\forall y \in \mathbb{R}$, $\widetilde{F}(y) = \int_a^b \widetilde{f}(x)(y)dx = \vee\{\wedge\{\widetilde{f}(x)(g(x))|a \le x \le b\}|g \in \mathcal{I}(a, b), y = \int_a^b g(t)dt\}$, where $\mathcal{I}(a, b)$ denotes the set of all integrable functions whose domain is $[a, b]$.

Definition 2.6.7 Let $\widetilde{f}$ be a fuzzy-valued function with level sets $[f^-(a, x), f^+(\alpha, x)]$ such that $f^-(a, \_)$ and $f^+(\alpha, \_)$ are integrable functions on the interval $[a, b]$. Let $\widetilde{I} \in \widetilde{P}(\mathbb{R})$ be defined by $\forall y \in \mathbb{R}$,

$$\widetilde{I}(y) = \begin{cases} \vee\{\alpha \in (0, 1]| \int_a^b f^-(\alpha, x)dx \le y \le \int_a^b \widetilde{f}(\alpha, x)dx\}, \\ 0 \text{ otherwise} \end{cases}$$

Theorem 2.6.8 [6] Let $\widetilde{I}$ be defined as in Definition 2.6.7. Then $\widetilde{I} = \widetilde{F}$.

Definition 2.6.9 [[3], p. 12] Let $f : A \to \mathbb{R}$. We say that f is differentiable at $x_0 \in A$ if there exists $L \in \mathbb{R}^*$ such that for every nonzero infinitesimal ε, we have

$$\frac{f^*(x_0 + \varepsilon) - f^*(x_0)}{\varepsilon} \approx L.$$

If so, we define the derivative of f at x_0 to be the standard part of L, $f'(x_0) = st(L)$.

Theorem 2.6.10 Let $\widetilde{f}$ be a function of $\mathbb{D}$ into $\widetilde{P}(\mathbb{R})$ such that $\widetilde{f}(x)$ is a fuzzy number for all x in $\mathbb{D}$. Suppose that $\forall x \in \mathbb{D}$, $\forall \alpha \in [0, 1]$, $\widetilde{f}(x)_\alpha$ is a closed bounded interval. Then there exist unique functions f^-, f^+ of $[0, 1] \times \mathbb{D}$ into $\mathbb{R}$ such that

1. $\forall x \in \mathbb{D}$, $f^-(\_, x)$ $(f^+(\_, x))$ is a nondecreasing (nonincreasing) function of α.
2. $\forall (\alpha, x) \in [0, 1] \times \mathbb{D}$, $f^-(\alpha, x) \le f^+(\alpha, x)$.
3. $\forall (\alpha, x) \in [0, 1] \times \mathbb{D}$, $f(x)_\alpha = [f^-(\alpha, x), f^+(\alpha, x)]$.
4. $\forall x \in \mathbb{D}$, $f^-(1, x) = f^+(1, x)$.

Theorem 2.6.11 Let g and h be functions of $[0, 1] \times \mathbb{D}$ into $\mathbb{R}$ such that $\forall x \in \mathbb{D}$, $g(\_, x)$ $(h(\_, x))$ is a nondecreasing (nonincreasing) function of α and $\forall(\alpha, x) \in [0, 1] \times \mathbb{D}$, $g(\alpha, x) \leq h(\alpha, x)$. Let $\widetilde{f}$ be the function $\mathbb{D} \times \mathbb{R}$ into $[0, 1]$ defined as follows: $\forall(x, y) \in \mathbb{D} \times \mathbb{R}$,

$$\widetilde{f}(x, y) = \begin{cases} \vee\{\beta \in [0, 1] | y \in g(\beta, x), h(\beta, x)]\} \text{ if } y \in [g(0, x), h(0, x)]\} \\ 0 \text{ otherwise.} \end{cases}$$

If $\forall x \in \mathbb{D}$, $g(\_, x)$ and $h(\_, x)$ are continuous from the left, then $\widetilde{f}(x)_\alpha = [g(\alpha, x), h(\alpha, x)] \forall \alpha \in [0, 1]$.

Theorem 2.6.12 Let g and h be functions of $[0, 1] \times \mathbb{D}$ into $\mathbb{R}$ such that $\forall x \in \mathbb{D}$, $g(\_, x)$ $(h(\_, x))$ is a nondecreasing (nonincreasing) function of α and $\forall(\alpha, x) \in [0, 1] \times \mathbb{D}$, $g(\alpha, x) \leq h(\alpha, x)$. Suppose there exists a function $\widetilde{f}$ of $[0, 1] \times \mathbb{D}$ into $[0, 1]$ such that $\forall \alpha \in [0, 1]$, $\widetilde{f}(x)_\alpha = [g(\alpha, x), h(\alpha, x)]$. Then $\forall x \in \mathbb{D}$, $g(\_, x)$ and $h(\_, x)$ are continuous functions of α from the left. Furthermore,

$$\widetilde{f}(x)(y) = \begin{cases} \vee\{\beta \in [0, 1] | y \in g(\beta, x), h(\beta, x)]\} \text{ if } y \in [g(0, x), h(0, x)]\} \\ 0 \text{ otherwise.} \end{cases}$$

Let $x \in \mathbb{D}$ and hold x fixed. Let $f(x)^*$ be the natural extension of $f(x)$ to $\mathbb{R}^*$. ($f(x)$ is a fuzzy number so $f(x) : \mathbb{R} \to [0, 1]$. We assume $f(x)$ is a continuous fuzzy number throughout.) Thus $f(x)^* : \mathbb{R}^* \to [0, 1]^*$ since $f((x)^*([(r_1, r_2, ...)\}]_U = [(f(x)(r_1), f(x)(r_2), ...)]_U$ and $f(x)(r_i) \in [0, 1], i = 1, 2, ...$ Note that since $f(x)(r_i) \geq 0, [(f(x)(r_1), f(x)(r_2), ...)]_U \geq [(0, 0, ...)]_U = 0$.

Write $f^-(x)$ for $f^-(\_, x)$ and $f^+(x)$ for $f^+(\_, x)$. Then $f^-(x)$ and $f^+(x)$ map $[0, 1]$ into $\mathbb{R}$. Let $f^-(x)^*$ and $f^+(x)^*$ be the natural extensions of $f^-(x)$ and $f^+(x)$ to mappings of $[0, 1]^*$ to $\mathbb{R}^*$, respectively. Let $\alpha \in [0, 1]$. Then $f(x)_\alpha = [f^-(\alpha, x), f^+(\alpha, x)]$. Since $f(x)$ is a fuzzy number, there exists $a, b \in R$ such that $a \leq b$ and $f(x)(y) = 0$ if $y \notin (a, b)$. (a and b are dependent on x.) Now $a \leq f^-(x)(\alpha) \leq f^+(x)(\alpha) \leq b$ for $\alpha > 0$.

Proposition 2.6.13 Let $a, b, c \in \mathbb{R}^*$. Then $a \lessapprox c \lessapprox b$ if and only if $st(a) \leq st(c) \leq st(b)$.

Proof We have that $a \approx c \Leftrightarrow st(a) = st(c)$. Also, $a < c \Leftrightarrow st(a) < st(c)$ or ($st(a) = st(c)$ and $nst(a) < nst(c)$). Similar arguments hold for c and b. □

Proposition 2.6.14 Let $x \in \mathbb{D}$. Let $f(x)$ be a fuzzy number and $f(x)^*$ its natural extension to $\mathbb{R}^*$. Then for all $\alpha \in [0, 1]$, $f^-(x)^*(\alpha^*) \lessapprox y \lessapprox f^+(x)^*(\alpha^*)$ if and only if $f^-(\alpha, x) \leq st(y) \leq f^+(\alpha, x)$.

Proof Since $f^-(x)$ and $f^+(x)$ are continuous on $[0,1]$, $f^-(x)^*$ and $f^+(x)^*$ are micro-continuous on $[0,1]^*$. Now $f^-(x)^*(\alpha^*) \approx f^-(x)(\alpha)$ and $f^+(x)^*(\alpha^*) \approx f^+(x)(\alpha)$. Also, $st(f^-(x)^*(\alpha^*)) = f^-(x)(\alpha)$ and $st(f^+(x)^*(\alpha^*)) = f^+(x)(\alpha)$.

□

2.7 Relativity

This section relies heavily on [2] and [6]. We next make a connection to the theory of relativity. Let $q : \mathbb{D} \times \mathbb{R} \to [0,1]$. For example, let $\mathbb{D}$ be a time interval $[a,b]$ and let r be distance, [2]. Then $q(t,r)$ is the intensity with which an object travels r units in time t. We next consider $q^* : \mathbb{D}^* \times \mathbb{R}^* \to [0,1]^*$. The following Proposition shows that q^* can be considered to be the natural extension of q to $(\mathbb{D} \times \mathbb{R})^*$.

Proposition 2.7.1 Define f from $\mathbb{D}^* \times \mathbb{R}^*$ into $(\mathbb{D} \times \mathbb{R})^*$ by for all $([(t_1,t_2,...)]_{\mathcal{U}}, [(r_1,r_2,...)]_{\mathcal{U}}) \in \mathbb{D}^* \times \mathbb{R}^*$,

$$f(([(t_1,t_2,...)]_{\mathcal{U}}, [(r_1,r_2,...)]_{\mathcal{U}})) = [((t_1,r_1),(t_2,r_2),...)]_{\mathcal{U}}.$$

Then f is a one-to-one function of $\mathbb{D}^* \times \mathbb{R}^*$ onto $(\mathbb{D} \times \mathbb{R})^*$.

Proof We have

$$\begin{aligned}
([(t_1,t_2,...)]_{\mathcal{U}}, [(r_1,r_2,...)]_{\mathcal{U}}) = ([(t_1',t_2',...)]_{\mathcal{U}}, [(r_1',r_2',...)]_{\mathcal{U}}) &\Leftrightarrow \\
[(t_1,t_2,...)]_{\mathcal{U}} = [(t_1',t_2',...)]_{\mathcal{U}} \text{ and } [(r_1,r_2,...)]_{\mathcal{U}} = [(r_1',r_2',...)]_{\mathcal{U}} &\Leftrightarrow \\
(t_1,t_2,...) \approx (t_1',t_2',...) \text{ and } (r_1,r_2,...) \approx (r_1',r_2',...) &\Leftrightarrow \\
\{i|t_i = t_i'\} \in \mathcal{U} \text{ and } \{j|r_j = r_j'\} \in \mathcal{U}.&
\end{aligned}$$

Now $\{k|t_k = t_k' \text{ and } r_k = r_k'\} = \{i|t_i = t_i'\} \cap \{j|r_j = r_j'\} \in \mathcal{U}$.
Also,

$$\begin{aligned}
[((t_1,r_1),(t_2,r_2),...)]_{\mathcal{U}} = [((t_1',r_1'),(t_2',r_2'),...)]_{\mathcal{U}} &\Leftrightarrow \\
((t_1,r_1),(t_2,r_2),...) \approx ((t_1',r_1'),(t_2',r_2'),...) &\Leftrightarrow \\
\{k|(t_k,r_k) = (t_k',r_k')\} \in \mathcal{U} &\Leftrightarrow \\
\{k|t_k = t_k' \text{ and } r_k = r_k'\} \in \mathcal{U}.&
\end{aligned}$$

Thus f is a one-to-one function of $\mathbb{D}^* \times \mathbb{R}^*$ onto $(\mathbb{D} \times \mathbb{R})^*$. □

Let $l : \mathbb{D} \times \mathbb{R} \to [0,1]$ and $l^* : \mathbb{D}^* \times \mathbb{R}^* \to [0,1]^*$, where $l(t,r)$ is the intensity of the velocity $r \in \mathbb{R}$ of a particle at time $t \in \mathbb{D}$.

In ([2], p. 10), it is stated that Newton's approach created a schism in the philosophy of mathematical modeling. One group of scientists believed that there exists actual real world entities that can be characterized in terms of infinitesimal measures of time, mass, volume, charge, and the like. Another group assumed that such terms are auxiliary in character and do not correspond to objective reality. The mathematical model called *the nonstandard physical world* {i.e. NSP-world) uses the corrected theory of the infinitesimally small and infinitely large, with other techniques, along with a new physical language theory of correspondence.

In ([2], p. 27), it is stated that experiments show that for small time intervals $[a, b]$ the Galilean theory of average velocities suffices to give accurate information relative to the compositions of such velocities. Let there be an internal function $q : [a, b]^* \to \mathbb{R}^*$ where q represents the NSP-world distance function. Also, let non negative and internal $l : [a, b]^* \to \mathbb{R}^*$ be a function that yields the NSP-world velocity of the electromagnetic propagation at and time $t \in [a, b]^*$. As usual $\mu(t)$ denotes the monad of standard time $\mu(t)$, where "t" is an absolute NSP-world "time" parameter.

The general and correct methods of infinitesimal modeling state that, within the internal portion of the NSP-worlds, two measures m_1 and m_2 are *indistinguishable* for dt (i.e., infinitely close of order one) (notation $m_1 \sim m_2$) if and only if $0 \neq dt \in \mu(0)$, ($\mu(0)$ the set of infinitesimals)

$$\frac{m_1}{dt} - \frac{m_2}{dt} \in \mu(0).$$

Intuitively, indistinguishable in this sense means that, although within the NSP-world the two measures are only equivalent and not necessarily equal, the *first level* (or first-order) effects these measures represent over dt are indistinguishable within the N-world (i.e., they appear to be equal.)

In [2], some continuity conditions are placed on q and l. It is argued that for each $t \in [a, b]$ and $t' \in \mu(t) \cap [a, b]^*$,

$$\frac{q(t')}{t'} - \frac{q(t)}{t} \in \mu(0) \text{ and } l(t') - l(t) \in \mu(0)$$

The above expressions give relations between nonstandard time $t' \in \mu(t)$ and the standard time t. Recall that if $x, y \in \mathbb{R}^*$, then $x \approx y$ if and only if $x - y \in \mu(0)$. It thus follows that for each $dt \in \mu(0)$ such that $t + dt \in \mu(t) \cap [a, b]^*$,

$$\frac{q(t+dt)}{t+dt} \approx \frac{q(t)}{t},$$
$$l(t+dt) + \frac{q(t+dt)}{t+dt} \approx l(t) + \frac{q(t)}{t}. \tag{2.1}$$

Hence

$$(l(t+dt)+\frac{q(t+dt)}{t+dt})dt \sim (l(t)+\frac{q(t)}{t})dt$$

Thus, it follows from [2] that

$$q(t+dt)-q(t) \sim (l(t+dt)+\frac{q(t+dt)}{t+dt})dt$$

and

$$q(t+dt)-q(t) \sim (l(t)+\frac{q(t)}{t})dt. \tag{2.2}$$

It is stated in [2] that Expression 2.1 is the basic result that will lead to conclusions relative to the Special Theory of relativity. In order to find out exactly what standard functions will satisfy 2.2, let arbitrary $t_1 \in [a,b]$ be the standard time at which electromagnetic propagation from position F_1. Next, the definition of $\sim$, yields

$$\frac{s^*(t+dt)-s(t)}{dt} \approx l(t)+\frac{s(t)}{t}. \tag{2.3}$$

Note that l is microcontinuous on $[a,b]^*$. For each $t \in [a,b]$, the value of $l(t)$ is limited. Hence let $st(l(t)) = v(t) \in \mathbb{R}$. From Theorem 1.1 in [2] or 7.6 in [3], v is continuous on $[a,b]$. Now 2.3 may be rewritten as

$$(\frac{d(s(t)/t)}{dt})^* = \frac{v^*(t)}{t}, \tag{2.4}$$

where all functions 2.4 are $*$-continuous on $[a,b]^*$. Consequently, we may apply the *-integral to both sides of 2.4. Now 2.4 implies that for $t \in [a,b]$

$$\frac{s(t)}{t} =^* \int_{t_1}^{t} \frac{v^*(x)}{x}dx$$

where for $t_1 \in [a,b]$, $s(t_1)$ has been initialized to be zero.

We next provide a possible extension of these results to nonstandard fuzzy numbers. We define a nonstandard fuzzy number to be the natural extension of a fuzzy number to $\mathbb{R}^*$ into $[0,1]^*$. Consider the function l above. Let l be a function of $[a,b]^*$ into the set of nonstandard fuzzy numbers. Then for all $t' \in [a,b]^*$ and $r' \in \mathbb{R}_{fin}, l(t',r') \in [0,1]^*$ is the intensity with the velocity is r' at time t'. Define $l(t',r') \wedge l(t,r)$ to be the intensity with which $r'-r \in \mu(0)$. Note that if $l(t',r') = 1$ and $l(t,r) = 1$, then the intensity with which $r'-r \in \mu(0)$ equals 1. Similar, interpretations can be given to the other equations given above. For example, let $v(t) = q(t)/t$ and consider v be a function of $[a,b]^*$ into the set of nonstandard fuzzy numbers.

Consider the sum in 2.1. Let $v : [a, b] \times \mathbb{R} \rightarrow [0, 1]$ be such that for all $t \in [a, b]$, $v(t)$ is a fuzzy number. Let v^* be the natural extension of v. Then $v^* : [a, b]^* \times \mathbb{R}^* \rightarrow [0, 1]^*$. Consider two such v_1, v_2. Define $v_1^* + v_2^* : [a, b]^* \times \mathbb{R}_{fin} \rightarrow [0, 1]^*$ as follows: For all $t' \in [a, b]^*$ and $r' \in \mathbb{R}_{fin}$,

$$(v_1^* + v_2^*)(t_1')(r') = \vee\{st(v_1(t', r_1') \wedge v_2(t', r_2'))|r' = r_1' + r_2'; r_1', r_2' \in \mathbb{R}_{fin}\}$$
$$+nst(r_1') \vee nst(r_2').$$

Consider the definition of $m_1 \sim m_2$. Let $m_1 = [(1, \frac{1}{4}, \frac{1}{9}, ...)]_U$ and $m_2 = [(1, \frac{1}{2}, \frac{1}{3}, ...)]_U$. Then $m_1 \approx m_2$. Let $t = [(1, \frac{1}{2}, \frac{1}{3}, ...)]_U$. Then $t \in \mu(0)$ and $\frac{m_1}{t} = [(1, \frac{1}{2}, \frac{1}{3}, ...)]_U$ and $\frac{m_2}{t} = [(1, 1, 1, ...)]_U$. Thus it's not the case that $\frac{m_1}{t} \approx \frac{m_2}{t}$. Hence it is not the case that $m_1 \sim m_2$.

Recall that $s(t)$ is the distance traveled at time t so $s : [a, b] \times \mathbb{R} \rightarrow [0, 1]$ gives the intensity that the distance traveled at time t is r, $s(t, r) \in [0, 1]$. Also $v(t)$ is the velocity of a particle at time t so $v : [a, b] \times \mathbb{R} \rightarrow [0, 1]$ gives the intensity that a particle is traveling r at time t., $v(t, r) \in [0, 1]$. Hence $(s(t)/t)(r)$ is the intensity that the velocity at time t is r. Also $l(t)(r)$, is the intensity that the velocity is r at time t. Thus $l(t) + s(t)/t$ is the sum of two fuzzy numbers which we define as follows: Given $t \in [a, b]$,

$$(l(t) + \frac{s(t)}{t})(r) = \vee\{l(t)(r_1) \wedge \frac{s(t)}{t}(r_2)|r = r_1 + r_2, r_1, r_2 \in \mathbb{R}\}$$

for all $r \in \mathbb{R}$.

Let f be integrable on the interval $[a, b]$. For all $t \in [a, b]$, define the function F of $[a, b] \rightarrow \mathbb{R}$ by for all $t \in [a, b]$, $F(t) = \int_a^t f(x)dx$. Let f^* and F^* be the natural extensions of f and F to $\mathbb{R}$, respectively. Define $\int_a^t f^*(x)dx$ to be $F^*(t)$.

2.8 The Nonstandard Interval $]^-0, 1^+[$

We now define another nonstandard unit interval. We follow [10] in the presentation. In [10] the relation $\lessapprox$ is defined on $\mathbb{R}^*$ by for all $x, y \in \mathbb{R}^*$, $x \lessapprox y$ if and only if $x < y$ or $x \approx y$. However, note that it is possible for $m_1, m_2 \in \mu(0)$, $m_1 \neq m_2$, with $m_1 < m_2$, where $<$ is the order of $\mathbb{R}^*$. Yet $m_1 \approx m_2$. Thus when we write $x \lessapprox y$, we mean that if it is not the case that $x \approx y$, then $x < y$.

Let $x = s_1 + m_1$ and $y = s_2 + m_2$, where $s_1, s_2 \in \mathbb{R}$ and $m_1, m_2 \in \mu(0)$. If $s_1 = s_2$ and $m_1 < m_2$, then $x < y$ and yet $st(x) = st(y)$. That is, $x < y$ does not imply $st(x) < st(y)$. However $st(x) < st(y)$ implies $x < y$.

The nonstandard unit interval is defined as follows:

$$]^{-}0, 1^{+}[= \{x \in \mathbb{R}^{*} | 0 \lessapprox x \lessapprox 1\}.$$

Proposition 2.8.1 $]^{-}0, 1^{+}[= [0, 1] + \mu(0)$.

Proof Clearly, $[0, 1] + \mu(0) \subseteq]^{-}0, 1^{+}[$. Let $x \in]^{-}0, 1^{+}[$. Suppose $0 \approx x$. Then $x \in \mu(0)$. Suppose $x \approx 1$. Then $1 - x \in \mu(0)$ and so $x \in 1 + \mu(0)$. Suppose x is such that neither $0 \approx x$ nor $x \approx 1$. Then $0 < x$ and $x < 1$. Thus $x = r + m$, for some $r \in \mathbb{R}$ and $m \in \mu(0)$ with $0 < r < 1$. □

We note that $]^{-}0, 1^{+}[$ is not the natural extension of $[0, 1]$ to $\mathbb{R}^{*}$. See also [10].

A function from a set X into $]^{-}0, 1^{+}[$ is called a **nonstandard fuzzy subset** of X. Let σ be a nonstandard fuzzy subset of a set X. Then $st \circ \sigma$ is a fuzzy subset of X. Now $(st \circ \sigma)(x) > 0 \Leftrightarrow st(\sigma(x)) > 0 \Leftrightarrow \sigma(x) \notin \mu(0)$. Now $\text{Supp}(st \circ \sigma) = \{x \in X | (st \circ \sigma)(x) > 0\} = \{x \in X | \sigma(x) \notin \mu(0)\}$. We call $\{x \in X | \sigma(x) \notin \mu(0)\}$ the **nonstandard support** of σ and write $\text{NSupp}(\sigma) = \{x \in X | \sigma(x) \notin \mu(0)\}$. That is, $\text{Supp}(st \circ \sigma) = \text{NSupp}(\sigma)$.

Let $t \in]^{-}0, 1^{+}[$ and $x \in X$. Then $\sigma(x) \gtrapprox t \Leftrightarrow st(\sigma(x)) \geq st(t)$. Hence $x \in \sigma_t \Leftrightarrow x \in (st \circ \sigma)_{st(t)}$. That is, $\sigma_t = (st \circ \sigma)_{st(t)}$.

Consider $a + m, b + m', c + m''$, where $a, b, c \in R$ and $m, m', m'' \in \mu(0)$.

$(*) a + m \gtrapprox b + m' \Leftrightarrow a + m > b + m'$ or $a + m \approx b + m' \Leftrightarrow a > b$ or $a = b \Leftrightarrow a \geq b$.

$(**) b + m' \lessapprox a + m \lessapprox c + m'' \Leftrightarrow b < a$ or $b = a$ and $a < c$ or $a = c \Leftrightarrow b \leq a \leq c$.

Definition 2.8.2 Let $I \subseteq \mathbb{R}_{fin}$. Then I is called a **closed nonstandard interval** if there exists $s_1, s_2 \in \mathbb{R}_{fin}$ such that $I = \{x \in \mathbb{R}_{fin} | s_1 \lessapprox x \lessapprox s_2\}$.

Definition 2.8.3 Let $\sigma : \mathbb{R} \to]^{-}0, 1^{+}[$. Then σ is called a **nonstandard fuzzy number** if the following conditions hold:

(1) There exists a unique $r \in \mathbb{R}$ such that $\sigma(r) \approx 1$.
(2) For all $t \in]^{-}0, 1^{+}[$, $\sigma_t = \{x \in \mathbb{R} | \sigma(x) \gtrapprox t\}$ is a closed nonstandard interval.
(3) $\text{NSupp}(\sigma)$ is bounded.

Proposition 2.8.4 If σ is a nonstandard fuzzy number, then $st \circ \sigma$ is a fuzzy number. Conversely, if $\widehat{\sigma} : \mathbb{R} \to [0, 1]$ is a (standard) fuzzy number, then $\widehat{\sigma}$ can be extended to a nonstandard fuzzy number σ.

Let $\mathbb{D}$ be a nonempty subset of $\mathbb{R}$. Let $\mathcal{NFP}(\mathbb{R})$ denote the set of all nonstandard fuzzy subsets of $\mathbb{R}$.

Theorem 2.8.5 [6] Let $\widetilde{f}$ be a function of $\mathbb{D}$ into $\mathcal{NFP}(\mathbb{R})$ such that $\forall x \in \mathbb{D}$, $\widetilde{f}(x)$ is a nonstandard fuzzy number. Suppose $\forall x \in \mathbb{D}, \forall t \in]^-0, 1^+[$, $\widetilde{f}(x)_t$ is a closed nonstandard interval. Then there exists unique functions f^-, f^+ of $]^-0, 1^+[\times\mathbb{D}$ into $\mathbb{R}$ such that the following conditions hold:

(1) $\forall x \in \mathbb{D}$, $f^-(\_, x)$, $f^+\_, x)$ is a non decreasing function (non increasing function);

(2) $\forall(t, x) \in]^-0, 1^+[\times\mathbb{D}$, $f^-(t, x) \leq f^+(t, x)$;

(3) $\forall(t, x) \in]^-0, 1^+[\times\mathbb{D}$, $\widetilde{f}(x)_t = [f^-(t, x), f^+(t, x)]$;

(4) $\forall x \in \mathbb{D}$, $f^-(1, x) = f^+(1, x)$.

Theorem 2.8.6 [6] Let g and h be functions of $]^-0, 1^+[\times\mathbb{D}$ into $\mathbb{R}$ such that $\forall x \in \mathbb{D}$, $g(\_, x)$ and $h(\_, x)$ are non decreasing and non increasing functions of t, respectively, and $\forall(t, x) \in]^-0, 1^+[\times\mathbb{D}$, $g(t, x) \leq h(t, x)$. Define the functions $\widehat{g}$ and $\widehat{h}$ of $[0, 1] \times \mathbb{D}$ into $\mathbb{R}$ by for all $(a, x) \in [0, 1] \times \mathbb{D}$, $\widehat{g}(a, x) = g(a, x)$ and $\widehat{h}(a, x) = h(a, x)$. Let $\widehat{f}$ be the function of $\mathbb{D} \times \mathbb{R}$ into $[0, 1]$ defined as follows: $\forall(x, y) \in \mathbb{D} \times \mathbb{R}$,

$$\widehat{f}(x, y) = \begin{cases} \sup\{\beta \in [0, 1] | y \in [\widehat{g}(\beta, x), \widehat{h}(\beta, x)] \\ \quad \text{if } y \in [\widehat{g}(0, x), \widehat{h}(0, x)], \\ \quad 0 \text{ otherwise.} \end{cases}$$

Define the function $\widetilde{f} : \mathbb{D} \times \mathbb{R} \to [^-0, 1^+[$ by $\forall(x, y) \in \mathbb{D} \times \mathbb{R}$, $\widetilde{f}(x, y) = \widehat{f}(x, y) + m$ for some $m \in M$. If $\forall x \in \mathbb{D}$, $\widehat{g}(\_, x)$ and $\widehat{h}(\_, x)$ are continuous from the left, then $\widetilde{f}(x)_t = [g(t, x), h(t, x)]$ for all $t \in]^-0, 1^+[$.

Theorem 2.8.7 [6] Let g and h be functions of $]^-0, 1^+[\times\mathbb{D}$ into $\mathbb{R}$ such that $\forall x \in \mathbb{D}$, $g(\_, x)$ and $h(\_, x)$ are nondecreasing and nonincreasing functions of t, respectively, and $\forall(t, x) \in]^-0, 1^+[\times\mathbb{D}$, $g(t, x) \leq h(t, x)$. Then the following conditions hold.

(1) Suppose there exists a function $\widehat{f}$ of $\mathbb{D} \times \mathbb{R}$ into $[0, 1]$ such that $\forall a \in [0, 1]$, $\widehat{f}(x)_a = [\widehat{g}(a, x), \widehat{h}(a, x)]$. Then $\forall x \in \mathbb{D}$, $\widehat{g}(\_, x)$ and $\widehat{h}(\_, x)$ are continuous from the left, and $\widehat{f}(x)(y) = \sup[\beta \in [0, 1] | y \in [\widehat{g}(\beta, x), \widehat{h}(\beta, x)]$ if $y \in [\widehat{g}(0, x), \widehat{h}(0, x)]$ and 0 otherwise.

(2) Suppose there exists a function $\widetilde{f}$ of $\mathbb{D} \times \mathbb{R}$ into $]^-0, 1^+[$ such that $\forall t \in]^-0, 1^+$, $\widetilde{f}(x)_t = [g(t, x), h(t, x)]$. Then $\widetilde{f}(x, y) \approx \widehat{f}(x, y)$ for all $(x, y) \in \mathbb{D} \times \mathbb{R}$.

Definition 2.8.8 Let $\widetilde{f}$ be a function of $\mathbb{D} \times \mathbb{R}$ into $]^-0, 1^+[$. Suppose that $\forall \alpha \in]^-0, 1^+[$, $f^-(\alpha, x)$ and $f^+(\alpha, x)$ are differentiable at x_0 and that

(1) $D_x f^-(\_, x_0)$ is a nondecreasing function of α and is continuous from the left in a.

(2) $D_x f^+(\_, x_0)$ is a nonincreasing function of α and is continuous from the left in a.

(3) $D_x f^-(\alpha, x) \le D_x f^+(\alpha, x)$ for all $\alpha \in]^-0, 1^+[$.

Then $\widetilde{f}$ is said to be **differentiable** at x_0 and we write $D_x \widetilde{f}(x_0)$. If $\widetilde{f}$ is differentiable at every point of $\mathbb{D}$, then $\widetilde{f}$ is said to be **differentiable** on $\mathbb{D}$.

Assume $\forall \alpha, \beta \in]^-0, 1^+[, \forall x \in \mathbb{D}, \alpha \approx \beta$ implies $f^-(\alpha, x) = f^-(\beta, x)$ and $f^+(\alpha, x) = f^+(\beta, x)$. Then $\forall \alpha \in]^-0, 1^+[, \forall x \in \mathbb{D}, f^-(\alpha, x) = f^-(st(\alpha), x) = \widehat{f}^-(st(\alpha), x)$ and $f^+(\alpha, x) = f^+(st(\alpha), x) = \widehat{f}^+(st(\alpha), x)$. For all.$(x, y) \in \mathbb{D} \times \mathbb{R}$, define $D_x \widetilde{f}(x, y) = D_x \widehat{f}(x, y)$, where we assume $\widehat{f}$ is differentiable in x.

2.9 Nonstandard Fuzzy Numbers

In the following, we do not assume that A^* is the natural extension of a function from R into [0, 1].

Let $A^* : \mathbb{R}^* \to]^-0, 1^+[$. Let $\alpha \in]^-0, 1^+[$. Then the α-**level set** of A^* is defined to be the set $\{x \in \mathbb{R}^* | A^*(x) \gtrapprox \alpha\}$ and is denoted by $A^{*\alpha}$. The **nonstandard support** of A^*, written, NSupp(A^*), is defined to be the set $\{x \in \mathbb{R}^* | A^*(x) \notin \mu(0)\}$.

We next consider a different definition of a nonstandard fuzzy number than that given in Definition 2.1.8

Definition 2.9.1 A^* is a **nonstandard fuzzy number** if the following properties hold:

(1) There exist $x \in \mathbb{R}^*$ such that $A^*(x) \approx 1$,

(2) $\forall \alpha \in]^-0, 1^+[$, there exists $c_\alpha, d_\alpha \in]^-0, 1^+[$ such that $c_\alpha \le d_\alpha$ and $A^{*\alpha} = \{x \in \mathbb{R}^* | c_\alpha \lessapprox x \lessapprox d_\alpha\}$.

(3) There exists $c, d \in \mathbb{R}_{fin}$ such that $c \le d$ and NSupp$(A^*) \subseteq \{x \in R^* | c \lessapprox x \lessapprox d\}$.

We note that $st \circ A^* : \mathbb{R}^* \to [0, 1]$.

We next consider $(st \circ A^*)|_\mathbb{R}$, where A^* is a nonstandard fuzzy number. We show $(st \circ A^*)|_\mathbb{R}$ is a fuzzy number.

(1′) Let $x \in \mathbb{R}^*$ be such that $A^*(x) \approx 1$. Then $(st \circ A^*)(x) = 1$.

(2′) Let $\alpha \in [0, 1]$ and c_α, d_α be as in (2) above. Now $A^{*\alpha} = \{x \in \mathbb{R}^* | c_\alpha \lessapprox x \lessapprox d_\alpha\}$. Thus $((st \circ A^*)|_\mathbb{R})^\alpha \subseteq \{x \in \mathbb{R}^* | c_\alpha \lessapprox x \lessapprox d_\alpha\}$. Let $x \in \mathbb{R}$. Now $c_\alpha \lessapprox x \lessapprox$

$d_\alpha \Leftrightarrow st(c_\alpha) \le x \le st(d_\alpha)$ since $x \in \mathbb{R}$. Hence $((st \circ A^*)|_{\mathbb{R}})^\alpha \subseteq \{x \in \mathbb{R}|st(c_\alpha) \le x \le st(d_\alpha)\}$. Let $y \in \{x \in \mathbb{R}|st(c_\alpha) \le x \le st(d_\alpha)\}$. Then $st(c_\alpha) \le y \le st(d_\alpha)$ and so $c_\alpha \lessapprox y \lessapprox d_\alpha$. Thus $y \in A^{*\alpha}$. Hence $A^*(y) \gtrapprox \alpha$ and so $(st \circ A^*)|_R(y) \ge \alpha$ since $\alpha \in \mathbb{R}$. Thus $y \in ((st \circ A^*)|_{\mathbb{R}})^\alpha$. Therefore, $((st \circ A^*)|_{\mathbb{R}})^\alpha = [st(c_\alpha), st(d_\alpha)]$.

(3′) Let $x \in \mathbb{R}$ and $\alpha \in [0, 1]$. Then $(st \circ A^*)|_{\mathbb{R}}(x) > 0 \Rightarrow (st \circ A^*)(x) > 0$ (since $x \in \mathbb{R}$) $\Rightarrow A^*(x) \notin M \Rightarrow x \in$NSupp$(A^*) \Rightarrow$Supp$(st \circ A^*) \subseteq [st(c), st(d)]$, where c, d are as in (3). Thus Supp$(st \circ A^*)|_{\mathbb{R}}$ is bounded.

Note that if A is a fuzzy number, then there exists $\omega_1, \omega_2 \in \mathbb{R}$ such that $\omega_1 \le \omega_2$ and $A(x) = 0$ for all $x \in (-\infty, \omega_1) \cup (\omega_2, \infty)$.

Theorem 2.9.2 [8] Let A be a fuzzy subset of $\mathbb{R}$. Then A is a fuzzy number if and only if there exists a closed interval $[a, b] \ne \emptyset$ such that

$$A(x) = \begin{cases} 1 \text{ if } x \in [a, b], \\ l(x) \text{ if } x \in (-\infty, a), \\ r(x) \text{ if } x \in (b, \infty), \end{cases}$$

where l is a function from $(-0, a)$ to $[0, 1]$ that is monotonic increasing, continuous from the right, and such that $l(x) = 0$ for $x \in (-\infty, \omega_1)$; r is a function from $(0, \infty)$ to $[0, 1]$ that is monotonic decreasing, continuous from the left, and such that $z(x) = 0$ for $x \in (\omega_2, \infty)$.

Corollary 2.9.3 Let A^* be a nonstandard fuzzy number on $\mathbb{R}^*$. Then there exists a closed interval $[a, b] \ne \emptyset$ such that

$$(st \circ A^*)|_{\mathbb{R}}(x) = \begin{cases} 1 \text{ if } x \in [a, b], \\ l(x) \text{ if } x \in (-\infty, a), \\ r(x) \text{ if } x \in (b, \infty), \end{cases}$$

where l is a function from $(-0, a)$ to $[0, 1]$ that is monotonic increasing, continuous from the right, and such that $l(x) = 0$ for $x \in (-\infty, \omega_1)$; r is a function from $(0, \infty)$ to $[0, 1]$ that is monotonic decreasing, continuous from the left, and such that $r(x) = 0$ for $x \in (\omega_2, \infty)$.

Proof The result follows since $(st \circ A^*)|_{\mathbb{R}}$ is a fuzzy number. □

References

1. Robinson, A.: Nonstandard Analysis, North-Holland Publishing Co. (1966)
2. Herrmann, R.A.: Nonstandard Analysis Applied to Special and General Relativity—The Theory of Infinitesimal Light-Clocks, Special arXiv:math/0312189v10 [math.GM], pp. 1–109 (2014)

3. Rayo, D.A.B.: Introduction to non-standard analysis, http://math.uchicago.edu REUPapers 2–15
4. Davis, I.: An introduction to nonstandard analysis, https://www.math.uchicago.edu VIGRE(2009)
5. Mordeson, J.N., Mathew, S.: Fuzzy mathematics and nonstandard analysis: applications to the theory of relativity. Trans. Fuzzy Sets Syst. **1**(1), 143–154 (2022)
6. Mordeson, J.N., Mathew, S., Binu, M.: *Applications of Mathematics of Uncertainty, Grand Challenges—Human Trafficking—Coronavirus—Biodiversity and Extinction*, Studies in Systems and Control vol. 391, Springer (2022)
7. Goldberg, I.: Lecture notes on nonstandard analysis, UCLA Summer School in Logic (2014)
8. Klir, G.J., Yuan, B.: Fuzzy Sets and Fuzzy Logic: Theory and Applications, Prentice Hall P T R Upper Saddle River, New Jersey 07458 (1995)
9. Staunton, E., Ryan, E., Supervisor.: Infinitesimals, Nonstandard Analysis and Applications to Finance, NUI Galway OE Gaillimh, National University of Ireland, Galway (2013)
10. Mordeson, J.N., Mathew, S.: Nonstandard fuzzy mathematics. Adv. Fuzzy Sets Syst. **27**, 35–52 (2022)

Chapter 3
Social Networks and Climate Change

The purpose of this chapter is to introduce the ideas from social network theory to model feedback processes in climate change. We first discuss negative feedback, positive feedback, and tipping points. We then introduce the basics of social networks and show how they can be applied to the study of feedback loops.

3.1 Feedback in the Climate System

The following three paragraphs are from [1]. Climate change feedbacks are important in the understanding of global warming because feedback processes amplify or diminish the effect of each climate forcing. They thus play an important part in determining the climate sensitivity and future climate state. Feedback in general is the process in which changing one quantity changes a second quantity, and the change in the second quantity in turn changes the first. Positive or reinforcing feedback amplifies the change in the first quantity while negative or balancing feedback reduces it.

The term forcing means a change which may push the climate system in the direction of warming or cooling. An example of climate forcing is increased atmospheric concentrations of greenhouse gases. Forcings are external to the climate system while feedbacks are internal. Feedbacks present internal processes of the system. Some feedbacks may act in relative isolation to the rest of the climate system, others may be tightly couples; hence it may be difficult to tell just now much a particular process contributes.

Forcing and feedbacks together determine how much and how fast climate changes. The main positive feedback in global warming is the tendency of warming to increase the amount of water vapor in the atmosphere, which in turn leads to further warming. The main cooling response comers from the Stefan-Boltzmann law, the amount of heat radiated from earth into space changes with the fourth power of the temperature of Earth's surface and atmosphere. It is typically not considered a feedback. Observations and modelling studies indicate that there is a net positive

J. N. Mordeson et al., *Fuzzy Graph Theory*, Studies in Fuzziness and Soft Computing 424, https://doi.org/10.1007/978-3-031-23108-7_3

feedback to warming. Large positive feedbacks can lead to effects that are abrupt or irreversible, depending upon the rate and magnitude of the climate change.

The following is taken from [2]. The cascading effects of climate change can have unforeseen consequences. These are the climate feedback loops that either amplify or reduce the effects of climate change.

Positive Feedback Loop: In a positive feedback loop, an initial warming triggers a feedback to amplify the effects of warming.

Negative Feedback Loop: Negative feedback loops reduce the effects of climate change.

Once you begin taking climate out of its balance, these positive and negative feedback loops start to kick in. Then they can eventually go beyond our ability to control it. Negative climate feedback loops have beneficial results. Instead of continued warming, they spark a favorable chain of events that lessen the severity of climate change. The following are examples of negative feedback mechanisms for climate change.

The following is taken from [2].

Negative Feedback Loops

1. Increased cloudiness reflects more incoming solar radiation.
   As ice sheets melt, this could increase cloudiness with more water vapor in the atmosphere. Because clouds reflect 1/3 of incoming solar radiation, there would be less heat absorption on earth's surface.
2. Higher rainfall from moisture in the atmosphere.
   Similarly, if there's more water held in the atmosphere, then higher water volume leads to more precipitation. This is because the atmosphere can retain more moisture with higher temperatures. The downside is that ocean circulation patterns would change and create an imbalance of where rainfall occurs.
3. Net primary productivity increase.
   As higher concentrations of CO_2 enter the atmosphere, plants have more material to photosynthesize. If you isolate a single plant in a laboratory, then adding CO_2 makes Earth greener for now. But this fertilization effect diminishes with time. However plants can't grow indefinitely with rising CO_2. This is because plants require other factors like nitrogen in the nutrient cycle. And if temperature rises, this can negatively influence plant growth.
4. Blackbody radiation.
   The energy released by Earth' is a function of temperature. If Earth's temperature increases, it raises the amount of outgoing radiation. Thus the more energy you add to Earth, the more energy it will emit. This concept is the Stefan-Boltzmann law which has an overall cooling effect.
5. Chemical weathering as a carbon dioxide sink.
   With more CO2 and water in the atmosphere, it increases carbonic acid which is just CO_2 and water. Chemical weathering in rocks is a sink atmospheric carbon dioxide. Thus it weakens the greenhouse effect and leads to cooling.

6. The ocean's solubility pump.
   The solubility pump refers to the oceans ability to transport carbon from its surface to the interior. The ocean serves an important role in regulating CO_2 by dissolving it in water.
   As ice sheets melt, carbon storage increases. Currently, oceans absorb 33% of CO_2 emitted to the atmosphere. Although this process cannot continue indefinitely, solubility pump efficiency depends on ocean circulation.
7. Lapse rate and altitude temperature.

Lapse rate to the change of temperature with altitude. Air expands higher in thew troposphere because there is less pressure. Conversely, air compresses lower in the troposphere because there is more pressure.

Climate models indicate that global warming will reduce the decreasing rate of temperatures with height. Overall, thus weakens the strength of the greenhouse effect.

Positive Feedback Loops

Positive climate feedback loops accumulate to a more harmful result with increased heating. The bad outweighs the good for climate scenarios by far. Methane release is the most devastating. It has the potential to cause a lethal chain of atmospheric heating.

1. Permafrost melt sparks methane release

In the Arctic tundra, permafrost melt will trigger methane release in the atmosphere. Because methane is a more potent greenhouse gas than CO_2.

This type of positive feedback loop could be a tipping point for our climate. Currently, there are only about 5 gigatons of methane in the atmosphere. However the amount of methane in the Arctic is in the hundreds of gigatons.

2. The removal of ice high albedo

Once the Arctic, Greenland and Antarctic ice sheets melt, water absorbs more heat. Because ice has a high albedo, it reflects 84% of incoming solar radiation.

But once we remove our protective shield water vapor reflects as low as 5% of solar radiation.

3. Ocean circulation patterns disruption

Once ice melts in the Arctic, it will start shifting deep ocean circulation patterns in the Gulf Stream. Currently, this circulation pattern relies on heavy salt water from the north to transport warm water to Great Britain.

Once ice sheets melt, it releases freshwater into the oceans. This disturbs this ocean conveyor belt by slowing downflow in the Atlantic Ocean.

4. Sea level rise

As the planet warms, ocean waters expand. Rising sea levels hit coastal cities the hardest. But another result will trigger further glacier calving, i.e., the process by which an iceberg breaks off from a shelf or glacier.

If you increase water volume, this could cause further chunks of ice to outpour into the oceans.

5. Rainforest drought and loss

Temperatures are projected to rise 2 to 6^0 by 2100. As a result of the warmer climate, this will result in larger evaporation losses.

Despite the possibility of more rainfall, unpredictable weather may result in less soil moisture. Drought due to a warmer means the loss of some of the most productive places in the world.

6. Wetland methane release

Wetlands are the largest natural source of methane in the world. Climate change is concerned with their health because heating can cause bogs to release methane.

The amount of methane production is dependent on a number of factors. For example, soil temperature, oxygen availability, and warmer environments all relate to climate change contributors.

7. More kindle for forest fires

Mid-latitude regions are poised to receive an imbalance in rainfall and increasing risk of drought. As a result, forest fires and desertification in forested regions will lower their ability to be carbon sinks.

Overall, this releases more carbon than forests can absorb into the atmosphere. Thus, this positive feedback loop causes further warming.

8. Gas hydrates in shallow water

In the shallow oceans, gas hydrates store enormous amounts of methane. They occur naturally throughout the world as forms of ice and methane,

Because we find them relatively shallow, they are particularly susceptible to warmer temperatures. And similar to melting permafrost, methane releases a potent greenhouse gas. Thus it causes further global warming.

In summary,climate feedback loops are a process in which an external factor, such as the release of heat-trapping greenhouse gases or the injection of aerosols into the atmosphere, cause a change in one part of the climate system that feedback and amplifies itself.

The Earth's climate is constantly changing due to such feedback loops. These loops can be either positive or negative. Positive feedback loops increase the rate of change in a particular system. Whereas negative feedback loops slow down or reverse changes caused by external factors.

3.2 Tipping Points

The following discussion is from [3]. The following possible tipping points are presented in [3].

A. Amazon rainforest

Frequent droughts

B. Arctic sea ice

Reduction in area

C. Atlantic circulation

In slowdown since 1950

D. Boreal forest

Fires and pests changing

F. Coral reefs

Large-scale die-offs

G. Greenland ice sheet

Ice loss accelerating

H. Permafrost

Thawing

I. West Antarctic ice sheet

Ice loss accelerating

J. Wilkes Basin

East Antarctica ice loss accelerating

It is stated that several cryosphere tipping points are dangerously close, but mitigating greenhouse-gas emissions could still slow down the inevitable accumulation of impacts and help us to adapt. Consider ice collapse. The Amundsen Sea embayment of West Antarctic might have passed a tipping point. Also, part of the East antarctic ice sheet - the Wilkes Basin- might be similarly unstable. The Greenland ice sheet is melting at an accelerating rate. Models suggest that the Greenland ice sheet could be doomed at 1.5°C of warming, which could happen as soon as 2030.

Climate change and other human activities risk triggering biosphere tipping points across a range of ecosystems and scales. Ocean heatwaves have led to mass coral bleaching and to the loss of half of the shallow-water corals on Australia's Great Barrier Reef. It is projected that 99% of tropical corals to be lost if global average temperature rises by 2°C owing to interactions between warming, ocean acidification and pollution. This would represent a profound loss of marine biodiversity and human livelihoods. Deforestation and climate change are destabilizing the Amazon—The world's largest rainforest. With the Arctic warming at least twice as quickly as the global average, the boreal forest in the subarctic is increasingly vulnerable. Already, warming has triggered large-scale insect disturbances and an increase in fires that have led to dieback of North American boreal forests, potentially turning some regions from a carbon sink to a carbon source. Permafrost across the Arctic is beginning to irreversibly thaw and release carbon dioxide and methane.

It is stated in [3] that the clearest emergency would be if we were approaching a global cascade of tipping points that led to a new, less habitable hothouse climate state. Interactions could happen through ocean and atmospheric circulation or through feedbacks that increase greenhouse-gas levels and global temperature. Alternatively, strong cloud feedbacks could cause global tipping points.

The following Table 3.1 presents the domino effect of the tipping points listed previously.

Table 3.1 Domino effect of the tipping points

	A	B	C	D	F	G	H	I	J
A									
B			→	→		→	→		
C	→							→	→
D									
F									
G			→						
H									
I									
J									

The following formula is presented in [3].

$$E = R \times U = p \times D \times \tau/T,$$

where E denotes emergency, R risk defined by insurers as probability p multiplied by damage D. Urgency U is defined in emergency situations as reaction time to an alert τ divided by the intervention time left to avoid a bad outcome T. The situation is an emergency if both risk and urgency are high. If the reaction time is longer than the intervention time left ($\tau/T > 1$), we have lost control.

3.3 Social Networks

The following is from ([4], p. 9). Our standard model describes a method by which persons weigh and integrate their own attitudes and the attitudes of others of an issue:

$$y_i^{(t+1)} = a_{ii} \sum_{j=1}^{n} w_{ij} y_j^{(t)} + (1 - a_{ii}) y_i^{(1)}$$

for each $i = 1, 2, ..., n$ and $t = 1, 2, ...$ The n group members' time t positions on an issue are $y_1^{(t)}, y_2^{(t)}, ..., y_n^{(t)}$, and these positions include their initial set of positions $y_1^{(1)}, y_2^{(1)}, ..., y_n^{(1)}$. Group members' individual susceptibilities to interpersonal influence are $a_{11}, a_{22}, ..., a_{nn}$, where $0 \leq a_{ii} \leq 1$ for all i. The relative interpersonal influence of each group member j on i is $w_{i1}, w_{12}, ..., w_{1n}$, where $0 \leq w_{oij} \leq 1$ for all i and j, $\sum_{j=1}^{n} w_{ij} = 1$, and $w_{ii} = 1 - a_{ii}$ for all i. Note that $\sum_{j=1}^{n} a_{ii} w_{ij} + (1 - a_{ii}) = 1$, so that i's attitude at time $t + 1$ is formed as a weighted average of the attitudes of others and self at time t, and i's initial position. If we assume that $0 \leq y_i^{(1)} \leq 1$ for $i = 1, 2, ..., n$, then it is easy to show by induction on t that $0 \leq y_i^{(t)} \leq 1, i = 1, 2, ..., n$.

The following is from ([4], p. 10). Individual preferences are allowed in persons' susceptibilities to interpersonal influence and in their profiles of accorded interpersonal influences. If a person i's susceptibility is $a_{ii} = 0$, then i's position on the issue does not change. If $a_{ii} = 1$, then i attaches no weight to his or her initial position, and i's initial position may be modified by the interpersonal influences of one or more other members of the group. The relative weights of others' positions, and the positions that they take on the issue at time t, determine the modification. If $0 < a_{ii} < 1$, then i's initial position on the issue has some weight in any modification of i' s position that occurs.

We introduce this social network concept to the modelling of feedback loops with respect to climate change. We consider both negative and positive feedbacks. We interpret $y_i^{(t)}$ to be the degree of strength of i at time t in effecting climate change. When dealing with negative feedback loops, $0 \leq y_i^{(t)} \leq 1$ and when considering positive feedback loops, $-1 \leq y_i^{(t)} \leq 0$. As for social networks, tipping points' individual susceptibilities to interpersonal change are $a_{11}, a_{22}, ..., a_{nn}$, where $0 \leq a_{ii} \leq 1$ for all i. The relative interpersonal influence of each member j on i is $w_{i1}, w_{12}, ..., w_{1n}$, where $0 \leq w_{oij} \leq 1$ for all i and j, $\sum_{j=1}^{n} w_{ij} = 1$, and $w_{ii} = 1 - a_{ii}$ for all i.

$$y_i^{(t+1)} = a_{ii} \sum_{j=1}^{n} w_{ij} y_j^{(t)} + (1 - a_{ii}) y_i^{(1)}, i = 1, 2, ..., n.$$

Let $t = 1$. Then

$$\begin{aligned} y_i^{(2)} - y_i^{(1)} &= a_{ii} \sum_{j=1}^{n} w_{ij} y_j^{(t)} + (1 - a_{ii}) y_i^{(1)} - y_i^{(1)} \\ &= a_{ii} \left(\sum_{j=1}^{n} w_{ij} y_j^{(1)} - y_i^{(1)} \right). \end{aligned}$$

Assume $a_{ii} > 0$. Then

$$\begin{aligned} y_i^{(2)} - y_i^{(1)} > 0 &\Leftrightarrow \sum_{j=1}^{n} w_{ij} y_j^{(1)} - y_i^{(1)} > 0 \\ &\Leftrightarrow \sum_{j=1}^{n} w_{ij} (y_j^{(1)} - y_i^{(1)}) > 0 \end{aligned} \tag{3.1}$$

since $\sum_{j=1}^{n} w_{ij} = 1$.

Suppose $t \geq 2$. Then

$$y_i^{(t+1)} - y_i^{(t)} = a_{ii} \sum_{j=1}^{n} w_{ij} y_j^{(t)} + (1 - a_{ii}) y_i^{(1)}$$
$$- \left[a_{ii} \sum_{j=1}^{n} w_{ij} y_j^{(t-1)} + (1 - a_{ii}) y_i^{(1)} \right]$$
$$= a_{ii} \left(\sum_{j=1}^{n} w_{ij} (y_j^{(t)} - y_j^{(t-1)}) \right).$$

Thus

$$y_i^{(t+1)} - y_i^{(t)} > 0 \Leftrightarrow \sum_{j=1}^{n} w_{ij} (y_j^{(t)} - y_j^{(t-1)}) > 0. \tag{3.2}$$

Suppose $y_i^{(2)} = a_{ii}(w_{ii} y_i^{(1)} + w_{ij} y_j^{(1)}) + (1 - a_{ii}) y_i^{(1)}$ with $w_{ij} > 0$ and $0 < a_{ii}$. Then

$$y_i^{(2)} - y_i^{(1)} = a_{ii}(w_{ii} y_i^{(1)} - y_i^{(1)} + w_{ij} y_j^{(1)}).$$

Now $w_{ii} + w_{ij} = 1$. Thus

$$y_i^{(2)} - y_i^{(1)} = a_{ii}(w_{ii} y_i^{(1)} - y_i^{(1)} + (1 - w_{ii}) y_j^{(1)})$$
$$= a_{ii}((w_{ii} - 1) y_i^{(1)} + (1 - w_{ii}) y_j^{(1)})$$
$$= a_{ii}(1 - w_{ii})(-y_i^{(1)} + y_j^{(1)})$$

Hence $y_i^{(2)} - y_i^{(1)} \geq 0$ if and only if $y_j^{(1)} \geq y_i^{(1)}$. Since $y_j^{(1)} < y_i^{(1)}$ is possible, $y_i^{(2)} - y_i^{(1)} < 0$ is possible. However, when we are considering only negative feedback and assuming the climate is such that $y_i^{(2)} - y_i^{(1)} \geq 0$, we assume the condition

$$y_i^{(2)} = y_i^{(1)} \vee [a_{ii} \sum_{j=1}^{n} w_{ij} y_j^{(1)} + (1 - a_{ii}) y_i^{(1)}]$$

for $i = 1, \ldots, n$.

Proposition 3.3.1 Suppose $y_i^{(2)} \geq y_i^{(1)}$ for $i = 1, \ldots, n$. Then $y_i^{(t+1)} \geq y_i^{(t)}$ for $i = 1, \ldots, n$ and $t = 1, 2, \ldots$.

Proof The condition for $t = 1$ is assumed true. Suppose $y_i^{(t+1)} \geq y_i^{(t)}$. Then

$$y_i^{(t+2)} - y_i^{(t+1)} = a_{ii} \sum_{j=1}^{n} w_{ij}(y_j^{(t+1)} - y_j^{(t)}) \geq 0$$

since $y_j^{(t+1)} - y_j^{(t)}$ for $j = 1, ..., n$, the induction hypothesis. □

Example 3.3.2 Consider a cycle with $n = 4$, $w_{11} = w_{22} = w_{33} = w_{44} = w_{14} = w_{21} = w_{32} = w_{42} = 0.5$.. Let $y_1 = 0.5$, $y_2 = 0.4$, $y_3 = 0.3$, and $y_4 = 0.2$. Consider for $i = 1, 2, 3$,

$$\begin{aligned}
&a_{i+1,i+1}(w_{i+1,i+1}y_{i+1}^{(1)} + w_{i+1,i}y_i^{(1)}) + (1 - a_{i+1,i+1})y_{i+1}^{(1)} \\
&= y_{i+1}^{(1)} + \frac{1}{4}(y_{i+1}^{(1)} + \frac{1}{4}y_i^{(1)}) - \frac{1}{2}y_{i+1}^{(1)} \\
&= y_{i+1}^{(1)} + \frac{1}{4}(y_i^{(1)} - y_{i+1}^{(1)}).
\end{aligned}$$

For $i = 1, 2, 3$, we have $y_{i+1}^{(1)} + \frac{1}{4}(y_i^{(1)} - y_{i+1}^{(1)}) = y_{i+1}^{(1)} + \frac{1}{4}(0.1)$ and so $y_i^{(2)} > y_i^{(1)}$. We have $y_1^{(1)} + \frac{1}{4}(y_4^{(1)} - y_1^{(1)}) = 0.5 + \frac{1}{4}(-0.3) < 0.5 = y_1^{(1)}$. Hence we use $y_1^{(2)} = y_1^{(1)} = 0.5$ in this case.

Example 3.3.3 Let $n = 4$. Let (1) denote More heat-trapping gases emitted, (2) Atmosphere warms, (3) Water evaporation, (4) Vapor. Assume $w_{13} = w_{12} = w_{23} = w_{24} = w_{31} = w_{34} = w_{41} = w_{42} = 0$. That is, we assume that we have a directed cycle, hence a feedback loop; (1) $\overset{w_{21}}{\rightarrow}$ (2) $\overset{w_{32}}{\rightarrow}$ (3) $\overset{w_{43}}{\rightarrow}$ (4) $\overset{w_{14}}{\rightarrow}$ (1). Consider (3.1). Then

$$y_1^{(2)} - y_1^{(1)} > 0 \Longleftrightarrow y_4^{(1)} > y_1^{(1)}.$$

By (3.2) with $t \geq 2$,

$$y_1^{(t+1)} - y_1^{(t)} > 0 \Leftrightarrow w_{11}(y_1^{(t)} - y_1^{(t-1)}) + w_{14}(y_4^{(t)} - y_4^{(t-1)}) > 0.$$

Similarly,

$$y_2^{(2)} - y_2^{(1)} > 0 \Longleftrightarrow y_1^{(1)} > y_2^{(1)}.$$

By (3.2) with $t \geq 2$,

$$y_2^{(t+1)} - y_2^{(t)} > 0 \Leftrightarrow w_{22}(y_2^{(t)} - y_2^{(t-1)}) + w_{21}(y_1^{(t)} - y_1^{(t-1)}) > 0.$$

Also,

$$y_3^{(2)} - y_3^{(1)} > 0 \Longleftrightarrow y_2^{(1)} > y_3^{(1)}.$$

By (3.2) with $t \geq 2$,

$$y_3^{(t+1)} - y_3^{(t)} > 0 \Leftrightarrow w_{33}(y_3^{(t)} - y_3^{(t-1)}) + w_{32}(y_2^{(t)} - y_2^{(t-1)}) > 0.$$

Also,

$$y_4^{(2)} - y_4^{(1)} > 0 \Longleftrightarrow y_3^{(1)} > y_4^{(1)}.$$

By (3.2) with $t \geq 2$,

$$y_4^{(t+1)} - y_4^{(t)} > 0 \Leftrightarrow w_{44}(y_4^{(t)} - y_4^{(t-1)}) + w_{43}(y_3^{(t)} - y_3^{(t-1)}) > 0.$$

We see that

$$\begin{aligned} y_4^{(2)} - y_4^{(1)} &> 0 \Rightarrow y_1^{(3)} - y_1^{(2)} > 0 \Rightarrow y_2^{(4)} - y_2^{(3)} > 0 \Rightarrow y_3^{(5)} - y_3^{(4)} > 0 \\ &\Rightarrow y_4^{(6)} - y_4^{(5)} > 0 \Rightarrow y_1^{(7)} - y_1^{(6)} > 0. \end{aligned} \tag{3.3}$$

If $y_i^{(1)} > y_{i+1}^{(1)}$, $i = 1$ or2 or 3 or $y_4^{(1)} > y_1^{(1)}$ can produce the implications in (3.3).

We can see that each time we have $y_4^{(t+1)} > y_4^{(t)}$, we 'add' w_{14} to (1), where by 'add' we mean an appropriate conorm. Hence (1) may strictly increase. At some point in time we may reach a tipping point.

Example 3.3.4 Let (1) denote Earth gets hotter, (2) Thawing tundra, Heat stressed forests, Methane hydrates, Warming oceans, (3) Release CO_2 and methane, (4) Arctic ice melt, and (5) Dark sea water absorbs light. Consider the two feedback loops:

$$(1) \overset{w_{21}}{\rightarrow} (2) \overset{w_{32}}{\rightarrow} (3) \overset{w_{13}}{\rightarrow} (1)$$

and

$$(1) \overset{w_{41}}{\rightarrow} (4) \overset{w_{54}}{\rightarrow} (5) \overset{w_{15}}{\rightarrow} (1).$$

Then

$$W = \begin{bmatrix} w_{11} & 0 & w_{13} & 0 & w_{15} \\ w_{21} & w_{22} & 0 & 0 & 0 \\ 0 & w_{32} & w_{33} & 0 & 0 \\ w_{41} & 0 & 0 & w_{44} & 0 \\ 0 & 0 & 0 & w_{54} & w_{55} \end{bmatrix}$$

and

$$y_1^{(t+1)} = a_{11}(w_{11}y_1^{(t)} + w_{13}y_1^{(t)} + w_{15}y_1^{(t)}) + (1 - a_{11})y_1^{(1)}. \tag{3.4}$$

Now (if $w_{11} = 0$),

$$W = \begin{bmatrix} w_{11}=0 & 0 & w_{13} & 0 & 0 \\ w_{21} & w_{22} & 0 & 0 & 0 \\ 0 & w_{32} & w_{33} & 0 & 0 \\ 0 & 0 & 0 & 0 & 0 \\ 0 & 0 & 0 & 0 & 0 \end{bmatrix} + \begin{bmatrix} w_{11}=0 & 0 & 0 & 0 & w_{15} \\ 0 & 0 & 0 & 0 & 0 \\ 0 & 0 & 0 & 0 & 0 \\ w_{41} & 0 & 0 & w_{44} & 0 \\ 0 & 0 & 0 & w_{54} & w_{55} \end{bmatrix}$$

Consider the two cycles separately. Then we have

$$y_1^{(t+1)} = a_{11}(w_{11}y_1^{(t)} + w_{13}y_3^{(t)}) + (1 - a_{11})y_1^{(1)} \tag{3.5}$$

for one cycle and

$$y_1^{(t+1)} = a_{11}(w_{11}y_1^{(t)} + w_{15}y_5^{(t)}) + (1 - a_{11})y_1^{(1)} \tag{3.6}$$

for the other.

Adding Eqs. (3.5) and (3.6) together we obtain

$$2y_1^{(t+1)} = a_{11}(2w_{11}y_1^{(t)} + w_{13}y_1^{(t)} + w_{15}y_1^{(t)}) + 2(1 - a_{11})y_1^{(1)} \tag{3.7}$$

Notice for W, $w_{11} + w_{13} + w_{15} = 1$ and that for (3.5)$w_{11} + w_{13} = 1$ and for (3.6)$w_{11} + w_{15} = 1$. Adding the latter two equations, we obtain $2w_{11} + w_{13} + w_{15} = 2$. Subtracting the two, we obtain $w_{11} = 1$. Thus even with $w_{11} = 0$, we have a problem with adding the two subcycles together to obtain (3.4).

For Eqs. (3.5) and (3.6),write

$$y_1^{\prime(t+1)} = a_{11}(w_{11}y_1^{(t)} + w_{13}y_3^{(t)}) + (1 - a_{11})y_1^{(1)} \tag{3.5$'$}$$

and

$$y_1^{\prime\prime(t+1)} = a_{11}(w_{11}y_1^{(t)} + w_{15}y_5^{(t)}) + (1 - a_{11})y_1^{(1)}. \tag{3.6$'$}$$

Then (3.7) becomes

$$y_1^{\prime(t+1)} + y_1^{\prime\prime(t+1)} = a_{11}(2w_{11}y_1^{(t)} + w_{13}y_1^{(t)} + w_{15}y_1^{(t)}) + 2(1 - a_{11})y_1^{(1)} \tag{3.7$'$}$$

We have $w_{13} = 1$ in (3.5′) and $w_{15} = 1$ in (3.6′). However $w_{13} + w_{15} = 1$ in (3.4). This approach will work if the two subcycles are vertex disjoint. For a definition of a subgraph, we might not want to require that weights add to 1.

3.4 Positive Feedback Loops

$$y_i^{(t+1)} = b_{ii}\sum_{j=1}^{n} z_{ij}y_j^{(t)} + (1 - b_{ii})y_i^{(1)}$$

Here the $0 \leq z_{ij} \leq 1$ and $\sum_{j=1}^{n} z_{ij} = 1$ for all $i = 1, ..., n$. Also, $b_{ii} = 1 - z_{ii}, i = 1, ..., n$. We assume $-1 \leq y_i^{(1)} \leq 0, i = 1, ..., n$.

Example 3.4.1 Consider the negative feedback loop:

$$(1) \overset{w_{21}}{\rightarrow} (2) \overset{w_{32}}{\rightarrow} (3) \overset{w_{13}}{\rightarrow} (1)$$

and the positive feedback loop

$$(1) \overset{z_{41}}{\rightarrow} (4) \overset{z_{54}}{\rightarrow} (5) \overset{z_{15}}{\rightarrow} (1).$$

Then

$$y_1^{(t+1)} = a_{11}(w_{11}y_1^{(t)} + w_{12}y_2^{(t)}) + (1 - a_{11})y_1^{(1)}$$

and

$$y_1^{\prime(t+1)} = b_{11}(z_{11}y_1^{\prime(t)} + z_{12}y_2^{\prime(t)}) + (1 - a_{11})y_1^{\prime(1)}.$$

Let $a_{11} = b_{11} = 1$. Then $w_{11} = z_{11} = 0$ and so $w_{12} = z_{15} = 1$. Hence

$$\begin{aligned} y_1^{(t+1)} + y_1^{\prime(t+1)} &= w_{12}y_2^{(t)} + z_{15}y_5^{\prime(t)} \\ &= y_2^{(t)} + y_5^{\prime(t)}. \end{aligned}$$

In this case, we have $y_2^{(1)} + y_5^{\prime(1)} > 0 \Rightarrow y_2^{(2)} + y_5^{\prime(2)} > 0 \Rightarrow y_2^{(3)} + y_5^{\prime(3)} > 0 \Rightarrow$ ....

We next consider the domino effect for the nine tipping points given previously. We have (Table 3.2)

Table 3.2 Domino effect for the given nine tipping points

$W =$

	A	B	C	D	F	G	H	I	J
A	$\frac{2}{3}$	0	w_{13}	0	0	0	0	0	0
B	0	1	0	0	0	0	0	0	0
C	0	w_{23}	$\frac{1}{3}$	0	0	w_{36}	0	0	0
D	0	w_{42}	0	$\frac{2}{3}$	0	0	0	0	0
F	0	0	0	0	1	0	0	0	0
G	0	w_{62}	0	0	0	$\frac{2}{3}$	0	0	0
H	0	w_{72}	0	0	0	0	$\frac{2}{3}$	0	0
I	0	0	w_{83}	0	0	0	0	$\frac{2}{3}$	0
J	0	0	w_{93}	0	0	0	0	0	$\frac{2}{3}$

Example 3.4.2 We consider y_3 in the following. From W above, we have (assuming $w_{23} = w_{36}$) that

$$y_3^{(t)} = \frac{2}{3}(w_{23}y_2^{(t-1)} + \frac{1}{3}y_3^{(t-1)} + w_{36}y_6^{(t-1)}) + \frac{1}{3}y_3^{(1)})$$
$$= \frac{2}{3}(\frac{1}{3}y_2^{(t-1)} + \frac{1}{3}y_3^{(t-1)} + \frac{1}{3}y_6y_6^{(t-1)}) + \frac{1}{3}y_3^{(1)}.$$

Thus

$$y_3^{(t+1)} - y_3^{(t)} = \frac{2}{3}(\frac{1}{3}y_2^{(t)} + \frac{1}{3}y_3^{(t)} + \frac{1}{3}y_6y_6^{(t)} - \frac{1}{3}y_2^{(t-1)} - \frac{1}{3}y_3^{(t-1)} - \frac{1}{3}y_6y_6^{(t-1)})$$
$$= \frac{2}{3}(\frac{1}{3}(y_3^{(t)} - y_3^{(t-1)} + y_2^{(t)} - y_2^{(t-1)} + y_6^{(t)} - y_6^{(t-1)})).$$

Hence

$$y_3^{(t+1)} - y_3^{(t)} > 0 \Leftrightarrow y_3^{(t)} - y_3^{(t-1)} + y_2^{(t)} - y_2^{(t-1)} + y_6^{(t)} - y_6^{(t-1)} > 0.$$

Since $a_{22} = 0$, we have that $y_2^{(t)} - y_2^{(t-1)} = 0$. Thus

$$y_3^{(t+1)} - y_3^{(t)} > 0 \Leftrightarrow y_3^{(t)} - y_3^{(t-1)} + y_6^{(t)} - y_6^{(t-1)} > 0.$$

Now y_6 is influence by y_2. Now y_2 remains the same over time and so $y_6^{(t)} - y_6^{(t-1)} = \frac{1}{3}(\frac{2}{3}(y_6^{(t-1)} - y_6^{(t-2)}))$. Hence if $y_6^{(2)} - y_6^{(1)} > 0$, then $y_6^{(t)} - y_6^{(t-1)} > 0$ for all t. Thus $y_3^{(t+1)} - y_3^{(t)}$ increases over time by an ever smaller amount (a factor of $\frac{2}{9}$).

Proposition 3.4.3 Let $S_i = \{j | w_{ij} > 0\}$. Let $S_i{}' = \{j \in S_i | y_j^{(t)} = y_j^{(t-1)}\}$, i.e., there is no k influencing j and let $S_i'' = \{j \in S_i | y_j^{(t)} > y_j^{(t-1)}\}$.

(1) If $S_i'' = \emptyset$, then $y_i^{(t+1)} = y_i^{(t)}$.

(2) If $S_i'' \neq \emptyset$, then $y_i^{(t+1)} > y_i^{(t)}$.

Proof The follows from our assumption that $y_i^{(t+1)} \geq y_i^{(t)}$. and so $S_i = S_i' \cup S_i''$. □

3.5 General Theory

The following is influenced by ([4], p. 9). Our standard model describes a mechanism by which persons weigh and integrate their own attitudes and the attitudes of others of an issue:

$$y_i^{(t+1)} = a_{ii} \sum_{j=1}^{n} w_{ij} y_j^{(t)} + (1 - a_{ii}) y_i^{(1)}$$

for each $i = 1, 2, ..., n$ and $t = 1, 2, ...$ The n group members' time t positions on an issue are $y_1^{(t)}, y_2^{(t)}, ..., y_n^{(t)}$, and these positions include their initial set of positions $y_1^{(1)}, y_2^{(1)}, ..., y_n^{(1)}$. Group members' individual susceptibilities to interpersonal influence are $a_{11}, a_{22}, ..., a_{nn}$, where $0 \leq a_{ii} \leq 1$ for all i. The relative interpersonal influence of each group member j on i is $w_{i1}, w_{12}, ..., w_{1n}$, where $0 \leq w_{oij} \leq 1$ for all i and j, $\sum_{j=1}^{n} w_{ij} = 1$, and $w_{ii} = 1 - a_{ii}$ for all i. Note that $\sum_{j=1}^{n} a_{ii} w_{ij} + (1 - a_{ii}) = 1$, so that i's attitude at time $t + 1$ is formed as a weighted average of the attitudes of others and self at time t, and $i's$ initial position. If we assume that $-1 \leq y_i^{(1)} \leq 1$ for $i = 1, 2, ..., n$, then it is easy to show by induction on t that $-1 \leq y_i^{(t)} \leq 1, i = 1, 2, ..., n$.

$-1 \leq y_i^{(1)} \leq 0$ is for positive feedback and $0 \leq y_i^{(1)} \leq 1$ is for negative feedback with respect to climate change terminology.

Example 3.5.1 Suppose

$$
\begin{aligned}
y_1^{(2)} &= a_{11}(w_{11}y_1^{(1)} + w_{1j}y_j^{(1)}) + (1 - a_{11})w_{1j}y_1^{(1)} \\
&= a_{11}(1 - a_{11})y_1^{(1)} + a_{11}w_{ij}y_j^{(1)} + (1 - a_{11})w_{1j}y_1^{(1)} (\text{since } w_{11} = 1 - a_{11}) \\
&= (1 + a_{11})(1 - a_{11})y_1^{(1)} + a_{11}w_{1j}y_j^{(1)} \\
&= (1 - a_{11}^2)y_1^{(1)} + a_{11}w_{1j}y_j^{(1)}.
\end{aligned}
$$

Then $y_1^{(2)} < 0$ if $(1 - a_{11}^2)y_1^{(1)} + a_{11}w_{1j}y_j^{(1)} < 0$. Assume $(1 - a_{11}^2) = a_{11}w_{1j}$. Then $y_1^{(2)} < 0$ if $y_1^{(1)} + y_j^{(1)} < 0$.

Example 3.5.2 Let $n = 3$. Consider the matrix M in ([3], p. 115). Consider the cycle $(1) \to (2) \to (3) \to (1)$. Suppose $y_1^{(1)} < 0$ and $y_3^{(1)} < 0$ and so (3.4) has a positive effect on (3.5) and (3.6) has a positive effect on (3.6). Also, (3.5) has a positive effect on (3.6). Hence in the reachability matrix $M = [m_{ij}]$, $m_{12} = m_{31} = -1$ and $m_{23} = 1$. Thus

$$
M = \begin{matrix} 1 & -1 & 0 \\ 0 & 1 & 1 \\ -1 & 0 & 1 \end{matrix}.
$$

Then

$$
M^2 = \begin{matrix} 1 & -2 & -1 \\ -1 & 1 & 2 \\ -2 & 1 & 1 \end{matrix}.
$$

Consider the 3, 2 entry of M^2. There is a path of length 2 from 3 to 2, as indicated by the number 1 in the 3, 2 location. However, both edges are negative. That is, we have lost the fact the edges were positive feedback edges.

The following results are from [5] and [6]. Matrix analysis is used to identify various aspect of group structure. These include redundancies [7], complete cycle [8], liaison persons [9], and cliques. The structural concepts are developed in [10, 11].

A digraph is said to be strongly connected (or strong) if for every pair of distinct points, x and y, there exists a directed path from x to y and one from y to x. A digraph is said to be unilaterally connected (or unilateral) if for every pair of points, x and y, there is a directed path from x to y or one from y to x. A digraph is called disconnected if the points can be divided into two sets with no line joining any point in one set with a point in the other set. A digraph is called weakly connected (or weak) if it is not disconnected. These connected definitions are inclusive since every strong digraph is unilateral or every unilateral digraph is weak. We use the term digraph and group interchangeably.

In order to distinguish between groups on the basis of the kind of connectedness, we require exclusiveness connected categories. These may be obtained as follows. Let U_3 be the collection of all strong digraphs. We define U_2 as the set of all unilateral digraphs. Similarly, we define U_1 as the set of all weak digraphs. Finally,U_0 is the collection of all disconnected digraphs. Then $U_3 \subseteq U_2 \subseteq U_1$ and $U_1 \cap U_0 = \emptyset$. Let $C_3 = U_3$, $C_2 = U_2 \backslash U_3$, $C_1 = U_1 \backslash U_2$, and $C_0 = U_0$. Clearly, any digraph belongs to exactly one of the categories, C_3, C_2, C_1, or C_0.

If D is a digraph and x is a vertex, then $D \backslash x$ is the graph obtained from D by deleting the point x and all lines which are either directed toward x or away from x. We say that x is a point of type P_{ij} if the digraph D with x present is in class C_i, but $D \backslash x$ (with x absent) is in class C_j. Since the four categories C_3, C_2, C_1, and C_0 are numbered in accordance with the convention that the higher the subscript the stronger the kind of connectedness of the digraph, we may utilize this convention to describe and characterize which points are strengthening. Thus a point of a digraph is a strengthening point if it is of type P_{ij} such that $i >, j$; it is a weakening point if $i < j$ and the point is called neutral if $i = j$. By strengthening and weakening group members, we mean those individuals coordinated with the strengthening and weakening points of the digraph that represents the structure of the group. Similarly, if a point is of type $O_{ij,}$ we speak of the corresponding individual as an (i, j) member. When we contrast D and $D \backslash x$, we assume that the lines between pairs of distinct points exist independently.

Theorem 3.5.3 [6] There are no $(1, 3)$ members in any group.

The weakening members of a group can be described further. All the possible kinds of members are:

A. The $(0, j)$ members for $j = 1, 2$, or 3

B. The $(1, 2)$ members

C. The $(2, 3)$ members.

This follows from Theorem which asserts there are no $(1, 3)$ members.

Let $x \in V$. Then x is called an **isolate** if $\nexists\, y \in V \backslash \{x\}$ such that either $(x, y) \in P$ or $(y, x) \in P$.

Corollary 3.5.4 The group consisting of exactly two isolates, i.e., $(\{x, y\}, \emptyset)$, has two $(0, 3)$ members. Any other disconnected group has at most on weakening member.

Corollary 3.5.5 A C_1 group has at most two $(1, 2)$ members.

The members x and z are $(1, 2)$ members in the following digraphs: $(\{x, y, z\}, \{(x, y), (z, y)\})$ and $(\{x, y, z\}, \{(y, x), (y, z)\})$.

Corollary 3.5.6 A C_2 group has at most two $(2, 3)$ members.
The members x and y are $(2, 3)$ in the digraph $(\{x, y\}, \{(x, y)\})$.

Theorem 3.5.7 [6] Any group has at most two weakening members.

3.6 Impacts on Humans

We consider the impact of climate change on humans in this section. The following is from [1].

1. Human Population Affluence Technology
2. Invasive Species Diseases
3. Drought
4. Ocean Ph
5. Forests
6. CO_2
7. Fires
8. Permafrost
9. Evaporation
10. Atmospheric Temperature
11. Coral, birds, polar bears, penguins
12. Storms
13. Floods
14. Atmospheric H_2O
15. Sea Level
16. Algai blooms
17. Ocean temp.
18. Salinity
19. H_2O Expansion
20. Glacial Ice
21. Open Sea
22. Albeda
23. Sea Ice.

Table 3.3 Impact of climate change on humans

	1	2	3	4	5	6	7	8	9	10	11	12	13	14	15	16	17	18	19	20	21	22	23
1					N	P																	
2	N																						
3	N				N																		
4																							
5						N																	
6				P																			
7					N	P																	
8					N	P																	
9			P																				
10		P					P	N	P		N	P					P			N			N
11																							
12													P	P									
13	N																						
14										P													
15													P	P									
16																							
17																			P				
18																	N						
19															P								
20															N								
21																						N	
22																							N
23											P										N		

The following Table 3.3 is determined from [1]. See also, Al Gore (2006), An inconvenient truth.

The positive feedback loops are

$$10 \to 7 \to 6 \to 10,$$
$$10 \to 12 \to 14 \to 10,$$
$$17 \to 19 \to 15 \to 14 \to 17$$

We also have $10 \to 2$, $10 \to 17$. We see that if we consider only positive feedbacks, controlling Atmospheric Temperature would be important. Note also that we have the negative feedbacks $10 \to 11$, $10 \to 20$, and $10 \to 21$.

We next consider the directed graph with the positive and negative feedback combined. Consider the vertex 16 Algai blooms. This a (1, 1) vertex. Consider the subgraph with Algai blooms deleted. This subgraph remains unilateral since there is no directed path 11 Coral, birds, polar bears, penguins to 22 Albedo or one from 22 to 11.

We assume there are only positive feedbacks in the following. Then $-1 \leq y_i^{(t)} \leq 0$ for all $i = 1, ..., n$ and $t = 1, 2, ...$ It is natural to assume that $y_j^{(t+1)} \leq y_j^{(t)}$ for $j = 1, ..., n$ and $t = 1, 2, \ldots$

Theorem 3.6.1 Assume that $a_{ii} > 0, i = 1, ..., n$.
(1) Then $y_i^{(2)} - y_i^{(1)} < 0$ if and only if $\sum_{j=1}^n w_{ij}(y_j^{(1)} - y_i^{(1)}) < 0$.
(2) Suppose $t \geq 2$. Then $y_i^{(t+1)} - y_i^{(t)} < 0$ if and only if there exists j such that $w_{ij}(y_j^{(t)} - y_j^{(t-1)}) < 0$.

Proof We have for $i = 1, ..., n$ and any $t \geq 1$ that

$$y_i^{(t+1)} = a_{ii} \sum_{j=1}^n w_{ij} y_j^{(t)} + (1 - a_{ii}) y_i^{(1)}. \tag{3.8}$$

(1) Hence

$$\begin{aligned} y_i^{(2)} &= a_{ii} \sum_{j=1}^n w_{ij} y_j^{(1)} + (1 - a_{ii}) y_i^{(1)} \\ &= a_{ii} \sum_{j=1}^n (w_{ij} y_j^{(1)} - y_i^{(1)}) + y_i^{(1)} \\ &= a_{ii} \sum_{j=1}^n w_{ij} (y_j^{(1)} - y_i^{(1)}) + y_i^{(1)} \end{aligned}$$

since $\sum_{j=1}^n w_{ij} = 1$. Thus

$$y_i^{(2)} - y_i^{(1)} = a_{ii} \sum_{j=1}^n w_{ij} (y_j^{(1)} - y_i^{(1)})$$

Hence $y_i^{(2)} - y_i^{(1)} < 0$ if and only if $\sum_{j=1}^n w_{ij}(y_j^{(1)} - y_i^{(1)}) < 0$.

(2) Suppose $t \geq 2$. Then by (3.8) it follows that

$$y_i^{(t+1)} - y_i^{(t)} = a_{ii} \sum_{j=1}^n w_{ij} (y_j^{(t)} - y_j^{(t-1)}).$$

Since $y_j^{(t)} - y_j^{(t-1)} \leq 0$ by assumption, $y_i^{(t+1)} - y_i^{(t)} < 0$ if and only if there exists j such that $w_{ij}(y_j^{(t)} - y_j^{(t-1)}) < 0$. □

3.7 Business, Ethics, and Global Climate Change

In this and the next section, we consider the opinion that global climate change is an ethical issue. We in particular, consider ethical issues concerning the obligationof business organizations in reducing carbon emissions. We use concepts from directed graph theory and mathematical logic in our presentation. Interesting background material can be found in [3, 12, 13].

Let $\overrightarrow{G} = (V, E)$ be a directed graph, where $E \subseteq V \times V$. Let $G = (V, \Gamma)$ be a fuzzy directed graph. Let G_k be the fuzzy subgraph $(V \backslash \{x\}, \Gamma')$ obtained from G by the removal of a point x, where for Y a subset of V, $\Gamma' = \Gamma|_Y$. Define the fuzzy subset Γ^{-1} of $V \times V$ by for all $(u, v) \in V \times V$, $\Gamma^{-1}(u, v) = \Gamma(v, u)$. Define the fuzzy subset Δ of $V \times V$ by for all $(u, v) \in V \times V$, $\Delta(u, v) = \Gamma(u, v) \vee \Gamma^{-1}(u, v)$. Let $\widehat{\Gamma}, \widehat{\Gamma^{-1}}$, and $\widehat{\Delta}$ denote the transitive closure of Γ, Γ^{-1}, and Δ, respectively, ([8], p. 24)

Let U_1 be the set of all strong digraphs, U_2 the set of all unilateral digraphs, and U_1 the set of all weak digraphs. Let U_0 denote the collection of all disconnected digraphs. Then $U_3 \subseteq U_2 \subseteq U_1$ and $U_0 \cap U_1 = \emptyset$. Let $\overrightarrow{G} = (V, E)$ be a directed graph, where $E \subseteq V \times V$. For $(u, v) \in E$, we sometimes write $\overrightarrow{uv}$ for (u, v).

Let μ be a fuzzy subset function of E. Define the function γ of $E \times E$ into $[0, 1]$ by $\forall((u, v), (x, y)) \in E \times E$, $\gamma((u, v), (x, y)) = \mu(u, v) \oplus \mu(x, y)$, where $\oplus$ is a function such as a t-conorm or a t-norm, or an aggregation operator.

In our case, we are mainly interested in the situation where $(u, v) \in E$ represents logical implication and $\mu(u, v)$ denotes a measure of severity. For example, $\mu(u, v)$ could be a measure of severity of the combined vulnerability or combined government response of countries u and v to human trafficking. In the former case, $\oplus$ might be a t-conorm and in the latter case $\oplus$ might be a t-norm. Another example, could be u denotes increase gas emissions, v denotes climate change, and w denotes poverty. Then $\mu(u, v)$ and $\mu(v, w)$ the denote severity of the implications $u \to v$ and $v \to w$ and $\gamma((u, v), (v, w))$ denotes the severity of u leading to w (hypothetical syllogism). Suppose $u \to w$ is already in the graph.

A directed path $\overrightarrow{P}$ is a sequence $(v_1, v_2), (v_2, v_3), \ldots, (v_{n-1}, v_n), (v_n, v_{n+1})$ of directed edges. Define $\gamma^*(\overrightarrow{P}) = \oplus_{i=1}^{n} \mu(v_i, v_{i+1})$. Define $\delta^*(\overrightarrow{G}) = \oplus_{(u,v) \in E} \mu(u, v)$.

Let $\overrightarrow{G_1} = (V, E_1)$ and $\overrightarrow{G_2} = (V, E_2)$ be directed fuzzy graphs of $\overrightarrow{G}$. Let $\overrightarrow{G_1} \cup \overrightarrow{G_2} = (V, E_1 \cup E_2)$, where $E_1, E_2 \subseteq E$. Then $\delta^*(\overrightarrow{G_1} \cup \overrightarrow{G_2}) = \oplus_{(u,v) \in E_1 \cup E_2} \mu(u, v)$. Thus $\mu(u, v)$ appears only once in $\oplus_{(u,v) \in E_1 \cup E_2} \mu(u, v)$ even if (u, v) is a member of both E_1 and E_2. Note that here we are assuming that $\mu(u, v)$ is the assigned edge value for $\overrightarrow{G}$, $\overrightarrow{G_1}$, and $\overrightarrow{G_2}$.

Let $\overrightarrow{G} = (V, E)$ be a directed graph, where $E \subseteq V \times V$. Let $G = (V, \Gamma)$ be a fuzzy directed graph. Let G_k be the fuzzy subgraph $(V \backslash \{x\}, \Gamma')$ obtained from G by the removal of a point x, where for Y a subset of V, $\Gamma' = \Gamma|_Y$. Define the fuzzy subset Γ^{-1} of $V \times V$ by for all $(u, v) \in V \times V$, $\Gamma^{-1}(u, v) = \Gamma(v, u)$. Define the fuzzy subset Δ of $V \times V$ by for all $(u, v) \in V \times V$, $\Delta(u, v) = \Gamma(u, v) \vee \Gamma^{-1}(u, v)$.

Let $\widehat{\Gamma}$, $\widehat{\Gamma^{-1}}$, and $\widehat{\Delta}$ denote the transitive closure of Γ, Γ^{-1}, and Δ, respectively, [[8], p. 24]

Let U_1 be the set of all strong digraphs, U_2 the set of all unilateral digraphs, and U_1 the set of all weak digraphs. Let U_0 denote the collection of all disconnected digraphs. Then $U_3 \subseteq U_2 \subseteq U_1$ and $U_0 \cap U_1 = \emptyset$.

Let $\overrightarrow{G} = (V, E)$ be a directed graph, where $E \subseteq V \times V$. For $(u, v) \in E$, we sometimes write $\overrightarrow{uv}$ for (u, v). Let μ be a fuzzy subset function of E. Define the function γ of $E \times E$ into $[0, 1]$ by $\forall((u, v), (x, y)) \in E \times E, \gamma((u, v), (x, y)) = \mu(u, v) \oplus \mu(x, y)$, where $\oplus$ is a function such as a t-conorm or a t-norm, or an aggregation operator.

In our case, we are mainly interested in the situation where $(u, v) \in E$ represents logical implication and $\mu(u, v)$ denotes a measure of severity. For example, $\mu(u, v)$ could be a measure of severity of the combined vulnerability or combined government response of countries u and v to human trafficking. In the former case, $\oplus$ might be a t-conorm and in the latter case $\oplus$ might be a t-norm. Another example, could be u denotes increase gas emissions, v denotes climate change, and w denotes poverty. Then $\mu(u, v)$ and $\mu(v, w)$ the denote severity of the implications $u \to v$ and $v \to w$ and $\gamma((u, v), (v, w))$ denotes the severity of u leading to w (hypothetical syllogism). Suppose $u \to w$ is already in the graph.

A directed path $\overrightarrow{P}$ is a sequence $(v_1, v_2), (v_2, v_3), \ldots, (v_{n-1}, v_n), (v_n, v_{n+1})$ of directed edges. Define $\gamma^*(\overrightarrow{P}) = \oplus_{i=1}^{n} \mu(v_i, v_{i+1})$. Define $\delta^*(\overrightarrow{G}) = \oplus_{(u,v) \in E} \mu(u, v)$.

Let $\overrightarrow{G_1} = (V, E_1)$ and $\overrightarrow{G_2} = (V, E_2)$ be directed fuzzy graphs of $\overrightarrow{G}$. Let $\overrightarrow{G_1} \cup \overrightarrow{G_2} = (V, E_1 \cup E_2)$, where $E_1, E_2 \subseteq E$. Then $\delta^*(\overrightarrow{G_1} \cup \overrightarrow{G_2}) = \oplus_{(u,v) \in E_1 \cup E_2} \mu(u, v)$. Thus $\mu(u, v)$ appears only once in $\oplus_{(u,v) \in E_1 \cup E_2} \mu(u, v)$ even if (u, v) is a member of both E_1 and E_2. Note that here we are assuming that $\mu(u, v)$ is the assigned edge value for $\overrightarrow{G}$, $\overrightarrow{G_1}$, and $\overrightarrow{G_2}$.

3.8 Application

A: atmosphere more than it can handle

I: increase in greenhouse gases

R: reduction in emissions needed

B: business causes emissions.

We consider, $B \to I \to A \to R$, where $\to$ means implies (logical implication).

One could also have $B \to I \to CC \to P \to HB \to R$, where CC means climate change, P means poverty, and HB means hurts business.

One could have other such sequences. Putting these sequences together gives what is called a directed graph. Some of these implications are stronger than others. Thus one can assign numbers between 0 and 1 to these arrows and in so doing place the study in a fuzzy logic setting.(fuzzy directed graphs). B is the origin and R is the destination.

Human trafficking, modern slavery, illegal immigration, global hunger, terrorism are made worse by climate change because climate change increases poverty. Climate change also causes other problems such as extinction, biodiversity, rising sea levels, melting ice, on and on.

Recall that

$(A \text{ and } A \to B) \to B$ is called modus ponens.

$(A \to B \text{ and } B \to C) \to (A \to C)$ is called hypothetical syllogism.

We next present implications of important issues that relate to R. See [4, 12–15].

B_1 : creates emissions to a large extent

B_2 : creates emissions to some extent

C : consumers

CP : consumer preferences

Sh : shareholders

St : stakeholders

HB : hurt business

DC : democratic countries

NDC : nondemocratic countries

CC : climate change

IE : increase in emissions

P : poverty

HT : human trafficking

II : illegal immigration

MS : modern slavery

GH : global hunger

T : terrorism

AI : anthrpopedic impact.

We next consider the degree to which the following implications hold (Table 3.4).

$B \to IE \to CC \to P \to HB \to R$

$C \to IE \to CC \to P \to HB \to R$

We could combine the effect of the two implications in several ways, e.g., t-norms, t-conornms, or aggregation operators. Using the conorm algebraic sum, we get for

Table 3.4 Degree of implications

	B		IE		CC		P		HB		R
μ		0.7		0.9		0.7		0.7		0.7	
	C		IE		CC		P		HB		R
μ		0.5		0.9		0.7		0.7		0.7	

B and C combined, $0.7 + 0.5 - (0.7)(0.5) = 0.85$. Then the aggregation operator average yields $\frac{1}{5}(0.85 + 0.9 + 0.7 + 0.7 + 0.7) = \frac{3.85}{5} = 0.77$.

In the following P_i denotes a particular path, $i = 1, 2, ..., 6$.

$P_1 : B \to^{0.7} IE \to^{0.9} CC \to^{0.9} AI \to^{0.9} R$

$P_2 : B \to^{0.7} IE \to^{0.9} CC \to^{0.7} P \to^{0.5} HT \to^{0.3} R$

$P_3 : B \to^{0.7} IE \to^{0.9} CC \to^{0.7} P \to^{0.7} II \to^{0.3} R$

$P_4 : B \to^{0.7} IE \to^{0.9} CC \to^{0.7} P \to^{0.5} MS \to^{0.3} R$

$P_5 : B \to^{0.7} IE \to^{0.9} CC \to^{0.7} P \to^{0.7} GH \to^{0.3} R$

$P_6 : B \to^{0.7} IE \to^{0.9} CC \to^{0.7} P \to^{0.5} T \to^{0.3} R$

We have $\gamma^*(P_1) = 0.9997$, $\gamma^*(P_2) = \gamma^*(P_4) = \gamma^*(P_6) = 0.99685$, and $\gamma^*(P_3) = \gamma^*(P_5) = 0.99811$.

$0.3 \oplus 0.3 \oplus 0.3 \oplus 0.3 \oplus 0.3 = 0.83$ is more accurate in measuring the over all affect of climate change.

For this application, we have $\widehat{\Gamma}(\overrightarrow{G}) = \widehat{\Gamma^{-1}}(\overrightarrow{G}) = \widehat{\Delta}(\overrightarrow{G})$ since $\Gamma(u, v) > 0$ implies $\Gamma^{-1}(u, v) = 0$. Now $\widehat{\Gamma}(\overrightarrow{G}) = 0.3$ and $\widehat{\Gamma}(\overrightarrow{G}_R) = 0.5$. Thus R is weakening, i.e., $\mu(G) < \mu(G_R)$.

Consider dominating sets, ([16], p. 81). It follows easily that $\{B, CC, P, R\}$ is a minimal dominating set. as is $\{B, CC, AI, P\}$. Also $\{B, CC, HT, II, MS, GH, T\}$ is a minimal dominating set as is $\{B, CC, P, HT\}$.

References

1. Climate change feedback, Wikiwand https://www.wikiwand.com/en/Climate_change_feedback, 1–16
2. Earth How, 15 Climate feedback loops and examples, atmosphere. Climate Change (2022)
3. Lenton, T.M., Rockstrom, J., Gaffney, O., Rahmstorf, S., Richardson, K., Steffen, W., Schelinhuber, H.J.: Climate tipping points—too risky to bet against. Nature **575**, 28 (2019). Nov
4. Friedkin, N.E., Johnsen, E.C.: Social Influence Network Theory, A Social Examination of Small Group Dynamics, Structural Analysis in the Social Sciences 33, Cambridge (2011)
5. Mordeson, J.N., Nair, P.S.: *Fuzzy Graphs and Fuzzy Hypergraphs*, Studies in Fuzziness and Soft Computing, vol. 46 Physica-Verlag, Heidelberg, New York (2000)
6. Ross, I.C., Harary, F.: A description of strengthening and weakening members of a group. Sociometry **22**, 139–147 (1959)
7. Ross, I.C., Harary, F.: On the determination of redundancies in sociomatric chains. Psychometrika **17**, 195–208 (1952)
8. Harary, F., Ross, I.C.: The number of complete cycles in a communication network. J. Soc. Psychol. **40**, 329–332 (1953)
9. Ross, I.C., Harary, F.: Indentification of liaison persons of an organization using structure matrix. Manage. Sci. **1**, 251–258 (1955)
10. Harary, F., Norman, R.Z.: Graph Theory as a Mathematical Model in Social Science, Ann Arbor. Institute for Social Research, Mich. (1953)
11. Harary, F., Norman, R.Z. Cartwright, D.: *Structural Models: An Introduction to the Theory of Directed Graphs,* Wiley, New York (1965)
12. Arnold, D.G., Bustos, K.: Business, ethics, and global climate change. Bus. Prof. Ethics J., **24**(1,2), 103–130 (Summer/Fall 2005)

13. Freeman, R.S.: Stakeholder theory of the modern corporation. In: Donaldson, T., Werhane, P.(eds.), Ethical Issues in Business, A Philosophical Approach, 7th Edition, pp. 38-48. Prentice Hall (2002)
14. Asaka, J.: Climate change–terrorism nexus? a preliminary review. Analysis of the literature, terrorism research initiative, terrorism research initiative. Perspect. Terrorism **15**, 81–92 (2021)
15. O'Connell, C.: From a Vicious to a Virtuous Circle. Antislavery, DCU, Executive Summary (2021). April
16. Mathew, S., Mordeson, J.N., Malik, D.S.: *Fuzzy Graph Theory*, Springer International Publishing, vol. 363 (2018)

Chapter 4
Climate Change and Consequences

There are many serious challenges in the world such as human trafficking, modern slavery, illegal immigration, global hunger, and terrorism. However, climate change may be the worst of all. Climate change creates poverty and consequently makes the above challenges worse.

In Sect. 4.1, we take the country rankings of the ND-Gain scores, [1], and those in [2] with respect to climate risk and determine their similarity. We found a strong similarity between the two rankings. We then find the similarity of the readiness and vulnerability rankings in [1] and [2].

In Sect. 4.2, we consider terrorism. Climate change and terrorism are highly connected, [3].

Before proceeding we review the concept of a similarity measure.

Let X be a set and let S be a function of $\mathcal{FP}(X) \times \mathcal{FP}(X)$ into $[0, 1]$. Then S is called a **fuzzy similarity measure** on $\mathcal{FP}(X)$ if the following properties holds: $\forall \mu, \nu, \rho \in \mathcal{FP}(X)$,

(1) $S(\mu, \nu) = S(\nu, \mu)$;
(2) $S(\mu, \nu) = 1$ if and only if $\mu = \nu$;
(3) If $\mu \subseteq \nu \subseteq \rho$, then $S(\mu, \rho) \leq S(\mu, \nu) \wedge S(\nu, \rho)$;
(4) If $S(\mu, \nu) = 0$, then $\forall x \in X, \mu(x) \wedge \nu(x) = 0$.

Suppose X is a finite set. Let A be a one-to-one function of A onto $\{1, 2, ..., n\}$. Then A is called a **ranking** of X . Define the fuzzy subset μ_A of X as follows: $\forall x \in X, \mu_A(x) = A(x)/n$. We wish to consider the similarity of two rankings of X by the use of similarity measures. Note that (4) of the previous definition holds vacuously for rankings.

Let μ_A and μ_B be fuzzy subsets of X associated with two rankings A and B of X, respectively. Then M and S are similarity measures.

J. N. Mordeson et al., *Fuzzy Graph Theory*, Studies in Fuzziness and Soft Computing 424, https://doi.org/10.1007/978-3-031-23108-7_4

$$M(\mu_A, \mu_B) = \frac{\sum_{x \in X} \mu_A(x) \wedge \mu_B(x)}{\sum_{x \in X} \mu_A(x) \vee \mu_B(x)},$$

$$S(\mu_A, \mu_B) = 1 - \frac{\sum_{x \in X} |\mu_A(x) - \mu_B(x)|}{\sum_{x \in X} (\mu_A(x) + \mu_B(x))}.$$

The following results are from [9].

(1) The smallest value that $M(\mu, \nu)$ can take on is $\frac{n+2}{3n+2}$ if n is even and $\frac{n+1}{3n-1}$ if n is odd.
(2) The smallest value that $S(\mu, \nu)$ can take on is $\frac{\frac{n}{2}+1}{n+1}$ if n is even and $\frac{1}{2} + \frac{1}{2n}$ if n is odd.

If $(M(\mu., \nu) - \frac{n+2}{3n+2})/(1 - \frac{n+2}{3n+2})$ or $(M(\mu, \nu) - \frac{n+1}{3n-1})/(1 - \frac{n+1}{3n-1})$ or $(S(\mu, \nu) - \frac{\frac{n}{2}+1}{n+1})/(1 - \frac{\frac{n}{2}+1}{n+1})$ or $(S(\mu, \nu) - (\frac{1}{2} + \frac{1}{2n)}))/(1 - (\frac{1}{2} + \frac{1}{2n)}))$ lie in the interval $[0.8, 1]$, $[0.6, 0.8)$, $[0.4, 0.6)$, $[0.2, 0.4)$, $[0, 0.2)$ we say the similarity is very strong, strong, medium, weak, very weak, respectively.

4.1 Climate Change

The ND-Gain Country Index summarizes a country's vulnerability to climate change and other global challenges in combination with its readiness to improve resilience. It aims to help governments, businesses and communities better prioritize investments for a more efficient response to the immediate global challenges ahead [1].

The Country Index is composed of two key dimensions of adaptation: vulnerability and readiness.

Vulnerability measures a country's exposure, sensitivity, and capacity to adapt to the negative effects of climate change. ND-GAIN measures overall vulnerability by considering six life-supporting sectors—food, water, ecosystem service, human habitat, and infrastructure.

Exposure: Degree to which a system is exposed to significant climate change from a biophysical perspective. It is a component of vulnerability independent of socioeconomic context. Exposure indicators are projected impacts for the coming decades and are therefore invariant overtime in ND-GAIN.

Sensitivity: Extent to which a country is dependent upon a sector negatively affected by climate hazards, or the proportion of the population particularly susceptible to a climate change hazard. A county's sensitivity can vary over time.

Adaptive Capacity: Availability of social resources for sector-specific adaptation. In some cases, these capacities reflect sustainable adaptation solutions. In other cases, they reflect capacities to put new, more sustainable adaptations into place. Adaptive capacity varies over time.

Readiness measures a country's to leverage investments and convert them to adaptation actions. ND-GAIN measures overall readiness by considering three components—economic readiness, governance, and social readiness.

Economic: The ability of a country's business environment to accept investment that could be applied to adaptation that reduces vulnerability (reduces sensitivity and improves adaptive capacity).

Governance: The institutional factors that enhance application of investment for adaptation.

Social: The factors such as social inequality, ICT infrastructure, education and innovation that enhance the mobility of investment and promote adaptation actions.

A country's ND-GAIN score is composed of a vulnerability score and a readiness score.

$$(\text{Readiness Indicators} - \text{Vulnerability Indicators} + 1) \times 50 = \text{GAIN}.$$

Climate vulnerability and adaptation are based on compiled indicators. Thirty-six indicators contribute to ND-GAIN's measure of vulnerability and nine indicators contribute to the measure of readiness.

$$\begin{aligned}\text{Vulnerability} = (&\text{Ecosystem} + \text{Food} + \text{Health} + \text{HumanHabitat} + \text{Infrastructure} \\ &+ \text{Water} + \text{AdaptiveCapacity} + \text{Exposure} + \text{Sensitivity})/9\end{aligned}$$

and

$$\text{Readiness} = (\text{Economics} + \text{Governance} + \text{Social Readiness})/3$$

In [2], 67 developed, emerging, and frontier market countries are ranked for their vulnerability to climate risks. Physical impacts, transition to low-carbon countries and the funds to respond to climate change are all key to this analysis (Table 4.1).

We have

$$M(\mu, \nu) = \frac{1996.5}{2559.5} = 0.78 \text{ and } S(\mu, \nu) = 1 - \frac{553}{4556} = 1 - 0.1214 = 0.8786.$$

Also, $n = 67$ is odd and so

$$\frac{M(\mu, \nu) - (67+1)/(3(67)-1)}{1 - (67+1)/(3(67)-1)} = \frac{0.78 - 0.34}{1 - 0.34} = \frac{0.44}{0.64} = 0.6875$$

and

$$\frac{S(\mu, \nu) - (0.5 + 0.0075)}{1 - (0.5 + 0.0075)} = \frac{0.8786 - 0.5075}{1 - 0.5075} = \frac{0.3711}{0.4925} = 0.7535.$$

Table 4.1 Climate risk

Country	Rank Fragile	Rank ND-GAIN
India	1	5
Pakistan	2	3
Philippines	3	6
Bangladesh	4	1
Oman	5	27
Sri Lanka	6	9
Columbia	7	14.5
Mexico	8	16
Kenya	9	4
S. Africa	10	12
Thailand	11	23
Israel	12	39
Lebanon	13	7
Vietnam	14	10
Nigeria	15	2
Morocco	16	21
Indonesia	17	11
Egypt	18	8
Brazil	19	13
Serbia	20	19.5
Malaysia	21	32
Peru	22	14.5
Bahrain	23	18
Saudi Arabia	24	28.5
Greece	25	36
China	26	26
Tunisia	27	22
Argentina	28	17
Australia	29	55.5
Mauritius	30	31
United Kingdom	31	59
Poland	32	43
Qatar	33	33
Czech Rep	34	46
Russia	35	38
Portugal	36	45
Kuwait	37	24
Jordan	38	19.5
United States	39	51

(continued)

Table 4.1 (continued)

Country	Rank Fragile	Rank ND-GAIN
Belgium	40	47
Kazakhstan	41	35
Japan	42	53
France	43	54
Hungary	44	34
Romania	45	25
Slovenia	46	50
Italy	47	37
Turkey	48	28.5
UAE	49	40
Croatia	50	30
Chile	51	42
Singapore	52	60
Spain	53	44
Germany	54	58
Lithuania	55	41
Netherland	56	52
S. Korea	57	55.5
Austria	58	61.5
Canada	59	57
Switzerland	60	64
Denmark	61	61.5
Ireland	62	49
New Zealand	63	66
Estonia	64	48
Norway	65	67
Sweden	66	63
Finland	67	65

We see that the similarity between the two rankings are very strong with respect to M and strong with respect to S.

We next consider readiness and vulnerability rankings (Table 4.2).

In the following, μ, ν, λ and ρ denotes the ranking for Fragile Readiness, ND-GAIN Readiness, Fragile Vulnerability, and ND-GAIN Vulnerability, respectively.

We have

$$M(\mu, \nu) = \frac{2064}{2492} = 0.828 \text{ and } S(\mu, \nu) = 1 - \frac{431}{4556} = 1 - 0.095 = 0.905.$$

Table 4.2 Readiness and vulnerability

Country	Fragile Readiness rank	ND-GAIN Readiness rank	Fragile Vulnerability rank	ND-GAIN Vulnerability rank
Kenya	1	5	9	2.5
Lebanon	2	4	13	17
Pakistan	3	3	2	2.5
Sri Lanka	4	15	6	8
Egypt	5	7	18	12
Brazil	6	8	19	27.5
Mexico	7	11	8	20
Bangladesh	8	2	4	1
Nigeria	9	1	15	5
India	10	9	1	4
Philippines	11	6	3	9
S. Africa	12	10	10	19
Tunesia	13	22	27	26
Indonesia	14	14	17	10
Morocco	15	18.5	16	29
Columbia	16	12.5	7	18
Peru	17	17	22	13
Jordan	18	16	38	30
Argentina	19	12.5	28	23
Vietnam	20	18.5	14	7
Serbia	21	23	20	16
Thailand	22	24.5	11	15
Greece	23	35	25	47.5
Italy	24	34	47	58
Mauritius	25	42	30	14
Croatia	26	27.5	50	33
Portugal	27	45	36	47.5
Turkey	28	26	48	39
Hungary	29	27.5	44	38
China	30	37	26	25
Malaysia	31	32	21	32
Russia	32	41	35	44
Romania	33	20	45	22
Spain	34	38	53	59.5
Israel	35	36	12	51.5
Lithuania	36	46	55	34
Bahrain	37	24.5	23	11
Kazakhstan	38	33	41	40

(continued)

Table 4.2 (continued)

Country	Fragile Readiness rank	ND-GAIN Readiness rank	Fragile Vulnerability rank	ND-GAIN Vulnerability rank
Japan	39	61	42	35
Poland	40	40	32	49.5
UK	41	54.5	31	59.5
France	42	53	43	57
Chile	43	39	51	49.5
Slovenia	44	49	46	53.5
Oman	45	29	5	21
Belgium	46	47	40	45.5
Czech Rep	47	43	34	53.5
Estonia	48	50	64	41
Qatar	49	31	33	36
Kuwait	50	21	37	31
Germany	51	56	54	62
Saudi Arabia	52	30	24	24
Austria	53	57	58	65
USA	54	52	39	45.5
Switzerland	55	58	60	66
Netherlands	56	54.5	56	43
UAE	57	44	49	37
Canada	58	51	59	56
Sweden	59	63	66	61
Singapore	60	67	52	27.5
Finland	61	64	67	63
Ireland	62	48	62	51.5
Denmark	63	65	61	42
S. Korea	64	62	57	6
Australia	65	59	29	55
New Zealand	66	60	63	64
Norway	67	66	65	67

Also, $n = 67$ is odd and so

$$\frac{M(\mu,\nu) - (67+1)/(3(67)-1)}{1-(67+1)/(3(67)-1)} = \frac{0.828 - 0.34}{1 - 0.34} = \frac{0.488}{0.64} = 0.762$$

and

$$\frac{S(\mu,\nu) - (0.5+0.0075)}{1-(0.5+0.0075)} = \frac{0.905 - 0.5075}{1 - 0.5075} = \frac{0.3975}{0.4925} = 0.807.$$

We have

$$M(\lambda, \rho) = \frac{1926}{2630} = 0.732 \text{ and } S(\lambda, \rho) = 1 - \frac{694}{4556} = 1 - 0.152 = 0.848.$$

Also, $n = 67$ is odd and so

$$\frac{M(\lambda, \rho) - (67+1)/(3(67)-1)}{1 - (67+1)/(3(67)-1)} = \frac{0.762 - 0.34}{1 - 0.34} = \frac{0.422}{0.64} = 0.659$$

and

$$\frac{S(\lambda, \rho) - (0.5 + 0.0075)}{1 - (0.5 + 0.0075)} = \frac{0.848 - 0.5075}{1 - 0.5075} = \frac{0.3405}{0.4925} = 0.691.$$

The similarity between the rankings is very strong.

4.2 Terrorism

Although climate change is considered the top international threat, many feel terrorism is also a major concern. The Intergovernmental Panel on Climate Change issued a report expressing serious concerns about the possible impacts of climate change. However global warming is just one of many concerns. Terrorism, specifically from the Islamic extremist group ISIS and cyberattacks are also seen by many as major security threats, [5].

Over the past several decades, scholars and policy makers have dedicated a considerable amount of time and other resources to understanding the connection between security and climate change. In [6], it is discussed that the climate change-terrorism nexus plays out in two ways, one a simple, indirect relationship and the other a complex, cyclical relationship.

First, because climate change acts as a threat multiplier, it can worsen existing social vulnerability if adaptation and/or mitigation measures are not in place to help reduce such vulnerability and/or build resilience. Social vulnerability has been linked to both the spread of terrorism as well as the likelihood that an individual may be recruited to join a terrorist group. For example, poverty-stricken youth have been known to be recruited to terrorism groups. Addressing poverty and climate change's impact on poor people's livelihoods are ways to reduce this recruiting, [3].

Second, the climate change-terrorism nexus has a complex, feedback loop relationship in which climate change drives and/or enables terrorism, which in turn drives climate change. This cycle is particularly concerning in pyro-terrorism, where terrorists use arson to terrorize a people and/or government for socio-political reasons. If pyro-terrorism were to be committed on a grand scale, it would undoubtedly contribute to climate change through emission of greenhouse gases. Considering the feedback-loop relationship between climate change and terrorism, the cycle would likely continue until some intervention disrupts it. With forest fires expected to get worse in terms of frequency and intensity as climate changes, vulnerability of forests to terrorist attacks remains a real concern for governments access across the globe, [3].

The Global Terrorism Index (GTI) is a report published annually by the Institute for Economics and Peace (IEP). The index provides a comprehensive summary of the key global trends and patterns in terrorism since 2000. It is an attempt to systematically rank nations of the world according to terrorist activity. It produces a composite score in order to provide an ordinal ranking of countries on the impact of terrorism. We take the ranking of the countries given in [7] and rerank them with respect to the region they are in.

The Global Peace Index (GPI) is a report produced by the Institute for Economics and Peace (IEP) which measures the relative position of nations and regions peacefulness. The GPI ranks 172 independent states and territories according to their levels of peacefulness. Since a high rank by GTI for a country means a high degree of terrorism and high rank by GPI means a high degree of peacefulness, we reverse the GPI rankings so that we can determine a similarity measure of the two rankings.

The German watch Global Climate Risk Index is an analysis based on one of the most reliable data sets available on the impacts of extreme weather events and associated socio-economic data. Its aim is to contextualize on-going climate policy debates—especially the international climate negotiations-looking at real-world impacts over the last year and the last 20 years. The index is not a comprehensive climate vulnerability scoring. It represents one important piece in the overall puzzle of climate-related impacts and the associated vulnerabilities. The index focuses on extreme weather events such as storms, floods and heat waves, but does not take into account important slow-onset processes such as rising sea levels, glacier melting or ocean warming and acidification.

In the following, μ denotes global terror and ν denotes global peace. From the rankings in [8] and [7], we determine the following rankings for the indicated regions. We then determine the similarity measures of global terror and global peace for each region.

In Table 4.3, the Global Peace ranking is the reverse of that given in [7] (Tables 4.4, 4.5, 4.6, 4.7, 4.8 and 4.9).

Global Terror—Global Peace

OECD

We have that $M(\mu, \nu) = \frac{500.5}{759.5} = 0.659$ and $S(\mu, \nu) = 1 - \frac{259}{1260} = 1 - 0.206 = 0.794$.

Now $n = 35$. Hence $\frac{n+1}{3n-1} = \frac{36}{104} = 0.346$. Thus

$$\frac{M(\mu, \nu) - 0.346}{1 - 0.346} = \frac{0.659 - 0.346}{0.654} = \frac{0.313}{0.654} = 0.479.$$

Table 4.3 OECD rankings

Country	Global terror rank	Global peace Reverse rank	Climate risk rank
Australia	18	21	2
Austria	22	30	9
Belgium	12.5	18	14.5
Canada	11	26	13
Chile	9	8	3
Czech Rep.	26	27	28
Denmark	20	33	17
Estonia	31	14	18
Finland	23	24	30
France	4	7	8
Germany	10	20	11
Greece	8	5	6
Hungary	24	19	22
Iceland	34	35	24
Ireland	15	28	32
Israel	5	2	33
Italy	12.5	12	5
Japan	19	25	1
Korea Rep.	25	6	12
Latvia	30	10	23
Lithuania	29	9	34
Luxembourg			xx
Mexico	7	3	10
Netherlands	17	17	29
New Zealand	6	34	16
Norway	21	23	27
Poland	28	16	20
Portugal	34	32	7
Slovak Rep.	32	15	19
Slovenia	34	31	26
Spain	16	13	4
Sweden	14	22	31
Switzerland	27	29	21
Turkey	1	1	14.5
U. K.	3	11	25
U. S.	2	4	

Table 4.4 East and South Asia

Country	Global terror rank	Global peace Reverse rank	Climate risk rank
Bangladesh	8	9	3
Bhutan	16.5	18	19
Brunei Darussalm			19
Cambodia	16.5	11	14
China	10	7	10
India	2	3	1
Indonesia	9	16	4
Korea, Dem. Rep.	16.5	1	
Lao PDR	12	14	13
Malaysia	11	17	15
Maldives			16
Mongolia	16.5	15	8
Myanmar	6	4	7
Nepal	7	10	2
Pakistan	1	2	5
Philippines	3	5	6
Singapore	16.5	19	17
Sri Lanka	4	8	9
Thailand	5	6	11
Timor Leste	16.5	12	19
Vietnam	13	13	12

Also, $\frac{1}{2} + \frac{1}{2n} = \frac{1}{2} + \frac{1}{70} = 0.514$. Hence

$$\frac{S(\mu, \nu) - 0.514}{1 - 0.514} = \frac{0.794 - 0.514}{1 - 0.514} = \frac{0.280}{0.486} = 0.576.$$

East and South Asia

We have that $M(\mu, \nu) = \frac{158}{222} = 0.712$ and $S(\mu, \nu) = 1 - \frac{64}{380} = 1 - 0.168 = 0.832$.
Now $n = 19$. Hence $\frac{n+1}{3n-1} = \frac{20}{56} = 0.357$. Thus

$$\frac{M(\mu, \nu) - 0.357}{1 - 0.357} = \frac{0.712 - 0.357}{0.643} = \frac{0.355}{0.643} = 0.552.$$

Also, $\frac{1}{2} + \frac{1}{2n} = \frac{1}{2} + \frac{1}{38} = 0.526$. Hence

$$\frac{S(\mu, \nu) - 0.526}{1 - 0.526} = \frac{0.832 - 0.526}{1 - 0.526} = \frac{0.306}{0.474} = 0.646.$$

Table 4.5 Eastern Europe and Central Asia

Country	Global terror rank	Global peace Reverse rank	Climate risk rank
Afghanistan	1	1	
Albania	7.5	17	17
Andorra			
Armenia	10	8	12
Azerbaijan	12	4	17
Belarus	20	5	17
Bosnia and Herzegovina	7.5	12	2
Bulgaria	13	20	8
Croatia			3
Cyprus	14	14	17
Georgia	9	10	6
Kazakhstan	6	13	17
Kyrgz Rep.	5	11	17
Liecherstan			
Malta			1
Moldova	16.5	15	17
Monaco			
Montenegro	11	16	11
North Macedonia	15	19	17
Romania	20	21	9
Russian Federation	3	2	4
San Marino			
Serbia	16.5	18	7
Tajikistan	4	7	10
Turkmenistan	20	6	
Ukraine	2	3	5
Uzbekistan	18	9	17

Eastern Europe and Central Asia

We have that $M(\mu, \nu) = \frac{180.5}{281.5} = 0.649$ and $S(\mu, \nu) = 1 - \frac{101}{462} = 1 - 0.219 = 0.781$.

Now $n = 21$. Hence $\frac{n+1}{3n-1} = \frac{22}{62} = 0.355$. Thus

$$\frac{M(\mu, \nu) - 0.355}{1 - 0.355} = \frac{0.641 - 0.355}{0.645} = \frac{0.286}{0.645} = 0.443.$$

Also, $\frac{1}{2} + \frac{1}{2n} = \frac{1}{2} + \frac{1}{42} = 0.524$. Hence

Table 4.6 Latin America and the Caribbean

Country	Global terror rank	Global peace Reverse rank	Climate risk rank
Antigua and Barbuda			22.5
Argentina	12	18	8
Bahamas and Barbados			22.5
Belize			22.5
Bolivia	3	8	1
Brazil	5	4	3
Columbia	1	2	4
Costa Rica	11	21	22.5
Cuba	20	12	
Dominica			11
Dominican Rep.	20	15	22.5
Ecuador	4	11	14
El Salvador	20	7	13
Grenada			22.5
Guatemala	13	6	9
Guyana	14	10	22.5
Haiti	7.5	9	7
Honduras	10	5	10
Jamaica	15	17	22.5
Nicaragua	7.5	3	6
Panama	18	19	15
Paraguay	6	16	2
Peru	9	13	5
St. Kitts and Nevis			22.5
St. Lucia			22.5
St. Vincent and the Grenadines			22.5
Suriname			22.5
Trinidad and Tobago	16	14	22.5
Uruguay	17	20	12
Venezuelia	2	1	22.5

$$\frac{S(\mu, \nu) - 0.524}{1 - 0.524} = \frac{0.781 - 0.524}{1 - 0.524} = \frac{0.267}{0.476} = 0.561.$$

Latin America and the Caribbean

We have that $M(\mu, \nu) = \frac{180.5}{281.5} = 0.641$ and $S(\mu, \nu) = 1 - \frac{101}{462} = 1 - 0.219 = 0.781$.

Now $n = 21$. Hence $\frac{n+1}{3n-1} = \frac{22}{62} = 0.355$. Thus

Table 4.7 Middle East and North Africa

Country	Global terror rank	Global peace Reverse rank	Climate risk rank
Algeria	10	9	10
Bahrain	11	10	13.5
Egypt	4	7	9
Iran, Islamic	7	6	1
Iraq	1	3	7
Jordan	9	14	13.5
Kuwait	12	16	13.5
Lebanon	14	5	2
Libya	5	4	5
Morocco	13	12	6
Oman	16.5	13	8
Qatar	15	17	13.5
Saudi Arabia	6	8	4
Syrian Arab Rep.	2	2	
Tunisia	8	11	13.5
United Arab Emirates	16.5	15	13.5
Yemen, Rep.	3	1	3

$$\frac{M(\mu, \nu) - 0.355}{1 - 0.355} = \frac{0.641 - 0.355}{0.645} = \frac{0.286}{0.645} = 0.443.$$

Also, $\frac{1}{2} + \frac{1}{2n} = \frac{1}{2} + \frac{1}{42} = 0.524$. Hence

$$\frac{S(\mu, \nu) - 0.524}{1 - 0.524} = \frac{0.781 - 0.524}{1 - 0.524} = \frac{0.257}{0.476} = 0.540.$$

Middle East and North Africa

We have that $M(\mu, \nu) = \frac{132}{174} = 0.767$ and $S(\mu, \nu) = 1 - \frac{42}{306} = 1 - 0.137 = 0.863$.

Now $n = 17$. Hence $\frac{n+1}{3n-1} = \frac{18}{50} = 0.360$. Thus

$$\frac{M(\mu, \nu) - 0.360}{1 - 0.360} = \frac{0.767 - 0.360}{0.640} = \frac{0.407}{0.640} = 0.636.$$

Also, $\frac{1}{2} + \frac{1}{2n} = \frac{1}{2} + \frac{1}{34} = 0.529$. Hence

$$\frac{S(\mu, \nu) - 0.529}{1 - 0.529} = \frac{0.863 - 0.529}{1 - 0.529} = \frac{0.334}{0.471} = 0.709.$$

Table 4.8 Sub-Saharan Africa

Country	Global terror rank	Global peace Reverse rank	Climate risk rank
Angola	17	29	8
Benin	26	25	40
Botswana	37.5	40	32
Burkino Faso	5	11	40
Burundi	14	14	17
Cabo Verde			40
Cameroon	6	8	22
Central African Rep.	8	2	23.5
Chad	13	13	40
Comoros			7
Congo Democratic Rep.	3	1	16
Congo Rep.	21	16	29
Cote d'Ivoire			34
Djibouti	32	22	26
Equatorial Guinea	37.5	35	
Erita	37.5	10	40
Eswatini	37.5	32	40
Gabon	24	26	40
Gambia, The	37.5	38	13
Ghana	22	41	14.5
Guinea	27	27	31
Guinea-Bissau	37.5	24	40
Kenya	10	18	10
Lesotho	31	21	20
Liberia		31	28
Madagascar	25	33	11
Malawi	23	36	3
Mali	4	6	27
Mauritania	37.5	17	18
Mauritius	37.5	42	3
Mozambique	7	23	1
Namibia	37.5	33	30
Niger	11	9	5
Nigeria	1	7	25
Rwanda	16	28	14.5
San Tome and Principe			
Senegal	28	34	23.5
Seychelles			40
Sierra Leone	29	39	40

(continued)

Table 4.8 (continued)

Country	Global terror rank	Global peace Reverse rank	Climate risk rank
Somalia	2	5	
South Africa	15	5	9
South Sudan	9	4	4
Sudan	12	3	6
Tanzania	19	37	21
Togo	37.5	19	40
Uganda	18	20	12
Zambia	30	30	19
Zimbabwe	20	12	2

Table 4.9 Oceania

Country	Global terror rank	Global peace Reverse rank	Climate risk rank
Fiji			75
Kiribati			130
Marshall Islands			130
Micronesia, Fed. Sts.			130
Nauru			
Palau			
Papua New Guinea	94		51
Samoa			130
Solomon Islands			130
Tonga			130
Tuvalu			130
Vanuatu			73

Sub-Saharan Africa

We have that $M(\mu, \nu) = \frac{752.5}{1053.5} = 0.714$ and $S(\mu, \nu) = 1 - \frac{301}{1806} = 1 - 0.167 = 0.833$.

Now $n = 42$. Hence $\frac{n+2}{3n+2} = \frac{44}{128} = 0.344$. Thus

$$\frac{M(\mu, \nu) - 0.344}{1 - 0.344} = \frac{0.714 - 0.344}{0.656} = \frac{0.370}{0.656} = 0.564.$$

Also, $(\frac{n}{2} + 1)/(n + 1) = \frac{22}{43} = 0.512$. Hence

$$\frac{S(\mu,\nu)-0.512}{1-0.512}=\frac{0.833-0.512}{1-0.512}=\frac{0.321}{0.488}=0.658.$$

Global Terror—Climate Risk

OECD

We have that $M(\nu,\rho)=\frac{428.5}{761.5}=0.563$ and $S(\nu,\rho)=1-\frac{333}{1190}=1-0.280=0.720$.

Now $n=34$. Hence $\frac{n+2}{3n+2}=\frac{36}{104}=0.346$. Thus

$$\frac{M(\nu,\rho)-0.346}{1-0.346}=\frac{0.563-0.346}{0.654}=\frac{0.217}{0.654}=0.332.$$

Also, $(\frac{n}{2}+1)/(n+1)=\frac{18}{35}=0.514$. Hence

$$\frac{S(\nu,\rho)-0.514}{1-0.514}=\frac{0.720-0.514}{1-0.514}=\frac{0.206}{0.486}=0.424.$$

East and South Asia

We have that $M(\mu,\nu)=\frac{144}{198}=0.727$ and $S(\mu,\nu)=1-\frac{54}{342}=1-0.158=0.842$.

Now $n=18$. Hence $\frac{n+2}{3n+2}=\frac{20}{56}=0.357$. Thus

$$\frac{M(\mu,\nu)-0.357}{1-0.357}=\frac{0.727-0.357}{0.643}=\frac{0.370}{0.643}=0.575.$$

Also, $(\frac{n}{2}+1)/(n+1)=\frac{10}{19}=0.526$ Hence

$$\frac{S(\mu,\nu)-0.526}{1-0.526}=\frac{0.842-0.526}{1-0.526}=\frac{0.316}{0.474}=0.667.$$

Eastern Europe and Central Asia

We have that $M(\mu,\nu)=\frac{145.5}{2834.5}=0.620$ and $S(\mu,\nu)=1-\frac{89}{380}=1-0.234=0.766$.

Now $n=19$. Hence $\frac{n+1}{3n-1}=\frac{20}{56}=0.357$. Thus

$$\frac{M(\mu,\nu)-0.357}{1-0.357}=\frac{0.641-0.357}{0.643}=\frac{0.284}{0.643}=0.442.$$

Also, $\frac{1}{2}+\frac{1}{2n}=\frac{1}{2}+\frac{1}{38}=0.526$. Hence

$$\frac{S(\mu,\nu)-0.526}{1-0.526}=\frac{0.781-0.526}{1-0.526}=\frac{0.255}{0.474}=0.538.$$

Latin America and the Caribbean

We have that $M(\mu, \nu) = \frac{168.5}{251.5} = 0.670$ and $S(\mu, \nu) = 1 - \frac{83.5}{420} = 1 - 0.199 = 0.801$.

Now $n = 20$. Hence $\frac{n+2}{3n+2} = \frac{22}{62} = 0.355$. Thus

$$\frac{M(\mu, \nu) - 0.355}{1 - 0.355} = \frac{0.670 - 0.355}{0.645} = \frac{0.315}{0.645} = 0.488.$$

Also, $(\frac{n}{2} + 1)/(n + 1) = \frac{11}{21} = 0.524$. Hence

$$\frac{S(\mu, \nu) - 0.524}{1 - 0.524} = \frac{0.801 - 0.524}{1 - 0.524} = \frac{0.277}{0.476} = 0.582.$$

Middle East and North Africa

We have that $M(\mu, \nu) = \frac{103}{169} = 0.609$ and $S(\mu, \nu) = 1 - \frac{69}{272} = 1 - 0.254 = 0.746$.

Now $n = 16$. Hence $\frac{n+2}{3n+2} = \frac{18}{50} = 0.360$. Thus

$$\frac{M(\mu, \nu) - 0.360}{1 - 0.360} = \frac{0.609 - 0.360}{0.640} = \frac{0.249}{0.640} = 0.389.$$

Also, $(\frac{n}{2} + 1)/(n + 1) = \frac{9}{17} = 0.529$. Hence

$$\frac{S(\mu, \nu) - 0.529}{1 - 0.529} = \frac{0.746 - 0.529}{1 - 0.529} = \frac{0.215}{0.471} = 0.456.$$

Sub-Saharan Africa

We have that $M(\mu, \nu) = \frac{626.5}{1013.5} = 0.618$ and $S(\mu, \nu) = 1 - \frac{376}{1640} = 1 - 0.229 = 0.771$.

Now $n = 40$. Hence $\frac{n+2}{3n+2} = \frac{42}{122} = 0.344$. Thus

$$\frac{M(\mu, \nu) - 0.344}{1 - 0.344} = \frac{0.618 - 0.344}{0.656} = \frac{0.274}{0.656} = 0.418.$$

Also, $(\frac{n}{2} + 1)/(n + 1) = \frac{21}{41} = 0.512$. Hence

$$\frac{S(\mu, \nu) - 0.512}{1 - 0.512} = \frac{0.771 - 0.512}{1 - 0.512} = \frac{0.259}{0.488} = 0.531.$$

References

1. ND-GAIN Country Index, University of Notre Dame, Notre Dame Global Adaptation Initiative, 721 Flanner Hall, Notre Dame, Indiana 46556, Copyright 2021
2. Paun, A., Acton, L., Chan, W.S.: Fragile Planet. Scoring Climate Risks Around the World, Climate Change Global (2018)
3. Asaka, J.: Climate change—Terrorism Nexus? A preliminary review/analysis of the literature, terrorism research initiative, terrorism research initiative. Perspect. Terrorism **15**, 81–92 (2021)
4. Mordeson, J.N., Mathew, S.: Fuzzy mathematics and nonstandard analysis: applications to the theory of relativity. Trans. Fuzzy Sets Syst. **1**(1), 143–154 (2022)
5. Poushter, J., Huang, C.: Climate Change Still Seen as the Top Global Threat, but Cyberattacks a Rising Concern, Pew Research Center (2021)
6. Asaka, J.: Climate Change and Terrorism, New Security Beat https://www.newsecurity.org/2021/04/climate-change-terrorism, April 14, 2021
7. Global Peace Index: Measuring Peace in a Complex World. Institute for Economics and Peace (2021)
8. Global Terrorism Index: Measuring the Impact of Terrorism. Institute for Economics and Peace (2020)
9. Mordeson, J.N., Mathew, S., Binu, M.: Applications of Mathematics of Uncertainty, Grand Challenges—Human Trafficking—Coronavirus—Biodiversity and Extinction. Studies in Systems and Control, vol. 391. Springer (2022)

Chapter 5
Fuzzy Soft Semigraphs and Graph Structures

In [1], Molodtsov presented a new concept called soft set theory. This is an approach to model vagueness and uncertainty. It considers the elements of a universe with respect to a given set of parameters. In this chapter, we apply these concepts to semigraphs. Our purpose is to apply them to problems concerning human trafficking. We also use the notion of a time intuitionistic fuzzy set introduced in [2], to model forest fires as a new technique to study human trafficking.

5.1 Fuzzy Soft Sets

We let V denote a universal set and $\mathcal{FP}(V)$ the fuzzy power set of V in the following. Let A be a set, called the set of parameters, and let A_1 and A_2 be subsets of A.

Definition 5.1.1 A pair (γ, A_1) is called a **fuzzy soft set** over V if $\gamma : A_1 \to \mathcal{FP}(V)$.

Definition 5.1.2 [3] The pair (V, A) denotes the collection of all fuzzy soft subsets on V with attributes from A. It is called a **fuzzy soft class**.

Example 5.1.3 Let $V = \{$Columbia, Costa Rica, El Salvador, Guatemala, Mexico, United States, Ecuador, Honduras, Nicaragua, Dominican Republic, Panama$\}$. Let $A = \{$Government Response, Vulnerability$\}$. Let g denote government response and v denote vulnerability. Define $\gamma : A \to \mathcal{FP}(V)$, where $\gamma(g)$ and $\gamma(v)$ are defined in Table 5.1, the entries of which are from [4, 5].

Definition 5.1.4 For two fuzzy soft sets (γ_1, A_1) and (γ_2, A_2) in a fuzzy soft class (V, A), we say (γ_1, A_1) is a **fuzzy soft subset** of (γ_2, A_2) if $A_1 \subseteq A_2$ and for all $x \in A_1$, $\gamma_1(x) \leq \gamma_2(x)$ and is written as $(\gamma_1, A_1) \subseteq (\gamma_2, A_2)$.

J. N. Mordeson et al., *Fuzzy Graph Theory*, Studies in Fuzziness and Soft Computing 424, https://doi.org/10.1007/978-3-031-23108-7_5

Table 5.1 Attribute values

	$\gamma(g)$	$\gamma(v)$
Columbia	0.53	0.42
Costa Rica	0.55	0.27
El Salvador	0.43	0.36
Guatemala	0.56	0.42
Mexico	0.62	0.47
United States	0.88	0.23
Ecuador	0.53	0.42
Honduras	0.39	0.43
Nicaragua	0.59	0.35
Dominican Republic	0.63	0.39
Panama	0.48	0.33

Definition 5.1.5 Let c denote a fuzzy complement. The **complement** of a fuzzy soft set (γ_1, A_1) is denoted by $(\gamma_1, A_1)^c$ and is defined by $(\gamma_1, A_1)^c = (\gamma_1^c, A_1)$, where $\gamma_1^c : A_1 \to \mathcal{P}(V)$ is a mapping given by $\gamma_1^c(x) = (\gamma_1(x))^c$ for all $x \in A_1$.

Definition 5.1.6 [3] The **union** of two fuzzy soft sets (γ_1, A_1) and (γ_2, A_2) in a soft class (V, A) is a fuzzy soft set (γ_3, A_3), where $A_3 = A_1 \cup A_2$ and for all $x \in A$,

$$\gamma_3(x) = \begin{cases} \gamma_1(x) \text{ if } x \in A_1 \backslash A_2, \\ \gamma_2(x) \text{ if } x \in A_2 \backslash A_1, \\ \gamma_1(x) \vee \gamma_2(x) \text{ if } x \in A_1 \cap A_2, \end{cases}$$

and is written as $(\gamma_1, A_1) \cup (\gamma_2, A_2)$.

Definition 5.1.7 [3] Let (γ_1, A_1) and (γ_2, A_2) be two fuzzy soft sets in a soft class (V, A) such that $A_1 \cap A_2 \neq \emptyset$. Then the **intersection** of (γ_1, A_1) and (γ_2, A_2) is a fuzzy soft set (γ_3, A_3), where $A_3 = A_1 \cap A_2$ and for all $x \in A_3$, $\gamma_3(x) = \gamma_1(x) \wedge \gamma_2(x)$. We write $(\gamma_3, A_3) = (\gamma_1, A_1) \wedge (\gamma_2, A_2)$.

Proposition 5.1.8 Let (γ_1, A_1), (γ_2, A_2) and (γ_3, A_3) be fuzzy soft sets in a soft class (V, A). Then the following properties hold:

(1) $(\gamma_1, A_1) \vee (\gamma_2, A_2) = (\gamma_2, A_2) \vee (\gamma_1, A_1)$;

(2) $(\gamma_1, A_1) \vee ((\gamma_2, A_2) \vee (\gamma_3, A_3)) = ((\gamma_1, A_1) \vee (\gamma_2, A_2)) \vee (\gamma_3, A_3)$;

(3) $(\gamma_1, A_1) \subseteq (\gamma_1, A_1) \vee (\gamma_2, A_2)$;

(4) $(\gamma_1, A_1) \subseteq (\gamma_2, A_2) \Rightarrow (\gamma_1, A_1) \vee (\gamma_2, A_2) = (\gamma_2, A_2)$.

Proposition 5.1.9 Let (γ_1, A_1), (γ_2, A_2) and (γ_3, A_3) be fuzzy soft sets in a soft class (V, A). Then the following properties hold:

(1) $(\gamma_1, A_1) \wedge (\gamma_1, A_1) = (\gamma_1, A_1)$;

(2) $(\gamma_1, A_1) \wedge (\gamma_2, A_2) = (\gamma_2, A_2) \wedge (\gamma_1, A_1)$;

(3) $(\gamma_1, A_1) \wedge (\gamma_2, A_2) \subseteq (\gamma_1, A_1)$;

(4) $(\gamma_1, A_1) \subseteq (\gamma_2, A_2) \Rightarrow (\gamma_1, A_1) \wedge (\gamma_2, A_2) = (\gamma_1, A_1)$;

(5) $(\gamma_1, A_1) \wedge ((\gamma_2, A_2) \wedge (\gamma_3, A_3)) = ((\gamma_1, A_1) \wedge (\gamma_2, A_2)) \wedge (\gamma_3, A_3)$.

Theorem 5.1.10 [3] Let c be a fuzzy complement and (γ_1, A_1) and (γ_2, A_2) be soft fuzzy sets in (V, A). Then the following properties hold.

(1) $(\gamma_1, A_1)^c \wedge (\gamma_2, A_2)^c = ((\gamma_1, A_1) \vee (\gamma_2, A_2))^c$;

(2) $(\gamma_1, A_1)^c \vee (\gamma_2, A_2)^c = ((\gamma_1, A_1) \wedge (\gamma_2, A_2))^c$.

We refer the reader to [6] for further discussion.

5.2 Semigraphs

The notion of a semigraph was introduced in [7].

A **semigraph** is a pair (V, X), where V is a nonempty set whose elements are called vertices and X is a set of n-tuples, called **edges** of G, of distinct vertices, satisfying the following conditions:

(1) Any two edges have at most one vertex in common.

(2) Two edges $(u_1, u_2, ..., u_n)$ and $(v_1, v_2, ..., v_m)$ are considered to be equal if and only if $m = n$ and either $u_i = v_i$ for $i = 1, ..., n$ or $u_i = v_{n-i+1}$ for $i = 1, ..., n$.

Let $G = (V, X)$ be a semigraph and $E = (v_1, ..., v_n)$ be an edge of G. Then v_1 and v_n are called the **end vertices** of E and v_i, $i = 2, ..., n-1$ are called the **middle vertices** (or m-vertices) of E. If a vertex v of a semigraph G appears only as an end vertex then it is called an **end vertex**. If a vertex v is only a middle vertex then it is a **middle vertex** or an **m-vertex** while a vertex v is called a **middle-cum-end vertex** or **(m, e)-vertex** if it is a middle vertex of some edge and an end vertex of some other edge. A subedge of an edge $E = (v_1, ..., v_n)$ is a k-tuple $E' = (v_{i_1}, ..., v_{i_k})$, where $1 \leq i_1 < i_2 < ... < i_k \leq n$. We say that the subedge E' is induced by the set of vertices $\{v_{i_1}, ..., v_{i_k}\}$. A **partial edge** of E is a $(j - i + 1)$-tuple $E(v_i, v_j) = (v_i, v_{i+1}, ..., v_j)$, where $1 \leq i < j \leq n$.

$G' = (V', X')$ is a **partial semigraph** of a semigraph G if the edges of G' are partial edges of G. Two vertices u and v in a semigraph G are said to be **adjacent** if they belong to the same edge. If u and v are adjacent and consecutive in order then they are said to be **consecutively adjacent**, u and v are said to be **e-adjacent** if they are the end vertices of an edge and **$1e$-adjacent** if both the vertices u and v belong to the same edge and at least one of them is an end vertex of that edge.

It follows that a subedge e' of an edge e is a partial edge if and only if any two consecutive vertices in e' are also consecutive vertices of e. Note that an edge is a subedge (partial edge) of itself and a proper subedge (partial edge) is not an edge of G. For otherwise, it would contradict the condition on edges that two edges have at most one vertex in common.

5.3 Soft Semigraphs

The results in this section are taken from [8].

Definition 5.3.1 Let V be the set of vertices of a semigraph G. Consider a subset V_1 of V. Then a partial edge formed by some or all vertices of V_1 is said to be a **maximum partial edge** or **mp edge** if it is not a partial edge of any other partial edge formed by some or all vertices of V_1.

Definition 5.3.2 Let $G = (V, X)$ be a semigraph having vertex set V and edge set X. Let X_p be the collection of all partial edges of the semigraph G and A be a nonempty set. Let $R \subseteq A \times V$ be an arbitrary relation from A to V. We define a mapping Q from A to $\mathcal{P}(V)$ by $Q(x) = \{y \in V | xRy\}, \forall x \in A$, where $\mathcal{P}(V)$ denotes the power set of V. Then the pair (Q, A) is a soft set over V. Also define a mapping W from A to $\mathcal{P}(X_p)$ by $W(x) = \{\text{mp edges}(Q(x))\}$, where mp edge $Q(x)$ denotes the set of all mp edges that can be formed by some or all vertices of $Q(x)$. The pair (W, A) is a soft set over X_p. We define the soft semigraph as follows:

Definition 5.3.3 The 4-tuple $G = (G^*, Q, W, A)$ is called a **soft semigraph** of G^* if the following conditions hold:

(1) $G^* = (V, X)$ is a semigraph having a vertex set V and edge set E;

(2) A is the set of parameters which is nonempty;

(3) (Q, A) is a soft set over V;

(4) (W, A) is a soft set over X_p;

(5) $H(a) = (Q(a), W(a))$ is a partial semigraph of G^* for all $a \in A$.

Example 5.3.4 Let $V = \{v_1, v_2, v_3, v_4, u_2, u_3, u_4\}$ and $X = \{(v_1, v_2, v_3, v_4), (v_1, u_2, u_3, u_4), (v_2, u_2)\}$. Let $A_1 = \{v_2\}$. Define R by $xRy \Leftrightarrow x = y$ or x and y are adjacent. Define Q_1 by $Q_1(x) = \{y \in V | xRy\}$. Then $Q_1(v_2) = \{v_1, v_2, v_3, v_4, u_2\}$. Let $W_1(x) = \{\text{mp edges}(Q_1(x))\}, x \in A_1$. Then $W_1(v_2) = \{(v_1, v_2, v_3, v_4), (v_2, u_2)\}$. Let $A_2 = \{u_2\}$. Using R, $Q_2(u_2) = \{u_1, u_2, u_3, u_4, v_2\}$. Let $W_2(x) = \{\text{mp edges}(Q_2(x))\}, x \in A_2$. Then $W_2(u_2) = \{(v_1, u_2, u_3, u_4), (v_2, u_2)\}$.

In this example, we assume we started with one edge (route), $(u_4, u_3, u_2, v_1, v_2, v_3, v_4)$. We select a vertex, say v_1,and vertices consecutively adjacent to v_1, say u_2, v_2. We then break $(u_4, u_3, u_2, v_1, v_2, v_3, v_4)$ at v_1 into two edges (routes). We will later compare (u_2, v_1, v_2) and (u_2, v_2) in a human trafficking setting.

Definition 5.3.5 Let $G^* = (V, X)$ be a semigraph and $G = (G^*, Q, W, A)$ be a soft semigraph of G^* which is also given by $\{H(x)|x \in A\}$. Then the partial semigraph $H(x)$ corresponding to any $x \in A$ is called a ***p*-part** of the soft semigraph G.

Definition 5.3.6 An edge in a soft semigraph G of G^* is called an ***f*-edge**. It may be a partial edge of some edges in G^* or an edge in G^*.

Definition 5.3.7 Let $G^* = (V, X)$ be a semigraph and let $G_1 = (G^*, Q_1, W_1, A_1)$ and $G_2 = (G^*, Q_2, W_2, A_2)$ be two soft semigraphs of G^*. Then the **extended union** of G_1 and G_2 is defined as $G_1 \cup_E G_2 = G = (G^*, Q, W, A)$, where $A = A_1 \cup A_2$ and for all $u \in A$,

$$Q(u) = \begin{cases} Q_1(u) \text{ if } u \in A_1 \backslash A_2, \\ Q_2(u) \text{ if } u \in A_2 \backslash A_1, \\ Q_1(u) \cup Q_2(u) \text{ if } u \in A_1 \cap A_2, \end{cases}$$

and

$$W(u) = \begin{cases} W_1(u) \text{ if } u \in A_1 \backslash A_2, \\ W_2(u) \text{ if } a \in A_2 \backslash A_1, \\ \{\text{mp edges}(Q_1(u) \cup Q_2(u))\} \text{ if } u \in A_1 \cap A_2. \end{cases}$$

If $H(u) = (Q(u), W(u)$ for all $u \in A$, then $G_1 \cup_E G_2 = \{H(u)|u \in A\}$.

In the previous definition, the relations for G_1 and G_2 may be different, ([8], Example 4.1).

Example 5.3.8 Let $V = \{v_0, v_1, v_2, v_3, v_4, v_5, v_6, v_7, v_8\}$ and $E = \{(v_0, v_1, v_2, v_3, v_4), (v_1, v_5, v_6), (v_3, v_8, v_7), (v_6, v_7)\}$. Let $A_1 = \{v_2, v_7\}$ and $A_2 = \{v_1, v_7\}$. Let R_1 be defined by xR_1y if and only if $x = y$ or x and y are adjacent. Let R_2 be defined by xR_2y if and only if $x = y$ or x and y are consecutively adjacent. Then $Q_1(v_2) = \{v_0, v_1, v_2, v_3, v_4\}$ and $Q_1(v_7) = \{v_6, v_7, v_8, v_3\}$. Also, $Q_2(v_1) = \{v_0, v_1, v_2, v_5\}$ and $Q_2(v_7) = \{v_6, v_7, v_8\}$. Thus $W_1(v_2) = \{(v_0, v_1, v_2, v_3, v_4)\}$ and $W_1(v_7) = \{(v_6, v_7), (v_7, v_8, v_3)\}$. Also $W_2(v_1) = \{(v_0, v_1, v_2), (v_1, v_5)\}$ and $W_2(v_7) = \{(v_6, v_7), (v_7, v_8)\}$.

We see that (Q_1, A_1) is a soft set over V and (W_1, A_1) is a soft set over X_p. Here $H_1(v_2) = (Q_1(v_2), W_1(v_2))$ and $H_1(v_7) = (Q_1(v_7), W_1(v_7))$ are partial semigraphs of G^*. Hence $G_1 = \{H_1(v_2), H_1(v_7)\}$ is a soft semigraph of $G^* = (V, X)$. Similarly, $G_2 = \{H_2(v_1), H_2(v_7)\}$ is a soft semigraph of G^*.

We next consider the extended union of G_1 and G_2. Let $A = A_1 \cup A_2$. We have $Q(v_2) = \{v_0, v_1, v_2, v_3, v_4\}$, $Q(v_1) = \{v_0, v_1, v_2, v_5\}$, and $Q(v_7)=\{v_6, v_7, v_8, v_3\} \cup \{v_6, v_7, v_8\} = \{v_6, v_7, v_8, v_3\}$. We also have $W(v_2) = \{(v_0, v_1, v_2, v_3, v_4)\}$, $W(v_1) = \{(v_0, v_1, v_2), (v_1, v_5)\}$, and $W(v_7) = \{(v_6, v_7), (v_7, v_8, v_3)\}$.

Theorem 5.3.9 Let $G^* = (V, X)$ be a semigraph having vertex set V and edge set E. Also let $G_1=(G^*, Q_1, W_1, A_1)$ and $G_2 = (G^*, Q_2, W_2, A_2)$ be two soft semigraphs of G^*. Then their extended union $G_1 \cup_E G_2$ is a soft semigraph of G^*.

Proof From the definition of extended union, we have in $G_1 \cup_E G_2$ that $A = A_1 \cup A_2$ is a parameter set, Q is a function from A to $\mathcal{P}(V)$, and W is a function from A to $\mathcal{P}(X_p)$. Here (Q, A) is a soft set over V and (W, A) is a soft set over X_p. Let $u \in A_1 \backslash A_2$. Then the coprresponding p-part $H(u)$ of $G_1 \cup G_2$ is $H(u) = (Q_1(u), W_1(u))$. This is a partial semigraph of G^* since G_1 is a soft semigraph of G^*. If $u \in A_2 \backslash A_1$, the corresponding p-part of $G_1 \cup_E G_2$ is $H(u) = (Q_2(u), W_2(u))$ This is a partial semigraph of G^* since G_2 is a soft semigraph og G^*. Let $u \in A_1 \cap A_2$. Then the corresponding p-part of $G_1 \cup_E G_2$ is $H(u) = (Q(u), W(u))$, where $Q(u) = Q_1(u) \cup Q_2(u)$ and $W(u) = \{\text{mp edges } (Q_1(u) \cup Q_2(u))\}$. Now $Q_1(u) \cup Q_2(u) \subseteq V$ and each f-edge in $W(u)$ is a partial edge of G^* for all $u \in A_1 \cap A_2$. That is, $H(u) = (Q(u), W(u))$ is a partial semigraph of G^* for all $u \in A_1 \cap A_2$. That is, $G_1 \cup G_2 = (G^*, Q, W, A)$ is a soft semigraph of G^*. □

Definition 5.3.10 Let $G^* = (V, X)$ be a semigraph. Let $G_1 = (G^*, Q_1, W_1, A_1)$ and $G_2 = (G^*, Q_2, W_2, A_2)$ be soft semigraphs of G^* such that $Q_1(u) \cap Q_2(u) \neq \emptyset$ for all $u \in A_1 \cap A_2$. Then the **extended intersection** of G_1 and G_2 denoted by $G_1 \cap G_2 = G = (G^*, Q, W, A)$, where $A = A_1 \cup A_2$ and for all $u \in A$, is defined as follows:

$$Q(u) = \begin{cases} Q_1(u) \text{ if } u \in A_1 \backslash A_2, \\ Q_2(u) \text{ if } u \in A_2 \backslash A_1, \\ Q_1(u) \cap Q_2(u) \text{ if } u \in A_1 \cap A_2, \end{cases}$$

and

$$W(u) = \begin{cases} W_1(u) \text{ if } u \in A_1 \backslash A_2, \\ W_2(u) \text{ if } a \in A_2 \backslash A_1, \\ \{\text{mp edges}(Q_1(u) \cap Q_2(u))\} \text{ if } u \in A_1 \cap A_2. \end{cases}$$

If $H(u) = (Q(u), W(u)$ for all $u \in A$, then $G_1 \cap_E G_2 = \{H(u) | u \in A\}$.

Example 5.3.11 Consider the semigraph $G^* = (V, X)$ and its soft semigraphs G_1 and G_2 in Example 5.3.8 Then $Q_1(v_7) \cap Q_2(v_7) = \{v_6, v_7, v_8\} = Q(v_7)$ and $W(v_7) = \{\{v_6, v_7), (v_7, v_8)\}$.

Theorem 5.3.12 Let $G^* = (V, X)$ be a semigraph having vertex set V and edge set E. Also let $G_1 = (G^*, Q_1, W_1, A_1)$ and $G_2 = (G^*, Q_2, W_2, A_2)$ be two soft semigraphs of G^*. Then their extended intersection $G_1 \cap_E G_2$ is a soft semigraph of G^*.

Proof From the definition of extended intersection, we have in $G_1 \cap_E G_2$ that $A = A_1 \cup A_2$ is the parameter set, Q is a function from A intp $\mathcal{P}(V)$ and W is a function of A intp $\mathcal{P}(X_p)$. Now (Q, A) is a soft set over V and (W, A) is a soft set over X_p. Let $u \in A_1 \backslash A_2$. Then the corresponding p-part of $G_1 \cap G_2$ is $H(u) = (Q_1(u), W_1(u))$. This is a partial semigraph of G^* since G_1 is a soft semigraph of $G^{*.}$. Let $u \in A_2 \backslash A_1$. Then the corresponding p-part of $G_1 \cap G_2$ is $H(u) = (Q_2(u), W_2(u))$. This is a partial semigraph of G^* since G_2 is a soft semigraph of $G^{*.}$. Let $u \in A_1 \cap A_2$. Then the corresponding p-part of $G_1 \cap G_2$ is $H(u) = (Q(u), W(u))$. where $Q(u) = Q_1(u) \cap Q_2(u)$ and $W(u) = \{\text{mp edge } (Q_1(u) \cap Q_2(u))\}$. Now $Q_1(u) \cap Q_2(u) \subseteq V$ and each f-edge in $W(u)$ is a partial edge of G^* for all $u \in A_1 \cap A_2$. That is, $G_1 \cap G_2 = (G^*, Q, W, A)$ is a soft semigraph of G^*. □

Theorem 5.3.13 Let $G^* = (V, X)$ be a semigraph having vertex set V and edge set E. Also let $G_1 = (G^*, Q_1, W_1, A_1)$ and $G_2 = (G^*, Q_2, W_2, A_2)$ be two soft semigraphs of G^*. If $Q_1(u) \cap Q_2(u) \neq \emptyset$ for all $u \in A_1 \cap A_2$, then $G_1 \cap G_2$ is a soft partial semigraph of $G_1 \cup G_2$.

Proof By Theorems 5.3.9 and 5.3.12 we have that $G_1 \cup_E G_2$ and $G_1 \cap_E G_2$ are soft semigraphs of G^*. Assume that $G_1 \cup_E G_2 = G_{EU} = \{(G^*, Q_{EU}, W_{EU}, A_{EU})$ and $G_1 \cap_E G_2 = G_{EI} = \{(G^*, Q_{EI}, W_{EI}, A_{EI})$. By the definitions of extended union and extended intersection of two soft semigraphs, $A_{EU} = A_{EI} = A_1 \cup A_2$. Therefore we have $A_{EI} \subseteq A_{EU}$.

(1) If $u \in A_1 \backslash A_2$, then the corresponding p-parts $H_{EI} = (Q_{EI}(u), W_{EI}(u))$ and $H_{EU} = (Q_{EU}(u), W_{EU}(u))$ of G_{EI} and G_{EU},respectively are equal to $H_1(u) = (Q_1(u), W_1(u))$. That is, $H_{EI}(u)$ is a partial semigraph of $H_{EU}(u)$ for all $u \in A_1 \backslash A_2$ since both p-parts are identical.

(2) If $u \in A_2 \backslash A_1$, the corresponding p-parts $H_{EI}(u) = (Q_{EI}(u), W_{EI}(u))$ and $H_{EU} = (Q_{EU}(u), W_{EU}(u))$ of G_{EI} and G_{EU}, respectively, are equal to $H_2(u) = (Q_2(u), W_2(u))$. That is, $H_{EI}(u)$ is a partial semigraph of $H_{EU}(u)$ for all $u \in A_2 \backslash A_1$ since both parts are identical.

(3) If $u \in A_1 \cap A_2$, $H_{EI}(u)$=$(Q_{EI}(u), W_{EI}(u))$, where $Q_{EI}(u) = Q_1(u) \cap Q_2(u)$ and $W_{EI}(u) = \{\text{mp edges } (Q_1(u) \cap Q_2(u))\}$ and $H_{EU} = (Q_{EU}(u), W_{EU}(u)$, where $Q_{EU}(u) = Q_1(u) \cap Q_2(u)$ and $W_{EU} = \{\text{mp edges } (Q_1(u) \cup Q_2(u))\}$. Clearly, $Q_{EI}(u) \subseteq Q_{EU}(u)$ and each f-edge in $W_{EI}(u)$ is a partial edge of an f-edge in $W_{EU}(u)$. Thus H_{EI} is a partial semigraph of $H_{EU}(u)$ for all $u \in A_1 \cap A_2$. That is, we have

   (i) $A_{EI} \subseteq A_{EU}$;

   (ii) For all $u \in A_{EI}$, $H_{EI}(u)$=$(Q_{EI}, W_{EI}(u))$ is a partial semigraph of $H_{EU} = (Q_{EU}(u), W_{EU}(u))$.

Hence $G_1 \cap_E G_2$ is a soft partial semigraph of $G_1 \cup_E G_2$. □

5.4 Fuzzy Soft Semigraphs

In this section, we place the human trafficking problem in a fuzzy soft semigraph setting. However, we first place the work in [2] concerning forest fires into a fuzzy soft semigraph setting. Wildfires pose serious problems for the world. One new approach to model wildfires is the Game Method for Modeling (GMM). The GMM has been approved as an appropriate tool for modelling of wildfire propagations, [2]. In [2], the authors proposed the introduction of temporal, rather than ordinary, intuitionistic fuzzy pairs (TIFPs) in order to evaluate the impact of the wildfire and investigate the basic properties of TIFPs.

Let $T = \{t_1, ...t_n\}$ be a fixed time-scale. Let U be a universal set. In [2] the set,

$$A(T) = \{((x, t), \mu_A(x, t), \nu_A(x, t))|x \in U, t \in T|$$

is defined as a Temporal Intuitionistic Fuzzy Set (TIFS), where $\mu_A(x, t), \nu_A(x, t)) \in [0, 1]$ and $0 \leq \mu_A(x, t) + \nu_A(x, t) \leq 1$ for all $x \in U, t \in T$.

In [2], the concept of a TIFP was defined by: $x(t) = (a(t), b(t))$, where $a, b : T \to [0, 1]$ and $a(t) + b(t) \leq 1$ for all $t \in T$. The following theorems were presented in [2].

The area of land in question is divided into cells. For the i-th iteration, the TIFP $(\mu(i), \nu(i))$ was determined, where $\mu(i)$ was the degree of totally burned areas (the number of totally burned cells divided by the number of all cells) and the degree of the yet unaffected area (the number of the unaffected cells divided by the number of all cells) for the whole considered area at that time-moment. The remaining intuitionistic fuzzy degree of hesitation (uncertainty), which is equal to the complement of these two degrees relative to 1, i.e., $\pi(i) = 1 - \mu(i) - \nu(i)$, corresponds to the number of currently burning cells of the area divided by the number of all cells. Clearly, before the start of the fire, the unaffected area is represented as the TIFP $(\mu(0), \nu(0) = (0, 1)$, meaning that none of the land has burned, At the final iteration when the whole area has been burned the respective TIFP's value is $(1, 0)$. We also note that the strength of the wind was taken into consideration, no wind, mild wind, strong wind.

In what follows, a, b, c, d can be thought of as functions of T into $[0, 1]$. Then the results apply to TIFP approach in [2]. Define $-$ on $[0, 1] \times [0, 1]$ by for all $a, b \in [0, 1]$, $-(a, b) = (b, a)$. Then $-$ is called **negation** in [9]. Let f_1 and f_2 be functions of $T \times T$ into $[0, 1]$. Define $g_1, g_2 : (T \times T) \times (T \times T) \to [0, 1]$ as follows: for all $(a, b), (c, d) \in [0, 1] \times [0, 1]$,

$$(a, b)g_1(c, d) = (f_1(a, c), f_2(b, d)),$$
$$(a, b)g_2(c, d) = (f_2(a, c), f_1(b, d)).$$

One possible example for f_1 and f_2 could be minimum and maximum, respectively, and minimum and maximum on pairs for g_1 and g_2, respectively.

Theorem 5.4.1 [2] Let $a, b, c, d \in [0, 1]$. Then

(1) $(a, b)g_1(c, d) = -(-(a, b)g_2 - (c, d))$;

(2) $(a, b)g_2(c, d) = -(-(a, b)g_1 - (c, d))$.

Proof (1) $-(-(a, b)g_2 - (c, d)) = -((b, a)g_2(d, c)) = -(-(-(b, a)g_1 - (d, c))) = (a, b)g_1(c, d)$.

(2) The proof here follows as in (1).

In the following, we let $\otimes_1 = g_1, \oplus_1 g_2$ and $\otimes = f_1, \oplus = f_2$. □

Corollary 5.4.2 Let $a, b, c, d \in [0, 1]$. Then

(1) $(a, b) \otimes_1 (c, d) = -(-(a, b) \oplus_1 -(c, d))$;

(2) $(a, b) \oplus_1 (c, d) = -(-(a, b) \otimes_1 -(c, d))$.

Let c denote a fuzzy complement. Define $c_1 : [0, 1] \times [0, 1] \to [0, 1]$ by $(a, b)^{c_1} = (a^c, b^c)$. Then c_1 is called a **complement** on pairs with respect to c.

Theorem 5.4.3 Let c, $\otimes$, and $\oplus$ be dual. Then the following properties hold.

(1) $(a, b) \otimes_1 (c, d) = ((a, b)^{c_1} \oplus_1 (c, d)^{c_1})^{c_1}$;

(2) $(a, b) \oplus_1 (c, d) = ((a, b)^{c_1} \otimes_1 (c, d)^{c_1})^{c_1}$;

Proof (1) We have

$$\begin{aligned}((a,b)^{c_1} \oplus_1 (c,d)^{c_1})^{c_1} &= ((a^c,b^c) \oplus_1 (c^c,d^c))^{c_1}\\ &= (a^c \oplus c^c, b^c \otimes d^c)^{c_1}\\ &= ((a^c \oplus c^c)^c, (b^c \otimes d^c)^c)\\ &= (a \otimes c, b \oplus d)\\ &= (a,b) \otimes_1 (c,d).\end{aligned}$$

(2) The proof here is similar to that of (1). □

Proposition 5.4.4 Let $a, b \in [0, 1]$. Then $-((a,b)^{c_1}) = (-(a,b))^{c_1}$.

Proof $-((a,b)^{c_1}) = -(a^c, b^c) = (b^c, a^c)$ and $(-(a,b))^{c_1} = (b,a)^{c_1} = (b^c, a^c)$. □

The modal operators of “necessity" and “possibility" were defined in [2] as follows:

$$\Box(a,b) = (a, 1-a),$$
$$\Diamond(a,b) = (1-b.b).$$

Theorem 5.4.5 ([2], Theorem 2, p. 6) Let $a, b \in [0, 1]$. Then the following properties hold.

(1) $\Box(a,b) = -(\Diamond - (a,b))$;

(2) $\Diamond(a,b = -(\Box - (a,b))$.

See also ([2], Theorem 3).

For placing the TIFP into a fuzzy soft set setting, we could consider T to be an n-tuple with t_0 and t_n as the end vertices and $t_1, \ldots, t_{n-1}$ the middle vertices. The universe would be T and the set of parameters $A = \{\text{no wind, mild wind, strong wind}\}$.

Example 5.4.6 We consider one of the main routes in the Americas with respect to human trafficking. It was taken from [5]. Let $A = \{g, v\}$ be the set of attributes. Let (σ, μ) be a fuzzy subgraph of a graph $G = (V, E)$ and let (τ, ν) be a complementary fuzzy subgraph of G. The σ and τ values in the following table are normalized government response and vulnerability values of [5], respectively. The μ and ν values are determined by the product of the σ values and the algebraic sum of the τ values. The Susceptibility of trafficking from one country to another is defined.

Let $V = \{$Columbia, Costa Rica, El Salvador, Guatemala, Mexico, United States, Ecuador, Honduras, Nicaragua, Dominican Republic, Panama$\}$. Let $A = \{$Government Response, Vulnerability$\}$. Let g denote government response and v denote vulnerability. Define $\gamma : A \to \mathcal{FP}(V)$, where $\gamma(g) : V \to [0, 1]$ and $\gamma(v) : V \to [0, 1]$ are defined in the following table. The combined government response of two countries is determined by taking the product of the $\gamma(g)$ values for the two countries and the combined vulnerability is determined by taking the algebraic sum of the two $\gamma(v)$ values of the two countries. The susceptibility for trafficking between the two countries is determine by the formula $\frac{1}{\mu(C_1,C_2)} + \frac{1}{1-\nu(C_1,C_2)}$, where $\mu(C_1, C_2) = \gamma(g)(C_1)\gamma(g)(C_2)$ and $\nu(C_1, C_2) = \gamma(v)(C_1) + \gamma(v)(C_2) - \gamma(v)(C_1) + \gamma(v)(C_2)$. The last column of the following table is determined by taking the sum of the susceptibiity entries as t increases, i.e., as those trafficked move from the origin country along the route from country to country to the destination country.

The average susceptibility of a route is defined by $\frac{1}{n}$ Susc(Origin, Desitnation), where n is the number of edges in the route. In the above example, $n = 5$ and so the average susceptibilty of the route in Table 5.2 is $\frac{29.38}{5} = 5.876$.

Table 5.2 Susceptibility values

	$\gamma(g)$	Combined $\gamma(g)$	$\gamma(v)$	Combined $\gamma(v)$	Susc	$\sum$
Columbia	0.53		0.42			
Col → C R		0.29		0.57	5.78	5.78
Costa Rica	0.55		0.27			
C R → El Sal		0.24		0.53	6.30	12.08
El Salvador	0.43		0.36			
El Sal → Guat		0.24		0.63	6.87	18.95
Guatemala	0.56		0.42			
Guat → Mex		0.34		0.69	6.17	25.12
Mexico	0.62		0.47			
Mex → U S		0.55		0.59	4.26	29.38
United States	0.88		0.23			

5.5 Soft Fuzzy Sets

In this section, we give regional rankings of countries with respect to three categories, social progress, fragility, and freedom. We place our work in a soft set theory setting. We also determine the similarity of the rankings using a similarity operation.

The Social Progress Index, [10], measures the extent to which countries provide for the social and environmental needs of their citizens. Fifty-four indicators in the areas of basic human needs, foundations of well-being, and opportunity to progress show the relative performance of nations. The SPI measures the well-being of a society by observing and environmental outcomes directly rather then the economic factors. The social and environmental factors include wellness (including health, shelter and sanitation), equality, inclusion, sustainability and personal freedom and safety.

The following is from [11]. *Freedom in the World* is an annual report on political rights and civil liberties, composed of numerical ratings and descriptive texts for each country and a select group of territories. The 2022 edition covers developments in 195 countries and 15 territories.

The Fragile States Index is a report released by the Fund for Peace and the magazine, Foreign Policy, [12]. The primary purpose of this report is to assess the vulnerability of sovereign states throughout the nation. All sovereign states that are members of the United Nations are included in the report when enough data ius available with the exception of several countries listed in [12]. The report uses indicators across four categories if a state is vulnerable to conflict or collapse. The categories that are assessed include: Cohesion, Economic, Political, and Social. There are 12 different indicators usee to determine the vulnerability of states. These factors include human rights, public services, demographic pressures, refugees and internally displaced persons, and security. The higher a state is ranked on the list, the more vulnerable it is.

Freedom of the World uses a two-tiered system consisting of scores and status. A country or territory is awarded 0 to 4 points for each of 10 political rights indicators and 15 civil liberties indicators, which take the form of questions; a score of 0 represents the smallest degree of freedom and 4 the greatest degree of freedom. The political rights questions are grouped into four categories: Electoral Process (3 questions), Political Pluralism and Participation (4), and Functioning of Government (3). The civil liberties questions are grouped into four subcategories: Freedom of Expression and Belief (4 questions), Associational and Organizational; Rights (3), Rule of Law (4), and Personal Autonomy and Individual Rights (4).

The combination of the overall score awarded for political rights and the overall score awarded for civil liberties, after being equally weighted, determines the status of Free, Partly Free, a Not Free given a country.

We take the ranks of the countries from [10–12] and rerank the countries with respect to regions. We then use a similarity operation to determine their similarity with respect the rankings given the Social Progress Index, the Fragile States Index, and the Freedom of the World Index.

We can let V be the set of countries and A={Social Progress, Fragile, Freedom}. Then γ(Social Progress):$V \to [0, 1]$.

We next use a fuzzy similarity operation, [13], to determine the similarity of the three rankings. In the following, when a ranking has missing data for a country, we delete the country for a different ranking even it has data for the ranking. After the deletion, we rerank.

Let μ denote the ranking for Social Progress, ν denote the ranking for Fragile, and ρ the ranking for Freedom. We use the fuzzy similarity operation,

$$M(\mu, \nu) = \frac{\sum_{i=1}^{n} \mu(C_i) \wedge \nu(C_i)}{\sum_{i=1}^{n} \mu(C_i) \vee \nu(C_i)}.$$

Let n be even. Then $\frac{n+2}{3n+2}$ is the smallest value $M(\mu, \nu)$ can be ([13], p. 14). Let n be odd. Then the smallest value $M(\mu, \nu)$ can be is $\frac{n+1}{3n-1}$. If one wishes to calculate a value for $M(\mu, \nu)$ in which the values are bounded below by 0, one can use the formulas $\frac{M(\mu,\nu)-(n+2)/(3n+2)}{1-(n+2)/(3n+2)}$ if n is even and $\frac{M(\mu,\nu)-(n+1)/(3n-1)}{1-(n+1)/(3n-1)}$ if n is odd.

The rankings in Tables 5.3, 5.4, 5.5, 5.6, 5.7 and 5.8 are used to determine the following similarity comparisons.

OECD

We have that $M(\mu, \nu) = \frac{609}{723} = 0.842$ with $n = 36$, $M(\mu, \rho) = \frac{589.5}{742.5} = 0.794$ with $n = 36$, $M(\nu, \rho) = \frac{585.5}{746.5} = 0.784$ with $n = 36$.

The smallest value $M(\mu, \nu)$ can have is $\frac{36+2}{108+2} = 0.345$ The smallest value $M(\mu, \rho)$ can have is $\frac{36+2}{108+2} = 0.345$. The smallest value $M(\nu, \rho)$ can have is $\frac{36+2}{108+2} = 0.345$.

$\frac{M(\mu,\nu)-0.345}{1-0.345} = \frac{0.842-0.345}{1-0.345} = \frac{0.497}{0.655} = 0.0.759$ and $\frac{M(\mu,\rho)-0.345}{1-0.345} = \frac{0.794-0.345}{1-0.345} = \frac{0.449}{0.655} = 0.685$ and $\frac{M(\nu,\rho)-0.345}{1-0.345} = \frac{0.784-0.345}{1-0.345} = \frac{0.439}{0.655} = 0.670$

East and South Asia

We have that $M(\mu, \nu) = \frac{180}{240} = 0.750$ with $n = 20$, $M(\mu, \rho) = \frac{156.5}{263.5} = 0.594$ with $n = 20$, $M(\nu, \rho) = \frac{162}{300} = 0.540$ with $n = 21$.

The smallest value $M(\mu, \nu)$ can have is $\frac{20+2}{60+2} = 0.355$. The smallest value $M(\mu, \rho)$ can have is $\frac{20+2}{60+2} = 0.355$ The smallest value $M(\nu, \rho)$ can have is $\frac{21+1}{63-1} = 0.355$.

Table 5.3 OECD

Country	Social progress	Fragile	Freedom
Australia	8	10	13.5
Austria	15	15	19.5
Belgium	16	17	11
Canada	7	8	5
Chile	31	31	16.5
Czech Rep.	25	21	21
Denmark	2	4	7.5
Estonia	24	26	16.5
Finland	3	1	2
France	18	18	26.5
Germany	11	13	16.5
Greece	26	33	29
Hungary	34	32	34
Iceland	9	5	16.5
Ireland	12	11	7.5
Israel	30	35	33
Italy	23	30	23.5
Japan	13	20	11
Korea Rep.	17	19	30.5
Latvia	32	29	28
Lithuania	29	22	26.5
Luxembourg	14	9	7.5
Mexico	35	34	35
Netherlands	10	12	7.5
N. Zealand	4	6	4
Norway	1	2	2
Poland	28	28	32
Portugal	21	14	13.5
Slovak Rep.	33	23	23.5
Slovenia	22	16	23.5
Spain	19	27	23.5
Sweden	5	7	2
Switzerland	6	3	11
Turkey	36	36	36
U. K.	20	24	19.5
U. S.	27	25	30.5

Table 5.4 East and South Asia

Country	Social progress	Fragile	Freedom
Bangladesh	3	4	13
Bhutan	11	13	5
Brunei Darussaiam		19	16
Cambodia	5	8	17
China	9	12	20.5
India	6	10	4
Indonesia	13	14	6
Korea, Dem. Rep.	20	20	2
Lao PDR	2	9	19
Malaysia	18	17	10
Maldives	15	15	12
Mongolia	16	18	1
Myanmar	4	2	20.5
Nepal	7	6	7
Pakistan	1	3	14
Philippines	10	7	8.5
Singapore	19	21	11
Sri Lanka	17	1	8.5
Thailand	14	11	15
Timor Leste	8	5	3
Vietnam	12	16	18

$\frac{M(\mu,\nu)-0.355}{1-0.355} = \frac{0.750-0.355}{1-0.355} = \frac{0.395}{0.645} = 0.612$ and $\frac{M(\mu,\rho)-0.355}{1-0.355} = \frac{0.594-0.355}{1-0.355} = \frac{0.239}{0.645} = 0.371$ and $\frac{M(\nu,\rho)-0.355}{1-0.355} = \frac{0.540-0355}{1-0.355} = \frac{0.186}{0.645} = 0.288$

Eastern Europe and Central Asia

We have that $M(\mu, \nu) = \frac{241}{311} = 0.775$ with $n = 23$, $M(\mu, \rho) = \frac{247}{305} = 0.810$ with $n = 23$, $M(\nu, \rho) = \frac{238.5}{318.5} = 0.761$ with $n = 23$.

The smallest value $M(\mu, \nu)$ can have is $\frac{23+1}{69-1} = 0.353$. The smallest value $M(\mu, \rho)$ can have is $\frac{23+1}{69-1} = 0.353$ The smallest value $M(\nu, \rho)$ can have is $\frac{23+1}{69-1} = 0.353$.

$\frac{M(\mu,\nu)-0.353}{1-0.353} = \frac{0.775-0.355}{1-0.355} = \frac{0.420}{0.645} = 0.651$ and $\frac{M(\mu,\rho)-0.355}{1-0.355} = \frac{0.810-0.355}{1-0.355} = \frac{0.455}{0.645} = 0.705$ and $\frac{M(\nu,\rho)-0.355}{1-0.355} = \frac{0.761-0.355}{1-0.355} = \frac{0.406}{0.645} = 0.629$

Table 5.5 Eastern Europe and Central Asia

Country	Social progress	Fragile	Freedom
Afghanistan	1	1	23
Albania	15	17	11
Andorra			2
Armenia	17	14	17
Azerbaijan	4	6	24
Belarus	18	13	25.5
Bosnia and Herzegovina	10	8	18
Bulgaria	20	20	9
Croatia	21	22	6
Cyprus	23	18	2
Georgia	14	7	16
Kazakhstan	9	16	20
Kyrgz Rep.	6	3	19
Liecheristan			4
Malta	22	23	5
Moldova	8	12	13.5
Monaco			7
Montenegro	13	19	11
North Macedonia	11	15	11
Romania	19	21	8
Russian Federation	7	5	21
San Marino			2
Serbia	16	11	13.5
Tajikistan	2	2	25.5
Turkmenistan	3	9	27
Ukraine	12	10	15
Uzbekistan	5	4	22

Latin America and the Caribbean

We have that $M(\mu, \nu) = \frac{246}{306} = 0.804$ with $n = 23$, $M(\mu, \rho) = \frac{244}{308} = 0.792$ with $n = 23$, $M(\nu, \rho) = \frac{306}{396} = 0.773$ with $n = 26$.
The smallest value $M(\mu, \nu)$ can have is $\frac{23+1}{69-1} = 0.353$. The smallest value $M(\mu, \rho)$ can have is $\frac{23+1}{69-1} = 0.353$ The smallest value $M(\nu, \rho)$ can have is $\frac{26+2}{78+2} = 0.350$. $\frac{M(\mu,\nu)-0.353}{1-0.353} = \frac{0.804-0.353}{1-0.353} = \frac{0.451}{0.647} = 0.697$ and $\frac{M(\mu,\rho)-0.353}{1-0.353} = \frac{0.792-0.353}{1-0.353} = \frac{0.439}{0.647} = 0.679$ and $\frac{M(\nu,\rho)-0.350}{1-0.350} = \frac{0.773-0.350}{1-0.350} = \frac{0.423}{0.647} = 0.654$

Table 5.6 Latin America and the Caribbean

Country	Social progress	Fragile	Freedom
Antigua and Barbuda		20	10
Argentina	20	24	11
Bahamas		22	
Barbados	19	23	2
Belize		16	9
Bolivia	7	7	21
Brazil	12	8	16.5
Columbia	13	6	23
Costa Rica	22	26	5.5
Cuba	10	19	30
Dominica			3
Dominican Rep.	8	15	20
Ecuador	16	9	19
El Salvador	6	10	24
Grenada		11	7.5
Guatemala	2	3	25
Guyana	5	13	16.5
Haiti	1	1	27
Honduras	3	5	26
Jamaica	15	18	14
Nicaragua	4	4	28
Panama	18	25	12
Paraguay	11	14	22
Peru	14	12	18
St. Kitts and Nevis			7.5
St. Lucia			4
St. Vincent and the Grenadines			5.5
Suriname	9	17	15
Trinidad and Tobago	17	21	13
Uruguay	21	27	1
Venezuela		2	29

Middle East and North Africa

We have that $M(\mu, \nu) = \frac{127.5}{178.5} = 0.714$ with $n = 17$, $M(\mu, \rho) = \frac{127}{179} = 0.709$ with $n = 17$, $M(\nu, \rho) = \frac{110}{196} = 0.561$ with $n = 17$.

Table 5.7 Middle East and North Africa

Country	Social progress	Fragile	Freedom
Algeria	8	9	6
Bahrain	4	13	13
Egypt	1	2.5	10
Iran	6	7	12
Iraq	2	4	7
Jordan	12	8	5
Kuwait	14	14	3.5
Lebanon	7	6	2
Libya		5	14.5
Morocco	5	10	3.5
Oman	11	15	9
Qatar	9	16	8
Saudi Arabia	3	11	16
Syrian Arab Rep.		2.5	17
Tunisia	13	12	1
United Arab Emirates	10	17	11
Yemen		1	14.5

The smallest value $M(\mu, \nu)$ can have is $\frac{17+1}{51-1} = 0.360$ The smallest value $M(\mu, \rho)$ can have is $\frac{17+1}{51-1} = 0.360$ The smallest value $M(\nu, \rho)$ can have is $\frac{17+1}{51-1} = 0.360$. $\frac{M(\mu,\nu)-0.360}{1-0.360} = \frac{0.714-0.360}{1-0.360} = \frac{0.354}{0.640} = 0.553$ and $\frac{M(\mu,\rho)-0.360}{1-0.360} = \frac{0.709-0.360}{1-0.360} = \frac{0.349}{0.640} = 0.545$ and $\frac{M(\nu,\rho)-0.360}{1-0.360} = \frac{0.561-0.360}{1-0.360} = \frac{0.201}{0.640} = 0.314$.

Sub-Saharan Africa

We have that $M(\mu, \nu) = \frac{921.5}{1240.5} = 0.743$ with $n = 46$, $M(\mu, \rho) = \frac{955}{1301} = 0.734$ with $n = 47$, and $M(\nu, \rho) = \frac{945}{1311} = 0.721$ with $n = 47$.

The smallest value $M(\mu, \nu)$ can have is $\frac{46+2}{138+2} = \frac{48}{140} = 0.343$. The smallest value $M(\mu, \rho)$ and $M(\nu, \rho)$ can be is $\frac{47+1}{141-1} = \frac{48}{140} = 0.343$. Now $\frac{M(\mu,\nu)-0.343}{1-0.343} = \frac{0.743-0.343}{1-0.343} = \frac{0.400}{0.637} = 0.628$ and $\frac{M(\mu,\rho)-0.343}{1-0.343} = \frac{0.734-0.343}{1-0.343} = \frac{0.391}{0.637} = 0.614$ and $\frac{M(\nu,\rho)-0.343}{1-0.343} = \frac{0.721-0.343}{1-0.343} = \frac{0.378}{0.637} = 0.593$.

The similarities range from low, 0.561, to high, 0.842.

Table 5.8 Sub-Saharan Africa

Country	Social progress	Fragile	Freedom
Angola	11	23	30
Benin	36	40	14
Botswana	43	47	7
Burkino Faso	20	25	15
Burundi	6	9	41
Cabo Verde	45	45	1
Cameroon	22	8	39.5
Central African Rep.	3	4	44
Chad	2	5.5	39.5
Comoros		34	23.5
Congo Dem. Rep.	8	3	36
Congo Rep,	13	17	37.5
Cote d'Ivoire	30	21	18
Djibouti	16	32	32
Equatorial Guinea	21	31	45
Erita	4	12.5	46
Eswatini	27	30	37.5
Ethiopia	17	15	33
Gabon	40	43	35
Gambia	34	33	
Ghana	41	46	3
Guinea	9	11	27
Guinea-Bissau	10	28.5	21
Kenya	38	19	19
Lesotho	31	36	11
Liberia	24	20	13
Madagascar	14	35	12
Malawi	32	28.5	9
Mali	12	39	29
Mauritania	18	22	25
Mauritius	46	49	2
Mozambique	19	18	21
Namibia	42	44	6
Niger	7	14	16.5
Nigeria	23	10	21
Rwanda	33	24	34

(continued)

Table 5.8 (continued)

Country	Social progress	Fragile	Freedom
Sao Tome Principe		41	
Senegal	39	38	8
Seychelles		48	4.5
Sierra Leone	25	27	10
Somalia	5	1	43
South Africa	44	42	4.5
South Sudan	1	2	47
Sudan	15	5.5	42
Tanzania	37	37	27
Togo	29	12.5	23.5
Uganda	28	16	27
Zambia	35	26	16.5
Zimbabwe	26	7	31

5.6 Fuzzy Semigraphs

Let V be a set with n elements, $n \geq 2$. Let $k \in N$ with $2 \leq k \leq n$ and let $P^k = \{(v_1, ..., v_k) | v_i \in V, i = 1, ..., k, |\{v_1, ..., v_k| = k\}, k = 2, ..., n$. Let $P = \cup_{k=2}^{n} P^k$.

Definition 5.6.1 Let $E = P$. Let $E' \subseteq E$. Suppose E' has the property that for all $(v_1, ..., v_k), (u_1, ..., u_j) \in E'$, with $\{v_1, ..., v_k\} \neq \{u_1, ..., u_j\}$, $|\{v_1, ..., v_k\} \cap \{u_1, ..., u_j\}| \leq 1$. Then $G = (V, E')$ is called a **directed semigraph**.

Note that in the previous definition, two edges $(u_1, u_2, ..., u_n)$ and $(v_1, v_2, ..., v_m)$ are considered to be equal if and only if $m = n$ and $u_i = v_i$ for $i = 1, ..., n$, but not $u_i = v_{n-i+1}$ for $i = 1, ..., n$.

We next determine the susceptibility of a route to human trafficking. The susceptibility is based on a country's vulnerability for and its government response to human trafficking. In [5], measurements of government response and vulnerability were provided for 181 countries. The data was normalized using the formula (number − minimum)/(maximum–minimum) and the Pearson correlation coefficient was used to determine the correlation between five types of government response values with the vulnerability values yielded a negative correlation. This is important because it shows that government response and vulnerability are opposites. The results can be found in [14]. Four routes through the Americas to the United States are examined. The indices of two of the measures agree on all four routes. it was shown in [14] that the route with the highest susceptibility has the Dominican Republic as its origin country.

Vulnerability Measure

(1) Government issues: Includes political instability, weapons access, women's physical security, rights for the disabled, political rights, and regulatory equality.

(2) Nourishment and access: Includes call phone availability, social "security net", undernourishment levels, access to clean water, tuberculosis rates, and the ability to borrow money.

(3) Inequality: Includes confidence in judicial system, violent crime, GINI coefficient (wealth inequality), ability to obtain emergency funds.

(4) Disenfranchised groups: Includes same sex rights and acceptance of immigrants and minorities.

(5) Effects of conflicts: Includes impact of terrorism, internal conflicts fought, and displaced persons.

Government Response

(1) Support for survivors: Survivors of slavery are supported to exit slavery and empowered to break the cycle of vulnerability.

(2) Criminal justice: Effective criminal justice responses are in place in every jurisdiction.

(3) Coordination: Effective and measurable national action plans are implemented and fully funded in every country.

(4) Response: Laws, policies, and programs address attitudes, social systems, and institutions that create vulnerability and enable slavery.

(5) Supply chains: Governments stop sourcing goods or services linked to modern slavery.

Definition 5.6.2 Let (σ, μ) be a fuzzy subgraph of $G = (V, \mathcal{E}')$ and let (τ, ν) be a complementary fuzzy subgraph of G. Let $V = \{v_1, ..., v_m\}$. Then $\frac{1}{\sigma(v_i)} + \frac{1}{1-\tau(v_i)}$ is called the **susceptibility** of v_i with respect to human trafficking. Let $P : v_1v_2, ..., v_{k-1}v_k$ be a path in G. Define $CRV = \sum_{i=1}^{k}(\frac{1}{\sigma(v_i)} + \frac{1}{1-\tau(v_i)})$. Then $CRP(P)$ is called the **susceptibility** of P with respect to human trafficking.

The following table provides a measure of the amount of flow from the country heading each row to the country heading each column (Table 5.9). The measure was provided in [4], where the terms (very) low, medium (very) high were used to describe the amount of flow. We assigned numbers 0.1, 0.3.0.5, 0.7, 0.9, respectively, to these terms in order to place the measures in a mathematics of uncertainty setting. We

Table 5.9 Flow amount

	Col	CR	DR	Ec	ES	Gu	Hon	Mex	Nic	Pan	US
Columbia		0.3		0.5						0.5	0.5
Costa Rica					0.5						
Dominican Rep.										0.5	0.3
Ecuador						0.3	0.3	0.3			
El Salvador						0.5		0.5			0.5
Guatemala					0.3			0.5			0.5
Honduras					0.5	0.5		0.5			0.3
Mexico											0.7
Nicaragua		0.5			0.3	0.5					0.3
Panama		0.3									0.3
US											

Table 5.10 Activity

	Col	CR	DR	Ecu	El S	Gu	Hon	Mex	Nic	Pan	US
Activity	0.65	0.8775	0.5	0.65	0.65	0.75	0.65	0.925	0.5	0.65	0.7

provide only those measures that correspond to the countries in the paths in Example 5.6.3 below.

The **activity**, $Act(C)$, of a country C with respect to human trafficking is defined to be the sum of amount of trafficking into the country plus the sum of the amount out of the country, where sum means algebraic sum. For example, $Act(\text{Costa Rica}) = (0.3 \oplus 0.5 \oplus 0.3) \oplus 0.5 = 0.8775$ (Table 5.10)

Example 5.6.3 It was determined in [5] and ([14], p. 216) that the main paths in the Americas that lead to the United States are the following:

$$\text{Columbia} \to \text{Costa Rica} \to \text{El Salvador} \to \text{Guatemala} \to \text{Mexico} \to \text{U. S.}$$
$$\text{Columbia} \to \text{Ecuador} \to \text{Honduras} \to \text{Guatemala} \to \text{Mexico} \to \text{U. S.}$$
$$\text{Nicaragua} \to \text{Costa Rica} \to \text{El Salvador} \to \text{Guatemala} \to \text{Mexico} \to \text{U. S.}$$
$$\text{Dom. Rep.} \to \text{Panama} \to \text{Costa Rica} \to \text{El Salvador} \to \text{Guatemala} \to \text{Mexico} \to \text{U. S.}$$

Let V denote the set of countries involved. These paths can be placed into a setting of a directed semigraph as follows:

$$E' = \{(\text{Columbia, Ecuador, Honduras, Mexico}), (\text{Columbia, Costa Rica, El Salvador, Guatemala}), (\text{Nicaragua, Costa Rica}), (\text{Dominican Rep., Panama, Costa Rica}), (\text{Mexico, U. S.})\}$$

It should be noted that this arrangement preserves all the paths that lead to the United States. We see that Costa Rica is in common with three of the paths and Mexico and Columbia each with two. Thus Costa Rica, Mexico, and Columbia are the countries whose presence may contribute the most to human trafficking.

Example 5.6.4 It was determined in [5] and ([13], p. 175) that an increasing number of people from Asia and Africa are seeking to enter the U. S. illegally over the Mexican border. The main paths are as follows:

China → Columbia → Guatemala → Mexico → United States
India → Guatemala → Mexico → United States
Ethiopia → S. Africa → Brazil → Ecuador → Mexico → United States
Somalia → UAE → Russia → Cuba → Columbia → Mexico → United States
Nigeria → Spain → Cuba → Columbia → Mexico → United States

Let V denote the set of countries involved. These paths can be placed into a setting of a directed semigraph as follows:

$$
\begin{aligned}
E' = \{&(\text{China, Columbia, Guatemala, Mexico, United States}),\\
&(\text{India, Guatemala}), (\text{Ethiopia, S. Africa, Brazil, Ecuador, Mexico}),\\
&(\text{Somalia, UAE, Russia, Cuba, Columbia}), (\text{Nigeria, Spain, Cuba})\}
\end{aligned}
$$

This arrangement preserves all paths to the United States. We see that Columbia, Guatemala, Cuba, Mexico each appear twice.

Definition 5.6.5 Let $G = (V, \mathcal{E}')$ be a semigraph. Let $\sigma : V \to [0, 1]$ and $\eta : \mathcal{E}' \to [0, 1]$. Then (σ, η) is called a **fuzzy semigraph** if $\forall e = (v_1, ..., v_k) \in \mathcal{E}', \eta(e) \leq \sigma(v_1) \wedge ... \wedge \sigma(v_k)$.

Definition 5.6.6 Let $G = (V, \mathcal{E}')$ be a semigraph and (V, S) a graph. Let $\sigma : V \to [0, 1]$, $\mu : S \to [0, 1]$, and $\eta : \mathcal{E}' \to [0, 1]$. Suppose (σ, μ) a fuzzy graph. Then (σ, η) is called a **fuzzy semigraph** if $\forall e = (v_1, ..., v_k) \in \mathcal{E}', \eta(e) \leq \mu(v_1v_2) \wedge ... \wedge \mu(v_{k-1}v_k)$.

One could also define intuitionistic fuzzy semigraphs.

Definition 5.6.7 Let $G = (V, E')$ be a directed semigraph. Let $\sigma : V \to [0, 1]$ and $\eta : E' \to [0, 1]$. Then (σ, η) is called a **fuzzy directed semigraph** if $\forall e = (v_1, ..., v_k) \in E', \eta(e) \leq \sigma(v_1) \wedge ... \wedge \sigma(v_k)$.

Table 5.11 Path to U. S.

	Col.		Costa R.		El Sal.		Guat.		Mex.		U. S.
σ	0.53		0.55		0.43		0.56		0.62		0.88
μ		0.29		0.24		0.24		0.34		0.55	
τ	0.42		0.27		0.36		0.42		0.47		0.23
ν		0.57		0.53		0.63		0.69		0.59	

Table 5.12 Path to U. S.

	Col.		Ecu.		Hond.		Guat.		Mex.		U. S.
σ	0.53		0.51		0.39		0.56		0.62		0.88
μ		0.27		0.20		0.22		0.35		0.55	
τ	0.42		0.35		0.43		0.42		0.47		0.23
ν		0.62		0.63		0.67		0.69		0.59	

Table 5.13 Path to U. S.

	Nic.		C. R.		El Sal.		Guat.		Mex.		U. S.
σ	0.59		0.55		0.43		0.56		0.62		0.88
μ		0.32		0.24		0.24		0.35		0.55	
τ	0.35		0.27		0.36		0.42		0.47		0.23
ν		0.52		0.53		0.63		0.69		0.59	

Table 5.14 Path to U. S.

	Dom. R.		Pan.		C. R.		El Sal.		Guat.		Mex.		U. S.
σ	0.63		0.48		0.55		0.43		0.56		0.62		0.88
μ		0.30		0.26		0.24		0.24		0.35		0.55	
τ	0.39		0.33		0.27		0.36		0.42		0.47		0.23
ν		0.59		0.51		0.53		0.63		0.69		0.59	

Definition 5.6.8 Let $G = (V, E')$ be a semigraph and (V, S) a graph. Let $\sigma : V \to [0, 1]$, $\mu : S \to [0, 1]$, and $\eta : E' \to [0, 1]$. Suppose (σ, μ) a fuzzy directed graph. Then (σ, η) is called a **fuzzy directed semigraph** if $\forall e = (v_1, ..., v_k) \in E'$, $\eta(e) \leq \mu(v_1 v_2) \wedge ... \wedge \mu(v_{k-1} v_k)$.

Consider Example 5.6.3 Let (σ, μ) be a fuzzy subgraph of $G = (V, \mathcal{E}')$ and let (τ, ν) be a complementary fuzzy subgraph of G. The σ and τ values in the following table are normalized government response and vulnerability values of ([14], p. 216), respectively. The μ and ν values are determined by the product of the σ values and the algebraic sum values of the τ values, respectively, in Tables 5.11, 5.12, 5.13 and 5.14.

5.7 Generalized Graph Structures

The notion of a graph structure was introduced in [15].

Definition 5.7.1 A graph structure $G = (V, R_1, ..., R_k)$ consists of a nonempty finite set V together with relations $R_!, ..., R_k$ on V which are mutually disjoint and such that R_i is symmetric and irreflexive, $i = 1, ..., k$.

If $(u, v) \in R_i$, then we write uv for (u, v). Since R_i is reflexive, $uv = vu$. If R_i is not reflexive, then we call G a **directed graph structure.** IfG is a directed graph structure and $(u, v) \in R_i$, then $uv \neq vu$.

Example 5.7.2 Consider Example 5.6.3 Let V denote the set of countries involved. The paths in Example 5.6.3 can be placed into a setting of a directed graph structure as follows:

R_1 = {(Columbia, Ecuador), (Ecuador, Honduras), (Honduras, Mexico), (Mexico, U. S.)},

R_2 = {(Columbia, Costa Rica), (Costa Rica, El Salvador), (El Salvador,Guatemala), (Guatemala, Mexico)},

R_3 = {((Nicaragua, Costa Rica)}, R_4 = {(Dominican Rep., Panama), (Panama, Costa Rica)},

It should be noted that this arrangement preserves all the paths that lead to the United States. We see that Costa Rica is in common with three of the paths and Mexico and Columbia each with two. Thus Costa Rica, Mexico, and Columbia are the countries whose presence may contribute the most to human trafficking.

Example 5.7.3 Consider Example 5.6.4. Let V denote the set of countries involved. The paths in Example 5.6.4 can be placed into a setting of a directed graph structure as follows:

R_1 = {(China, Columbia), (Columbia, Guatemala), (Guatemala, Mexico), (Mexico, U. S.)}

R_2 = {(India. Guatemala), (Guatemala, Mexico)}

R_3 = {(Ethiopia, S. Africa), (S. Africa, Brazil), (Brazil, Ecuador), (Ecuador, Mexico)}

R_4 = {(Somalia, UAE),(UAE, Russia), (Russia, Cuba), (Cuba, Columbia)}

R_5 = {(Nigeria, Spain), (Spain, Cuba)}

This arrangement preserves all paths to the United States. We see that Columbia, Guatemala, Cuba, Mexico each appear twice.

Let $G = (V, \mathcal{E}')$ be a semigraph. Let $m = |\mathcal{E}'|$. For each permutation $p_i = (v_{1i}, ..., v_{ki})$ in $\mathcal{E}'$, let $S_i = \{v_{1i}v_{2i}, v_{2i}v_{3i}, ..., v_{k-1,i}v_{ki}\}, i = 1, ..., m$. Then $S_i \cap S_j = \emptyset, i, j = 1, ..., m; i \neq j$. If we consider G to be $(V, S, ..., S_m)$, then G is a graph structure.

5.8 Fuzzy Graph Structures

Definition 5.8.1 Let $G = (V, R_1, R_2, ..., R_k)$ be a graph structure and $\sigma, \rho_1, \rho_2, ..., \rho_k$ be fuzzy subsets of $V, R_1, R_2, ..., R_k$, respectively, such that $\rho_i(x, y) \leq \sigma(x) \wedge \sigma(y)$ for all $x, y \in V$ and $i = 1, 2, ..., k$. Then $G^* = (\sigma, \rho_1, \rho_2, ..., \rho_k)$ is called a **fuzzy graph structure** of G.

Since the R_i are mutually disjoint, if we define $\mu(uv) = \rho_i(uv)$, where $uv \in R_i$, then μ is well-defined and in fact (σ, μ) is a fuzzy subgraph with $\mu = \cup_{i=1}^{k} \rho_i$. Also, (σ, ρ_i) is a subgraph (σ, μ). Consequently, known results from fuzzy graph theory hold automatically.

We next consider the Central Mediterranean Route. This refers to migration from North Africa to Italy and to a lesser extent, Malta., ([16], Table 5). The Central Mediterranean Route is the most frequently used route to Europe. A directed fuzzy incidence model was created in [16] for the traffic in the Mediterranean. Since most of deaths take place from the sea, we consider three major routes, namely, east, central, and western routes. In [16], a directed fuzzy incidence model was developed. The following is an example of a graph structure with respect to this situation.

Example 5.8.2 Let $V_1 = \{$Algiers, France, Spain, Melilla, Ceuta$\}$, $V_2 = \{$Alexandria, Malta, Lampedusa, Tripoli, Sousse, Italy$\}$, $V_3 = \{$Syria, Alexandria, Cyprus, Greece, Turkey$\}$. Let $R_1 = \{$(Algiers, France), (Algiers, Spain), (Melilla, France), (Melilla, Spain), (Ceuta, France), (Ceuta, Spain)$\}$, R_2=$\{$(Alexandria, Malta), (Malta, Lampedusa), (Malta, Libya), (Libya, Lampedusa), (Tripoli, Lampedusa), (Sousse, Italy)$\}$, $R_3 = \{$(Syria, Alexandria), (Syria, Turkey), (Alexandria,Cyprus), (Alexandria, Greece), (Turkey, Lampedusa)$\}$. Let $V = V_1 \cup V_2 \cup V_3$. Then (V, R_1, R_2, R_s) is a graph structure.

Definition 5.8.3 [7] Let $G_1 = (\sigma_1, \mu_1'\mu_2', ..., \mu_n')$ and $G_2 = (\sigma_2, \mu_1''\mu_2'', ..., \mu_n'')$ be two fuzzy-graph structures with underlying crisp graph structures of $G_1^* = (V_1, R_1'$ $R_2', ..., R_n')$ and $G_2^* = (V_2, R_1'', R_2'', ..., R_n'')$, respectively. Then $G_1 * G_2 = (\sigma, \mu_1, \mu_2, ..., \mu_n)$ is called a **maximal fuzzy graph** structure with underlying crisp graph structure $G^* = (V, R_1, R_2, ..., R_n)$, where $V = V_1 \times V_2$ and $R_i = \{(u_1, v_1)(u_2, v_2)| u_1 = u_2, v_1v_2 \in R_i''$ or $v_1 = v_2, u_1u_2 \in R_i'\}$ and σ and μ_i are defined as follows, $i = 1, ..., n$:

$$\sigma(u, v) = \sigma_1(u) \vee \sigma_2(v), \text{ where } (u, v) \in V_1 \times V_2$$

and

$$\mu_i((u_1, v_1)(u_2, v_2)) = \begin{cases} \sigma_1(u_1) \vee \mu_i''(v_1v_2), & \text{where } u_1 = u_{2.}, v_1v_2 \in R_i'', \\ \sigma_2(v_1) \vee \mu_i'(u_1u_2), & \text{where } v_1 = v_2, u_1u_2 \in R_i', \end{cases}$$

$i = 1, 2, ..., n$.

In the previous definition, we could write $\mu_i(u, v_1v_2)$ with $u \in V_1, v_1v_2 \in R_i''$ and $\mu_i(u_1u_2, v)$ with $v \in V_2, u_1u_2 \in R_i', i = 1, 2, ..., n$. This allows for the placement of maximal product into the area of fuzzy incidence graph structures.

Definition 5.8.4 A fuzzy graph structure $G = (\sigma, \mu_1, ..., \mu_n)$ is called μ_i**-strong** if $\mu_i(v_1v_2) = \sigma_1(v_1) \wedge \sigma_2(v_2)$ for all $v_1v_2 \in R_i$, where $i \in \{1, ..., n\}$. If G is u_i-strong for all $i \in \{1, ..., n\}$, then G is called a **strong fuzzy graph structure**.

Theorem 5.8.5 [17] The maximal product of two strong fuzzy graph structures is a strong fuzzy graph structure. □

Proof Let $G_1 = (\sigma_1, \mu_1'\mu_2', ..., \mu_n')$ and $G_2 = (\sigma_2, \mu_1''\mu_2'', ..., \mu_n'')$ be two fuzzy-graph structures. Then $\mu_i'(v_1v_2) = \sigma_1(v_1) \wedge \sigma_2(v_2)$ for any $v_1v_2 \in R_i'$ and $\mu_i''(u_1u_2) = \sigma_2(u_1) \wedge \sigma_2(u_1u_2) \in R_i'', i = 1, 2, ..., n$.

Suppose that $u_1 = u_2$ and $v_1v_2 \in R_i''$. Then

$$\begin{aligned}\mu_i((u_1, v_1)(u_2, v_2)) &= \sigma_1(u_1) \vee \mu_i''(v_1v_2) \\ &= \sigma_1(u_1) \vee (\sigma_2(v_1 \wedge \sigma_2(v_2)) \\ &= (\sigma_1(u_1) \vee \sigma_2(v_1)) \wedge (\sigma_1(u_1 \vee \sigma_2(v_2)) \\ &= \sigma(u_1, v_1) \wedge \sigma(u_2v_2).\end{aligned}$$

Suppose that $v_1 = v_2$ and $u_1u_2 \in R_i'$. Then

$$\begin{aligned}\mu_i((u_1, v_1)(u_2, v_2)) &= \sigma_2(v_1) \vee \mu_i'(u_1u_2) \\ &= \sigma_2(v_1) \vee (\sigma_1(u_1) \wedge \sigma_1(u_2)) \\ &= (\sigma_1(u_1) \vee \sigma_2(v_1)) \wedge (\sigma_1(u_2) \vee \sigma_2(v_1)) \\ &= \sigma(u_1, v_1) \wedge \sigma(u_2v_2).\end{aligned}$$

Thus $\mu_i((u_1, v_1)(u_2, v_2)) = \sigma(u_1, v_1) \wedge \sigma(u_2, v_2)$ for all edges. Hence $G = G_1 * G_2 = (\sigma, \mu_1, \mu_2, ..., \mu_n)$ is a strong fuzzy graph structure.

5.9 Fuzzy Incidence Graph Structures

We next consider fuzzy incidence graph structures as presented by Dinesh, [18, 19].

Let (V, E) be a graph and Let $I \in V \times E$. Then (V, E, I) is called an incidence graph. Let σ, μ, ψ be fuzzy subsets of V, E, I, respectively. If for all $uv \in E$, $\mu(uv) \leq \sigma(u) \wedge \sigma(v)$, and $\psi(u, uv) \leq \sigma(u) \wedge \mu(uv)$, then (σ, μ, ψ) is called a fuzzy incidence graph on (V, E, I).

Let $(V, R_1, ..., R_n)$ be a graph structure. Let $I_i = V \times R_i, i = 1, ..., n$. Then $(V, R_1, ..., R_n, I_1, ..., I_n)$ is called an **incidence graph structure**.

We next consider basic definitions for a fuzzy subgraph (σ, μ). Then these definitions hold for (σ, ρ_i), where $\rho_i(uv) = 0$ if $uv \notin R_i$. Note for $\mu = \cup_{i=1}^n \rho_i$, $\mu|_{R_i} = \rho_i$. If we consider a fuzzy path in (σ, ρ_i), we call the path a ρ_i-path. Consider a fuzzy incidence graph $(\sigma, \rho_1, ..., \rho_n, \psi_1, ..., \psi_n)$. We can consider $(\sigma, \rho_1, ..., \rho_n, \psi_1, ..., \psi_n)$ to be a fuzzy subgraph of $(V \times E, I)$, where $\sigma \times \mu$ is a fuzzy subset of $V \times E$, $E = \cup_{i=1}^n R_i$, and $\mu = \cup_{i=1}^n \rho_i$. However, we provide the basic definitions for fuzzy incidence graph structures.

Through out, $G = (V, R_1, ..., R_n, I_n, ..., I_n)$ denotes an incidence graph structure and $G^* = (\sigma, \rho_1, ..., \rho_n, \psi_1, ..., \psi_n)$ a fuzzy incidence subgraph of G. Since $uv = vu$, $(v, uv) = (v, vu)$ and so (v, uv) denotes flow from v to $vu = uv$. Recall (u, uv) denotes flow from u to uv. However, if we were working with directed graphs $uv \neq vu$, then (v, uv) denotes flow uv to v.

Definition 5.9.1 Let $xy \in \text{Supp}(\rho_i)$ in G^*. Then xy is called a ρ_i**-edge** of G^*. If $(x, xy), (y, xy) \in \text{Supp}(\psi_i)$, then $(x, xy), (y, xy)$ are called ψ_i**-pairs**.

Suppose that in the previous definition, we had stated: Let $xy \in \text{Supp}(\mu)$ in G^*. Then xy is called a **edge** of G^*. If $(x, xy), (y, xy) \in \text{Supp}(\psi)$, then $(x, xy), (y, xy)$ are called ψ**-pairs.** If $xy \in \text{Supp}(\mu)$, then $xy \in \text{Supp}(\rho_i)$ for some i, then xy is an ρ_i-edge since $\mu(xy) > 0$ implies $\rho_i(xy) > 0$. If $(x, xy), (y, xy) \in \text{Supp}(\psi)$, then $(x, xy), (y, xy) \in \text{Supp}(\psi_i)$. Thus $(x, xy), (y, xy)$ are ψ_i-pairs.

Definition 5.9.2 A sequence

$$v_0, (v_0, e_1), e_1, (v_1, e_1), v_1, (v_1, e_2), e_2, (v_2, e_2), v_2, ..., v_{n-1}, (v_{n-1}, e_n), e_n, (v_n, e_n), v_n$$

of vertices, edges, and ψ-pairs which are distinct except possibly for v_0, v_n such that $v_{j-1}v_j$ is an edge is called an μ**-path**. If all the edges are ρ_i-edges and all the pairs are ρ_i-pairs, then the sequence is called a ρ_i**-path.**

Definition 5.9.3 Two vertices u, v that are joined by a path are said to be **connected**. If u, v are joined by a ρ_i-path, they are called ρ_i**-connected.**

Definition 5.9.4 The **incidence strength** of a fuzzy incidence graph structure is defined to be $\wedge\{\psi(v_j, e_k) | (v_j, e_k) \in \text{Supp}(\psi)\}$. If all the $(v_j, e_k) \in \text{Supp}(\psi_i)$ for a fixed i, then the strength is called the ψ_i**-incidence strength**.

Example 5.9.5 $V = \{v_1, v_2, v_3, u_1, u_2\}$.Let $R_1 = \{v_1w, wv\}$ and $R_2 = \{u_1w, wu_2, u_2v\}$. Define $\rho_1(v_1w) = 0.2$, $\rho_1(wv) = 0.8$, $\rho_2(u_1w) = 0.7$, $\rho_2(wu) = 0.5$, $\rho_2(wv) = 0.5$. Then the ρ_2-strength of $u_1$to v is 0.5 while the ρ-strength of v_1 to v is 0.7.

Definition 5.9.6 Let G be an incidence graph structure and G^* be a fuzzy incidence graph structure on G. If $xy \in \text{Supp}(\rho_i)$, then xy is said to be an ρ_i**-edge**, $i = 1, ..., k$.

Definition 5.9.7 A ρ_i**-path** of a fuzzy graph structure G^* is a sequence of vertices, $x_0, x_1, ., , , x_n$ which are distinct (except possibly x_0 and x_n) such that $x_{j-1}x_j$ is a ρ_i-edge for $j = 1, 2, ..., n$.

Definition 5.9.8 Two vertices of a fuzzy graph structure G^*, joined by a ρ_i-path are said to be ρ_i**-connected**.

Since $(V, R_i, \sigma, \rho_i, \psi_i)$ is a fuzzy incidence graph, other definitions and results follow from immediately from known definitions and results.

Consider Example 5.7.3. We have the following tables, where σ represents government response and τ represents vulnerability, [5].

Definition 5.9.9 Let τ be a fuzzy subset of V and let ν be a fuzzy subset of E. Then (τ, ν) is called a **complementary fuzzy subgraph** of G if $\nu(xy) \geq \tau(x) \vee \tau(y)$ for all $xy \in E$. Let (τ, ν) be a complementary fuzzy subgraph of G and let ω be a fuzzy subset of $E' = \{(x, xy)|xy \in E\}$. Then (τ, ν, ω) is called a **complementary fuzzy incidence subgraph** of G if for all $(x, xy) \in E'$, $\omega(x, xy) \geq \tau(x) \vee \nu(xy)$.

Definition 5.9.10 Let (σ, μ) be a fuzzy subgraph of $G = (V, E)$ and let (τ, ν) be a complementary fuzzy subgraph of G. Let $u \in V$ be such that $\sigma(u) > 0$ and $\tau(u) < 1$. Then $\frac{1}{\sigma(u)} + \frac{1}{1-\tau(u)}$ is called the **susceptibility** of u with respect to human trafficking. Let $P : x_1, x_2, ..., x_{n-1}, x_n$ be a path in G. Define $S(P) = \sum_{i=1}^{n}(\frac{1}{\sigma(x_i)} + \frac{1}{1-\tau(x_i)})$. Then $S(P)$ is called the **susceptibility** of P with respect to human trafficking.

When combining the effect of the government response of two countries, say $\sigma(u)$ and $\sigma(v)$, we feel that the combined response should be less than $\sigma(u) \wedge \sigma(v)$ if both $\sigma(u)$ and $\sigma(v)$ are less than 1. Consequently, the t-norm product has been used. However, it might be argued than product is too severe. Hence we use a special kind of aggregation operator to determine the ψ values. In so doing, we have $\psi(u, uv) \leq \mu(uv) \wedge \sigma(u)$ for all pairs of distinct countries u and v, where $\mu(uv) = \sigma(u) \wedge \sigma(v)$. This follows since the smallest $\sigma(u) = 0.21$, where u = Cuba and so we choose $\lambda = 0.2$ for the following norm operation, ([20], p. 93),

$$h_\lambda(a, b) = \begin{cases} \lambda \wedge (a + b - ab) \text{ if } a, b \in [0, \lambda], \\ \lambda \vee (ab) \text{ if } a, b \in [\lambda, 1], \\ \lambda \text{ otherwise.} \end{cases}$$

Table 5.15 Norm operation

	China		Col.		Guat.		Mex.		U. S.
σ	0.36		0.53		0.56		0.57		0.82
ψ		0.2		0.3		0.32		0.47	
τ	0.45		0.42		0.42		0.43		0.18
ω		0.68		0.66		0.67		0.53	

Table 5.16 Norm operation

	India		Guat.		Mex.		U. S.
σ	0.46		0.56		0.57		0.82
ψ		0.26		0.32		0.47	
τ	0.53		0.42		0.43		0.18
ω		0.72		0.67		0.53	

Table 5.17 Norm operation

	Eth.		S. Afr.		Braz.		Ecu.		Mex.		U.S
σ	0.42		0.49		0.66		0.51		0.57		0.82
ψ		0.21		0.32		0.34		0.29		0.47	
τ	0.58		0.49		0.31		0.35		0.43		0.18
φ		0.72		0.65		0.55		0.63		0.53	

Table 5.18 Norm operation

	Som.		UAE		Rus.		Cuba		Col.		Mex.		U. S.
σ	0.28		0.56		0.30		0.21		0.53		0.57		0.82
ψ		0.2		0.2		0.2		0.2		0.3		0.47	
τ	0.72		0.26		0.42		0.32		0.42		0.43		0.18
ω		0.72		0.57		0.61		0.61		0.67		0.53	

We can use a similar argument for ω. Here the largest $\tau(c)$ value is 0.72 for $c =$ Somalia.

With $\lambda = 0.2$ for government response $\lambda = 0.72$ for vulnerability, we have Tables 5.15, 5.16, 5.17, 5.18 and 5.19.

Now $\psi(u, uv)$ can be thought of as the success in combatting human trafficking at u with respect to uv, while $\omega(u, uv)$ the failure of combatting human trafficking at u with respect to uv.

We use the length of a route (path) defined by Rosenfeld to provide measures determining the success in combatting human trafficking with respect to government response. The definition of ω-length of a path $P : x_0, x_1, ..., x_n$ is $\psi l(P) = \sum_{i=1}^{n} \frac{1}{\psi(x_{i-1}, x_i x_{i+1})}$.

Table 5.19 Norm operation

	Nigeria		Spain		Cuba		Col.		Mex.		U. S.
σ	0.44		0.71		0.21		0.53		0.57		0.82
ψ		0.31		0.2		0.2		0.3		0.47	
τ	0.56		0.20		0.32		0.42		0.43		0.18
ω		0.65		0.46		0.61		0.67		0.53	

The numbers in [5] provide high numbers if the vulnerability of a country is high. The standard complement of these numbers then provides high numbers if the vulnerability is low. Consequently, we are interested in the complement of ω. It follows that (τ, ν^c, ω^c) is a complementary fuzzy incidence graph. We have $\omega^c l(P) = \sum_{i=1}^{n} \frac{1}{\omega^c(x_{i-1}x_i)} = \sum_{i=1}^{n} \frac{1}{1-\omega(x_{i-1}x_i)}$.

Definition 5.9.11 Let (σ, μ) be a fuzzy subgraph of $G = (V, E)$. Let $P : x_0, x_1, ..., x_n$ be a path in G. Define $SI(P) = \sum_{i=1}^{n}(\frac{1}{\psi(x_{i-1}, x_i x_{i+1})} + \frac{1}{1-\omega(x_{i-1}x_i)})$. Then $SI(P)$ is called the **incidence susceptibility** of P with respect to human trafficking.

We provide the incidence susceptibility of the above paths.

$SI(\text{China, U. S.}) = \frac{1}{0.2} + \frac{1}{0.3} + \frac{1}{0.32} + \frac{1}{0.47} + \frac{1}{1-0.68} + \frac{1}{1-0.66} + \frac{1}{1-0.67} + \frac{1}{1-0.53} = 23.48$

$SI(\text{India, U. S.}) = \frac{1}{0.26} + \frac{1}{0.32} + \frac{1}{0.47} + \frac{1}{1-.072} + \frac{1}{1-0.67} + \frac{1}{1-0.53} = 17.84$

$SI(\text{Ethiopia, U. S.}) = \frac{1}{0.21} + \frac{1}{0.32} + \frac{1}{0.34} + \frac{1}{0.29} + \frac{1}{0.47} + \frac{1}{1-0.72} + \frac{1}{1-0.65} + \frac{1}{1-0.55} + \frac{1}{1-0.63} + \frac{1}{1-0.53} = 29.88$

$SI(\text{Somalia, U. S.}) = \frac{1}{0.2} + \frac{1}{0.2} + \frac{1}{0.2} + \frac{1}{0.2} + \frac{1}{0.3} + \frac{1}{0.47} + \frac{1}{1-0.72} + \frac{1}{1-0.57} + \frac{1}{1-0.61} + \frac{1}{1-0.61} + \frac{1}{1-0.67} + \frac{1}{1-0.53} = 41.64$

$SI(\text{Nigeria, U. S.}) = \frac{1}{0.31} + \frac{1}{0.2} + \frac{1}{0.2} + \frac{1}{0.3} + \frac{1}{0.47} + \frac{1}{1-0.65} + \frac{1}{1-0.46} + \frac{1}{1-0.61} + \frac{1}{1-0.67} + \frac{1}{1-0.53} = 31.12$

References

1. Molodtsov, D.: Soft set theory-first results. Comput. Math. Appl. **37**, 19–31 (1999)
2. Mavrov, D., Atanassova, V., Bureva, V., Roeva, O., Tavetkov, R., Zoteva, D., Sotirova, E., Atanassov, K., Alexandrov, A., Tsakov, H.: Application of game method for modelling and temporal intuitionistic fuzzy pairs to the forest fire spread in the presence of strong wind. Mathematics **10**, 1–17 (2022)
3. Maji, P.K., Biswas, R., Roy, A.R.: Fuzzy soft sets. J. Fuzzy Math. **9**, 589–602 (2001)
4. Trafficking in Persons: Global Patterns. The United Nations Office for Drugs and Crime, Trafficking in Persons Citation Index

5. Global Slavery Index 2016 (2017). http://www.globalslaveryondex.org/findings
6. Ahmad, B., Kharal, A.: On fuzzy soft sets. Adv. Fuzzy Syst. 1–6 (2009)
7. Sampathkumar, E.: Semigraphs. Res. Gate 1–16 (2020)
8. George, B., Thumbakara, H.K., Jose, J.: Soft semigraphs and some of their operations. New Mathematics and Natural Computation (to appear)
9. Hampiholi, P.R., Kaliwal, M.M.: Operations on semigraphs. Bull. Math. Sci. Appl. **18**, 11–22 (2017)
10. Social Progress Index: Wikipedia. http://en.wikipedia.org/Social_Progress_Index
11. Freedom House: Countries and Territories. https://freedomhouse.org/freedom-world/scores
12. World Population Review: Fragile States Index 2022. https://worldpopulationreview.com/country-rankings/fragile-state-index
13. Mordeson, J.N., Mathew, S., Binu, M.: Applications of mathematics of uncertainty, grand challenges—human trafficking—coronavirus—biodiversity and extinction. In: Studies in Systems and Control, vol. 391. Springer (2022)
14. Mordeson, J.N., Mathew, S.: Sustainable development goals: analysis by mathematics of uncertainty. In: Studies in Systems, Decision, and Control, vol. 299. Springer (2021)
15. Sampathkumar, E.: Generalized graph structures. Bull. Kerala Math. Assoc. **3**(2), 67–123 (2006)
16. www.independent.co.uk Independent, 6 charts and a map that shows where Europe's refugees are coming from and the perilous journey they are taking (2015)
17. Sitara, M., Akram, M., Bhatti, M.Y.: Fuzzy graph structures with application. Mathematics **7**(1), 63 (2019). https://doi.org/10.3390/math7010063
18. Dinesh, T.: Fuzzy incidence graph structures. Adv. Fuzzy Math. **125**, 21–30 (2020)
19. Dinesh, T., Ramakrishnan, T.V.: On generalized fuzzy graph structure. Appl. Math. Sci. **5**(4), 173–180 (2011)
20. Klir, G.J., Yuan, B.: Fuzzy Sets and Fuzzy Logic: Theory and Applications. Prentice Hall P T R Upper Saddle River, New Jersey (1995)

Chapter 6
Directed Fuzzy Incidence Graphs

This chapter presents directed fuzzy incidence graphs, a novel mathematical model for analyzing a wide range of stochastic networks. The theory of fuzzy graphs already includes the notion of fuzzy incidence graphs, which mainly focus on node-edge relationships. When working with systems of considerable external flow and support, such graphs are extremely useful. The ramping system on roads can be modeled using fuzzy incidence graphs, which makes it easier to manage unpredictable traffic flow between cities and highways. In a one-way traffic situation, however, such modeling will fall short of solving the problem. Since fuzzy incidence graphs lack a defined flow direction, two-way transport through the edges and incidence pairs can be included, which again complicates the system. Hence, we are in need of defining the direction in fuzzy incidence graphs.

Directed fuzzy incidence graphs (DFIG) are structures with a specified flow direction that can be used to evaluate the flow of a variety of unpredictable networks. Here, we develop such a model and investigate some of its connectivity aspects. Directed fuzzy incidence graph models help us to comprehend the influences of pairs on nodes and arcs where they meet. Our approach to connectivity in DFIG is extensively expanded by the use of legal and illegal flows so that a distinguished analysis of legal and illegal influences can be made. The contents of this chapter are from [1].

6.1 Directed Fuzzy Incidence Graphs (DFIG)

In this section, we develop the basic idea of directed fuzzy incidence graphs. If we consider an arc zx from z to x, the incidence (or influence) of z on zx is (z, zx), which is 1 and (x, zx) is 0. We generalize this concept in DFIG by incorporating the asymmetry of relations.

J. N. Mordeson et al., *Fuzzy Graph Theory*, Studies in Fuzziness and Soft Computing 424, https://doi.org/10.1007/978-3-031-23108-7_6

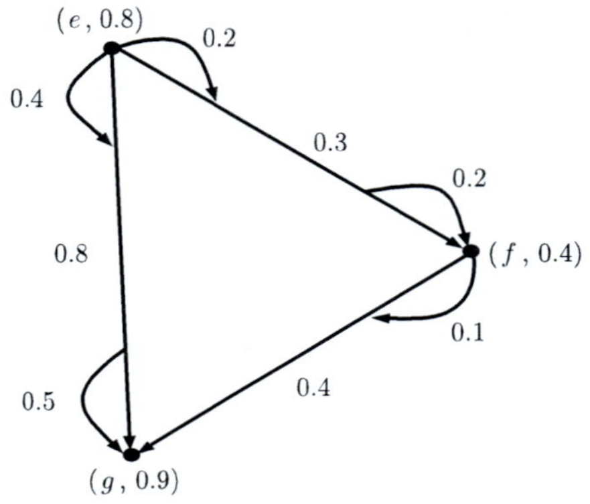

Fig. 6.1 Example of a DFIG

Definition 6.1.1 Let V be a set of points. A **directed fuzzy incidence graph** $\overrightarrow{G}_i$ is an ordered triple (σ, μ, ψ) where σ is a fuzzy subset of V, μ is a fuzzy subset of $V \times V$ such that $\mu(w, y) \leq \sigma(w) \wedge \sigma(y)$ for all $(w, y) \in V \times V$, where $\wedge$ denotes the minimum and ψ is a fuzzy subset of S where $S = (\sigma^* \times \mu^*) \cup (\mu^* \times \sigma^*)$ such that $\psi(w, wy) \leq \sigma(w) \wedge \mu(wy)$ for every $(w, wy) \in \sigma^* \times \mu^*$ and $\psi(wy, y) \leq \sigma(y) \wedge \mu(wy)$ for every $(wy, y) \in \mu^* \times \sigma^*$. It is denoted by $\overrightarrow{G}_i(\sigma, \mu, \psi)$.

It is important to note that the relations μ and ψ are defined over ordered sets. We call the members of σ^* as nodes, the members of μ^* as arcs, and the members of ψ^* as directed incidence pairs or d-pairs.

Example 6.1.2 Consider Fig. 6.1, which demonstrates a DFIG. It has node set $V = \{e, f, g\}$. Define σ, μ and ψ as $\sigma(e) = 0.8, \sigma(f) = 0.4, \sigma(g) = 0.9, \mu(ef) = 0.3, \mu(fg) = 0.4, \mu(eg) = 0.8, \psi(e, ef) = \psi(ef, f) = 0.2, \psi(e, eg) = 0.4, \psi(eg, g) = 0.5$ and $\psi(f, fg) = 0.1$. Take $\mu(ee) = \mu(ff) = \mu(gg) = \mu(fe) = \mu(gf) = \mu(ge) = \psi(fg, g) = 0$. Then $\overrightarrow{G}_i(\sigma, \mu, \psi)$ is an example of a DFIG.

For a DFIG, we ignore the possibility of loops like (z, zz) and (zz, z). Note that the d-pairs indicate the illegal part of the flow, through which one can reach the nodes as well as the arcs of the DFIG. For an arc zx, we let $\psi(z, zx)$ to represent the illegal inflow through zx and $\psi(zx, x)$ to represent the illegal outflow through zx. Subtracting the maximum of illegal inflow and illegal outflow through zx from the weight of zx yields the amount of regular flow through zx.

In a DFIG, we can have several types of paths, just as in fuzzy incidence graphs. However, the paths are limited in their directions. Incidence arcs in the paths follow the same direction of the flow. The existence of a path without incidence pairs is also possible in a DFIG. When a path has at least one directed incidence pair, it is called a directed path or a *di*-path. Consider the following definition, which gives the formal definition of a *di*-path.

Definition 6.1.3 Consider the DFIG $\overrightarrow{G_i}(\sigma, \mu, \psi)$. An $s-t$ directed incidence path $\overrightarrow{P}$ from the origin s to the destination t in $\overrightarrow{G_i}$ with both $s, t \in \sigma^*$ is a sequence given by $\overrightarrow{P} : s = w_1, (w_1, w_1w_2), w_1w_2, (w_1w_2, w_2), w_2, \ldots, w_{n-1}, (w_{n-1}, w_{n-1}w_n), w_{n-1}w_n, (w_{n-1}w_n, w_n), w_n = t$.

We remove a d-pair from the path if its ψ-value is zero. A *di*-path $\overrightarrow{P}$ is called a **regular path** if $\psi(w_i, w_iw_{i+1}) = \psi(w_iw_{i+1}, w_{i+1}) = 0$ for every $i = 1, 2, \ldots, (n-1)$ and is called an **irregular path** otherwise. If the origin coincides with the destination in a path, then they are called a regular or an irregular cycle. A regular path is the same as a directed path.

When we analyze Example 6.1.2, we see that there is a unique irregular path $\overrightarrow{Q} : k, (k, km), km, (km, m), m$ from k to m. But, we cannot identify any regular or irregular incidence paths from m to k. Moreover, $\overrightarrow{G_i}$ has no regular paths as every *di*-path includes some illegal pair.

The idea of connectivity demands a detailed study of both legal and illegal flow through the network. Next, we present the concepts of legal and illegal flow in a DFIG. To describe them, we have to define the strength of connectedness between two nodes. Take a look at the definitions below.

Definition 6.1.4 Consider a DFIG $\overrightarrow{G_i}(\sigma, \mu, \psi)$ and a directed incidence path $\overrightarrow{Q}$ in $\overrightarrow{G_i}$ given by $\overrightarrow{Q} : w_1, (w_1, w_1w_2), w_1w_2, (w_1w_2, w_2), w_2, \ldots, w_{n-1}, (w_{n-1}, w_{n-1} w_n), w_{n-1}w_n, (w_{n-1}w_n, w_n), w_n$. Then the **legal incidence strength (legal flow)** of $\overrightarrow{Q}$ is defined by

$$i_s(\overrightarrow{Q}) = \wedge\{\mu(w_iw_{i+1}) - \vee\{\psi(w_i, w_iw_{i+1}), \psi(w_iw_{i+1}, w_{i+1})\}, i = 1, 2, \ldots, (n-1)\}.$$

The **illegal incidence strength (illegal flow)** of $\overrightarrow{Q}$ is defined by $ii_s(\overrightarrow{Q}) = \wedge\{\mu(wy) : wy \in \overrightarrow{Q}\} - i_s(\overrightarrow{Q})$. A *di*-path $\overrightarrow{Q}$ is said to be a legal path if $i_s(\overrightarrow{Q}) > ii_s(\overrightarrow{Q})$, an illegal path if $i_s(\overrightarrow{Q}) < ii_s(\overrightarrow{Q})$, and an indeterminable path if $i_s(\overrightarrow{Q}) = ii_s(\overrightarrow{Q})$.

Example 6.1.5 Consider the DFIG given in Fig. 6.2, having four nodes e, f, g and h with 6 arcs and 8 d-pairs. Put $\sigma(e) = \sigma(f) = \sigma(g) = \sigma(h) = 1$. Define the μ values and ψ values as in the figure.

If we analyze the figure, we can see that the legal flow from e to f through the arc ef is $\mu(ef) - \vee\{\psi(e, ef), \psi(ef, f)\} = 0.7 - \vee\{0.2, 0.1\} = 0.5$. The difference $|\psi(k, km) - \psi(km, m)|$ is known as the incidence loss of the arc ef, and it is represented by $L_i(ef)$. Thus we get, $L_i(ef) = 0.2 - 0.1 = 0.1$. If we consider the $f-g$ *di*-path given by $f, (f, fg), fg, g$, then the flow entry to g is totally legal as the illegal outflow from f to fg is absent. Also, the flow from e to h is legal whereas

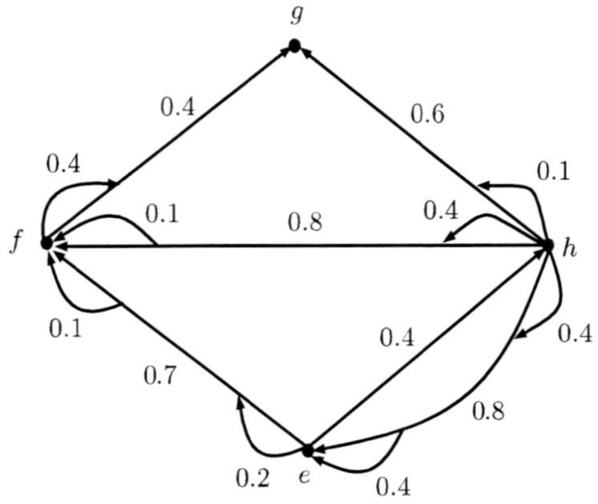

Fig. 6.2 Legal and illegal flows in a DFIG

a minor illegal to legal conversion through hg happens. Moreover, the arc he has no incidence loss.

Consider the two $e-g$ *di*-paths $\overrightarrow{P_1} : e, (e, ef), ef, (ef, f), f, (f, fg), fg, g$ and $\overrightarrow{P_2} : e, eh, h, (h, hg), hg, g$. The legal flow of $\overrightarrow{P_1}$ is given by $i_s(\overrightarrow{P_1}) = \wedge\{0.7 - \vee\{0.2, 0.1\}, 0.4 - \{0.4, 0\}\} = \wedge\{0.5, 0\} = 0$. Its illegal flow is $\wedge\{0.7, 0.4\} - 0 = 0.4$. Hence $\overrightarrow{P_1}$ is an illegal path. The legal flow of $\overrightarrow{P_2}$ is given by $i_s(\overrightarrow{P_2}) = 0.4$ and $ii_s(\overrightarrow{P_2}) = 0.4 - 0.4 = 0$. Thus $\overrightarrow{P_2}$ is legal.

The $h-f$ *di*-path $\overrightarrow{P_3} : h, (h, hf), hf, (hf, f), f$ is indeterminable because $i_s(\overrightarrow{P_3}) = 0.4 = ii_s(\overrightarrow{P_3})$.

It will be extremely helpful to deal with *di*-paths if we could establish an equivalent criterion for a legal path. The following proposition provides an equivalent criterion for a *di*-path to become legal. It is derived using the weights of arcs and *d*-pairs in the *di*-path.

Proposition 6.1.6 Let $\overrightarrow{G_i}(\sigma, \mu, \psi)$ be a DFIG. A directed incidence path $\overrightarrow{P}$ in $\overrightarrow{G_i}$ is legal if and only if $\text{Max}\{\psi(w, wy), \psi(wy, y)\} < \mu(wy) - \frac{1}{2}\text{Min}\{\mu(uv) : uv \in \overrightarrow{P}\}$, for every arc $wy \in \overrightarrow{P}$.

Proof Let $\overrightarrow{P}$ be a *di*-path in $\overrightarrow{G_i}$ and $s(\overrightarrow{P}) = \text{Min}\{\mu(uv) : uv \in \overrightarrow{P}\}$. Then we have to prove that $\overrightarrow{P}$ is legal if and only if $\text{Max}\{\psi(w, wy), \psi(wy, y)\} < \mu(wy) - \frac{1}{2}s(\overrightarrow{P})$. Suppose that $\overrightarrow{P}$ is legal. That is, $i_s(\overrightarrow{P}) > ii_s(\overrightarrow{P})$. That is,

$\Longleftrightarrow \wedge\{\mu(wy) - \vee\{\psi(w, wy), \psi(wy, y)\} : wy \in \overrightarrow{P}\} > s(\overrightarrow{P}) - \wedge\{\mu(wy) - \vee\{\psi(w, wy), \psi(wy, y)\} : wy \in \overrightarrow{P}\}$.

$\Longleftrightarrow \wedge\{\mu(wy) - \vee\{\psi(w, wy), \psi(wy, y)\} : wy \in \overrightarrow{P}\} > \frac{1}{2}s(\overrightarrow{P})$.

$\Longleftrightarrow \mu(wy) - \vee\{\psi(w, wy), \psi(wy, y)\} > \frac{1}{2}s(\overrightarrow{P})$, for every arc $wy \in \overrightarrow{P}$.

$\Longleftrightarrow \vee\{\psi(w, wy), \psi(wy, y)\} < \mu(wy) - \frac{1}{2}s(\overrightarrow{P})$, for every arc $wy \in \overrightarrow{P}$. □

Next, we characterize indeterminable paths of a DFIG.

Proposition 6.1.7 Let $\overrightarrow{P}$ be a *di*-path in a DFIG $\overrightarrow{G_i}(\sigma, \mu, \psi)$ and let $s(\overrightarrow{P}) = \text{Min}\{\mu(uv) : uv \in \overrightarrow{P}\}$. Then $\overrightarrow{P}$ is indeterminable if and only if there exists an arc wy in $\overrightarrow{P}$ such that $s(\overrightarrow{P}) = 2[\mu(wy) - \vee\{\psi(w, wy), \psi(wy, y)\}]$ and $2i_s(\overrightarrow{P}) = s(\overrightarrow{P})$.

Proof Assume the conditions given in the statement. Then,

$\mu(wy) - \vee\{\psi(w, wy), \psi(wy, y)\} = \frac{1}{2}s(\overrightarrow{P}) = i_s(\overrightarrow{P})$ for an arc wy in $\overrightarrow{P}$.

So we have

$$\begin{aligned} ii_s(\overrightarrow{P}) &= s(\overrightarrow{P}) - i_s(\overrightarrow{P}) \\ &= s(\overrightarrow{P}) - [\mu(wy) - \vee\{\psi(w, wy), \psi(wy, y)\}] \\ &= \mu(wy) - \vee\{\psi(w, wy), \psi(wy, y)\} \\ &= i_s(\overrightarrow{P}). \end{aligned}$$

Thus $\overrightarrow{P}$ is indeterminable.

Conversely, suppose that $i_s(\overrightarrow{P}) = ii_s(\overrightarrow{P})$. We have $ii_s(\overrightarrow{P}) = s(\overrightarrow{P}) - i_s(\overrightarrow{P})$. Thus $s(\overrightarrow{P}) = 2i_s(\overrightarrow{P})$. Let the minimum in the definition of $i_s(\overrightarrow{P})$ is attained by the arc wy. Then $i_s(\overrightarrow{P}) = \mu(wy) - \vee\{\psi(w, wy), \psi(wy, y)\}$. Therefore, $s(\overrightarrow{P}) = 2[\mu(wy) - \vee\{\psi(w, wy), \psi(wy, y)\}]$. □

Now, we look into legal, illegal, and pseudo connectivity ideas in a DFIG using legal and illegal flows.

Definition 6.1.8 A DFIG $\overrightarrow{G_i}(\sigma, \mu, \psi)$ is called **legally (illegally) connected** when there is a legal (illegal) di-path joining any two ordered pairs of nodes in σ^*.

"Legal connectedness" meets the characteristics of an equivalence relation in σ^*, where it partitions the vertex set V of $\overrightarrow{G_i}$ into equivalence classes $V_1, V_2, \ldots, V_j$. The **legal components** of $\overrightarrow{G_i}$ are nothing but the induced sub DFIG of $\overrightarrow{G_i}$, given by $\overrightarrow{G_i}(V_1), \overrightarrow{G_i}(V_2), \ldots, \overrightarrow{G_i}(V_j)$. So, if the number of legal components of $\overrightarrow{G_i}$ is one, we may conclude that $\overrightarrow{G_i}$ is legally connected. We may describe the illegal components of $\overrightarrow{G_i}$ in the same way.

When it comes to connectivity in a DFIG, we can observe that there are DFIGs that are neither legally nor illegally connected, as well as DFIGs that are both legally and illegally connected. As a result, we present the concept of "pseudo connectedness" in a DFIG.

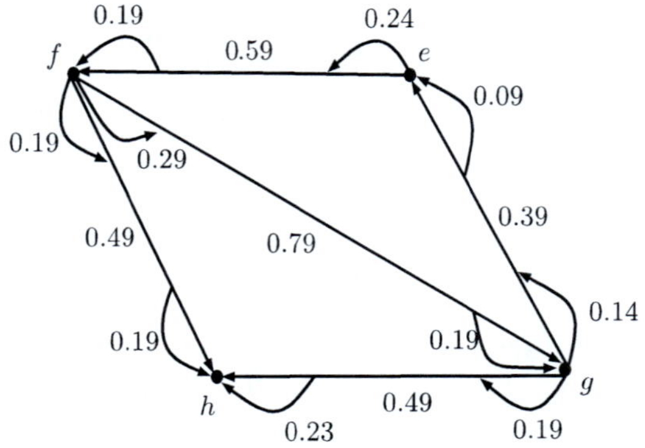

Fig. 6.3 Legally connected DFIG

Definition 6.1.9 A DFIG $\overrightarrow{G_i}(\sigma, \mu, \psi)$ is called **pseudo connected** when it is both legally and illegally connected.

We may now look at some examples of each of the above concepts to have a better idea of them. A legally connected *di*-graph is illustrated by the DFIG below.

Example 6.1.10 Consider the DFIG $\overrightarrow{G_i}(\sigma, \mu, \psi)$ given in Fig. 6.3. It has $V = \{e, f, g, h\}$ such that $\sigma(a) = 1$ for all $a \in V$. Let $\mu(ef) = 0.59$, $\mu(fh) = 0.49$, $\mu(hg) = 0.49$, $\mu(ge) = 0.39$, $\mu(fg) = 0.79$ and $\psi(e, ef)$=0.24, $\psi(ef, f) = 0.19$, $\psi(f, fh) = 0.19$, $\psi(fh, h) = 0.19$, $\psi(h, hg) = 0.23$, $\psi(hg, g) = 0.19$, $\psi(g, ge) = 0.14$, $\psi(ge, e) = 0.09$, $\psi(f, fg) = 0.29$ and $\psi(fg, g) = 0.19$.

Now, we find the legal and illegal incidence strength of *di*-paths between each ordered pair of nodes in σ^*. For $e, h \in \sigma^*$, $\overrightarrow{P} : e, (e, ef), ef, (ef, f), f, (f, fh), fh, (fh, h), h$ is the unique $e - h$ *di*-path with

$$i_s(\overrightarrow{P}) = \wedge\{\mu(ef) - \vee\{\psi(e, ef), \psi(ef, f)\}, \mu(fh) - \vee\{\psi(f, fh), \psi(fh, h)\}\}$$
$$= \wedge\{0.59 - \vee\{0.24, 0.19\}, 0.49 - \vee\{0.19, 0.19\}\} = \wedge\{0.35, 0.3\} = 0.3.$$
$$ii_s(\overrightarrow{P}) = \wedge\{\mu(ef), \mu(fh)\} - i_s(\overrightarrow{P}) = 0.49 - 0.3 = 0.19.$$

That is, $i_s(\overrightarrow{P}) > ii_s(\overrightarrow{P})$ implies $\overrightarrow{P}$ is a legal path from e to h.

Similarly, we calculate the legal and illegal incidence strength between rest of the pairs of nodes and tabulate them as follows.

Table 6.1 clearly shows that the given DFIG is legally connected. Because any two ordered pairs of σ^* have at least one legal path between them. It is also worth noting that there is no illegal path between any of the nodes in σ^*.

Following that, we'll look at an example of an illegally connected DFIG.

Table 6.1 Legal and illegal flow values

Ordered pair of nodes	Di-path	Legal flow (i_s)	Illegal flow (ii_s)	Legal/illegal
(e, f)	$e, (e, ef), ef, (ef, f), f$	0.35	0.24	Legal
(e, h)	$e, (e, ef), ef, (ef, f), f, (f, fh), fh, (fh, h), h$	0.3	0.19	Legal
(e, g)	$e, (e, ef), ef, (ef, f), f, (f, fg), fg, (fg, g), g$	0.35	0.24	Legal
	$e, (e, ef), ef, (ef, f), f, (f, fh), fh, (fh, h), h, (h, hg), hg, (hg, g), g$	0.26	0.23	Legal
(f, e)	$f, (f, fg), fg, (fg, g), g, (g, ge), ge, (ge, e), e$	0.25	0.14	Legal
	$f, (f, fh), fh, (fh, h), h, (h, hg), hg, (hg, g), g, (g, ge), ge, (ge, e), e$	0.25	0.14	Legal
(f, g)	$f, (f, fg), fg, (fg, g), g$	0.5	0.29	Legal
	$f, (f, fh), fh, (fh, h), h, (h, hg), hg, (hg, g), g$	0.26	0.23	Legal
(f, h)	$f, (f, fh), fh, (fh, h), h$	0.3	0.19	Legal
(g, e)	$g, (g, ge), ge, (ge, e), e$	0.25	0.14	Legal
(g, f)	$g, (g, ge), ge, (ge, e), e, (e, ef), ef, (ef, f), f$	0.25	0.14	Legal
(g, h)	$g, (g, ge), ge, (ge, e), e, (e, ef), ef, (ef, f), f, (f, fh), fh, (fh, h), h$	0.25	0.14	Legal
(h, e)	$h, (h, hg), hg, (hg, g), g, (g, ge), ge, (ge, e), e$	0.25	0.14	Legal
(h, f)	$h, (h, hg), hg, (hg, g), g, (g, ge), ge, (ge, e), e, (e, ef), ef, (ef, f), f$	0.25	0.14	Legal
(h, g)	$h, (h, hg), hg, (hg, g), g$	0.26	0.23	Legal

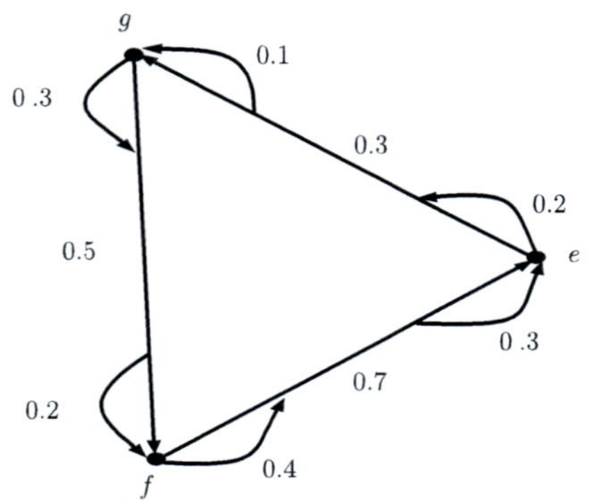

Fig. 6.4 Illegally connected DFIG

Table 6.2 Legal and illegal flow values

Ordered pair of nodes	Di-paths	Legal flow (i_s)	Illegal flow (ii_s)	Legal/illegal
(e, f)	$e, (e, eg), eg, (eg, g), g, (g, gf), gf, (gf, f), f$	0.1	0.2	Illegal
(e, g)	$e, (e, eg), eg, (eg, g), g$	0.1	0.2	Illegal
(f, e)	$f, (f, fe), fe, (fe, e), e$	0.3	0.4	Illegal
(f, g)	$f, (f, fe), fe, (fe, e), e, (e, eg), eg, (eg, g), g$	0.1	0.2	Illegal
(g, e)	$g, (g, gf), gf, (gf, f), f, (f, fe), fe, (fe, e), e$	0.2	0.3	Illegal
(g, f)	$g, (g, gf), gf, (gf, f), f$	0.2	0.3	Illegal

Example 6.1.11 Let the DFIG $\overrightarrow{G_i}(\sigma, \mu, \psi)$ of Fig. 6.4 has nodes e, f and g. Put σ-value 1 for all the three nodes. Define $\mu(eg) = 0.3$, $\mu(gf) = 0.5$, $\mu(fe) = 0.7$, $\psi(e, eg) = 0.2$, $\psi(e.g., g) = 0.1$, $\psi(g, gf) = 0.3$, $\psi(gf, f) = 0.2$, $\psi(f, fe) = 0.4$ and $\psi(fe, e) = 0.3$.

From the following table, we get the legal and illegal incidence strengths of *di*-paths between ordered pairs of nodes in σ^*.

Note from Table 6.2 that the given DFIG is illegally connected, and no legal paths exist between any pair of nodes in σ^*.

To sketch a pseudo connected DFIG, each node $x \in \sigma^*$ must have at least two *di*-paths starting from x and terminating at each of the other nodes in σ^* and at least two *di*-paths originating from each of the other nodes and terminating at x. So, the simplest example of a pseudo connected DFIG is shown below.

Example 6.1.12 Let e and f be the nodes of the DFIG $\overrightarrow{G_i}(\sigma, \mu, \psi)$ given in Fig. 6.5. Take $\sigma(e) = \sigma(f) = 1$.

We assign two parallel arcs from e to f and two parallel arcs from f to e. Then, each of the arcs form a *di*-path, given by

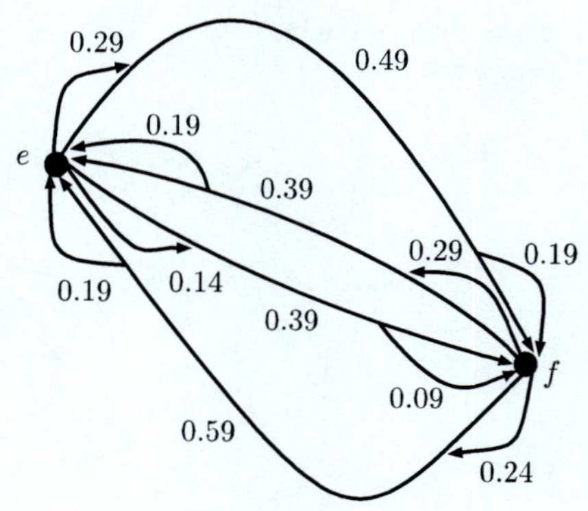

Fig. 6.5 Pseudo connected DFIG

$\overrightarrow{P_1}: e, (e, ef), ef, (ef, f), f$ with $i_s(\overrightarrow{P_1}) = 0.49 - \wedge\{0.29, 0.19\} = 0.2$ and $ii_s(\overrightarrow{P_1}) = 0.49 - 0.2 = 0.29$,
$\overrightarrow{P_2}: e, (e, ef), ef, (ef, f), f$ with $i_s(\overrightarrow{P_2}) = 0.39 - \wedge\{0.14, 0.09\} = 0.25$ and $ii_s(\overrightarrow{P_1}) = 0.39 - 0.25 = 0.14$,
$\overrightarrow{P_3}: f, (f, fe), fe, (fe, e), e$ with $i_s(\overrightarrow{P_3}) = 0.59 - \wedge\{0.24, 0.19\} = 0.35$ and $ii_s(\overrightarrow{P_3}) = 0.59 - 0.35 = 0.24$,
$\overrightarrow{P_4}: f, (f, fe), fe, (fe, e), e$ with $i_s(\overrightarrow{P_4}) = 0.39 - \wedge\{0.29, 0.19\} = 0.1$ and $ii_s(\overrightarrow{P_4}) = 0.39 - 0.1 = 0.29$.
Hence $\overrightarrow{P_2}$ and $\overrightarrow{P_3}$ are legal paths and $\overrightarrow{P_1}$ and $\overrightarrow{P_4}$ are illegal paths. Thus $\overrightarrow{G_i}$ is pseudo connected.

Our next goal is to create the concept of legal and illegal flow in a DFIG between any two nodes w and y. In a DFIG, we have already established the legal and illegal flow of a *di*-path, and we can use that to describe the legal and illegal flow between two nodes.

Definition 6.1.13 Let $\overrightarrow{G_i}(\sigma, \mu, \psi)$ be a DFIG and $w, y \in \sigma^*$. The **directed incidence connectivity** or **legal flow** between w and y is defined as

$$DICONN_{\overrightarrow{G_i}}(w, y) = \vee\{i_s(\overrightarrow{P}) : \overrightarrow{P} \text{ is a di-path between } w \text{ and } y\}.$$

The **directed illegal incidence connectivity** or **illegal flow** between w and y is defined by

$$DICONN^*_{\overrightarrow{G_i}}(w, y) = \vee\{ii_s(\overrightarrow{P}) : \overrightarrow{P} \text{ is a di-path between } w \text{ and } y\}.$$

A *di*-path $\overrightarrow{P}$ from w to y is called a **widest legal path** when $i_s(\overrightarrow{P}) = DICONN_{\overrightarrow{G_i}}(w, y)$ and a **widest illegal path** when $ii_s(\overrightarrow{P}) = DICONN^*_{\overrightarrow{G_i}}(w, y)$.

Example 6.1.14 Consider the DFIG $\overrightarrow{G_i}(\sigma, \mu, \psi)$ on four nodes e, f, g and h given in Fig. 6.6. Assign σ value 1 to each node. Also, define μ values and ψ values as in the figure. We use this DFIG to make a thorough idea on directed incidence connectivity and directed illegal incidence connectivity between two nodes.

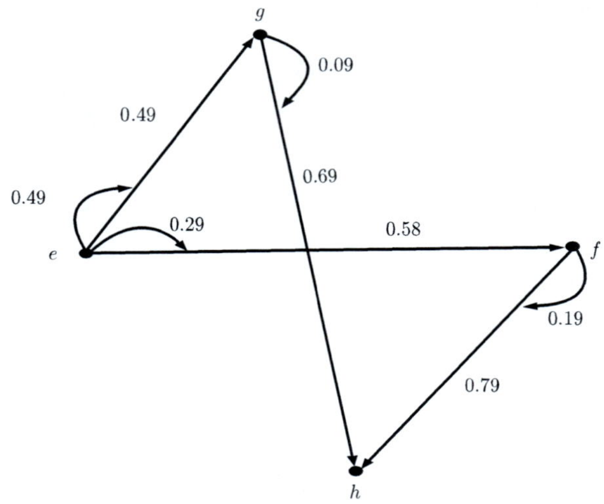

Fig. 6.6 DFIG given in Example 6.1.14

Consider $e, h \in \sigma^*$. Then there are two *di*-paths from e to h given by $\overrightarrow{P_1}$: $e, (e, ef), ef, (ef, f), f, (f, fh), fh, (fh, h), h$ and $\overrightarrow{P_2}$: $e, (e, eg), eg, (e.g., g), g, (g, gh), gh, (gh, h), h$. So, the legal flow through $\overrightarrow{P_1}$ and $\overrightarrow{P_2}$ is,

$$i_s(\overrightarrow{P_1}) = \wedge\{0.58 - 0.29, 0.79 - 0.19\} = 0.29.$$
$$i_s(\overrightarrow{P_2}) = \wedge\{0.49 - 0.49, 0.69 - 0.09\} = 0.$$

Hence, the directed incidence connectivity between e and h is obtained by $DICONN_{\overrightarrow{G_i}}(e, h) = \vee\{0.29, 0\} = 0.29$. Now, the illegal flow through $\overrightarrow{P_1}$ and $\overrightarrow{P_2}$ is given by

$$ii_s(\overrightarrow{P_1}) = \wedge\{0.58, 0.79\} - 0.29 = 0.29.$$
$$ii_s(\overrightarrow{P_2}) = \wedge\{0.49, 0.69\} - 0 = 0.49.$$

Thus, the directed illegal incidence connectivity between e and h is given by $DICONN^*_{\overrightarrow{G_i}}(e, h) = \vee\{0.29, 0.49\} = 0.49$.

Definition 6.1.15 Let $\overrightarrow{G_i}(\sigma, \mu, \psi)$ be a DFIG and $w, y \in \sigma^*$. Then (w, y) is said to be a **legal pair of nodes** if $DICONN_{\overrightarrow{G_i}}(w, y) > DICONN^*_{\overrightarrow{G_i}}(w, y)$ and an **illegal pair of nodes** if $DICONN_{\overrightarrow{G_i}}(w, y) < DICONN^*_{\overrightarrow{G_i}}(w, y)$. Also, (w, y) is called a **balanced flow pair of nodes** when $DICONN_{\overrightarrow{G_i}}(w, y) = DICONN^*_{\overrightarrow{G_i}}(w, y)$.

Example 6.1.16 Use Fig. 6.6. described in Example 6.1.14. Since $0.49 = DICONN^*_{\overrightarrow{G_i}}(e, h) > DICONN_{\overrightarrow{G_i}}(e, h) = 0.29$, clearly (e, h) is an illegal pair of nodes.

Now, we consider $e, f \in \sigma^*$. Then $DICONN^*_{\overrightarrow{G_i}}(e, f) = DICONN_{\overrightarrow{G_i}}(e, f) = 0.29$. So, (e, f) forms a balanced flow pair of nodes. Also, (f, h) forms a legal pair of nodes because $0.6 = DICONN_{\overrightarrow{G_i}}(f, h) > DICONN^*_{\overrightarrow{G_i}}(m, l) = 0.19$.

A DFIG subgraph can be defined in the same way as FIG can. However, in the case of subgraphs also, the relations are specified over ordered sets.

Definition 6.1.17 Let $\overrightarrow{G_i}(\sigma, \mu, \psi)$ be a DFIG. $\overrightarrow{H_i}(\tau, \nu, \chi)$ is said to be a **directed incidence subgraph** of $\overrightarrow{G_i}$ if $\tau(w) \leq \sigma(w)$ for all $w \in V$, $\nu(wy) \leq \mu(wy)$ for all $wy \in V \times V$ and $\chi(s, t) \leq \psi(s, t)$ for all $(s, t) \in \sigma^* \times \mu^* \cup \mu^* \times \sigma^*$. It is said to be a sub DFIG of $\overrightarrow{G_i}$ if $\tau(w) = \sigma(w)$ for all $w \in \tau^*$, $\nu(wy) = \mu(wy)$ for all $wy \in \nu^*$ and $\chi(s, t) = \psi(s, t)$ for all $(s, t) \in \tau^* \times \nu^* \cup \nu^* \times \tau^*$.

Like other graph structures, we cannot make a comparison between the legal flow in $\overrightarrow{H_i}$ with that in $\overrightarrow{G_i}$ between two nodes. To demonstrate this, we create a simple example of a DFIG.

Example 6.1.18 Let $\overrightarrow{G_i}(\sigma, \mu, \psi)$ be the DFIG shown in Fig. 6.7 having nodes e and f, each with weight 1. Define $\mu(ef) = 0.7$, $\psi(e, ef) = 0.3 = \psi(ef, f) = 0.3$. Then $DICONN_{\overrightarrow{G_i}}(e, f) = 0.7 - \vee\{0.3, 0.3\} = 0.4$.

Now, consider the directed incidence subgraph $\overrightarrow{H_i}(\tau, \nu, \chi)$ of $\overrightarrow{G_i}$ shown in Fig. 6.7a. It has $\tau(e) = 0.7$, $\tau(m) = 0.8$, $\nu(ef) = 0.5$, $\chi(e, ef) = \chi(ef, f) = 0.2$. Then $DICONN_{\overrightarrow{H_i}}(e, f) = 0.5 - \vee\{0.2, 0.2\} = 0.3$. That is, $DICONN_{\overrightarrow{G_i}}(e, f) > DICONN_{\overrightarrow{H_i}}(e, f)$.

Next, consider the directed incidence subgraph $\overrightarrow{K_i}(\tau, \nu, \chi)$ of $\overrightarrow{G_i}$ shown in Fig. 6.7b. It has $\tau(e) = 0.7$, $\tau(f) = 0.8$, $\nu(km) = 0.6$, $\chi(e, ef) = 0.09 = \chi(ef, f) = 0.09$. Then $DICONN_{\overrightarrow{K_i}}(e, f) = 0.6 - \vee\{0.09, 0.09\} = 0.51$. That is, $DICONN_{\overrightarrow{G_i}}(e, f) < DICONN_{\overrightarrow{K_i}}(e, f)$.

We also achieve equality on the legal flow between two nodes in the DFIG and its directed incidence subgraph since each DFIG is a directed incidence subgraph of itself.

If $\overrightarrow{H_i}(\tau, \nu, \chi)$ is a directed incidence subgraph of $\overrightarrow{G_i}(\sigma, \mu, \psi)$, then for $w, y \in \tau^*$, note that each $w - y$ *di*-path in $\overrightarrow{H_i}$ will also be a $w - y$ *di*-path in $\overrightarrow{G_i}$. So, we denote $i_s(\overrightarrow{P})_{G_i}$ as the legal flow of $\overrightarrow{P}$ in $\overrightarrow{G_i}$ whereas $i_s(\overrightarrow{P})_{H_i}$ as the legal flow of $\overrightarrow{P}$ in $\overrightarrow{H_i}$. In a similar manner, we can denote the illegal flow of $\overrightarrow{P}$ in $\overrightarrow{G_i}$ and $\overrightarrow{H_i}$ by $ii_s(\overrightarrow{P})_{G_i}$ and $ii_s(\overrightarrow{P})_{H_i}$, respectively.

Proposition 6.1.19 Let $\overrightarrow{G_i}(\sigma, \mu, \psi)$ be a DFIG and let $\overrightarrow{H_i}(\tau, \nu, \chi)$ be a directed incidence subgraph of $\overrightarrow{G_i}$. For $w, y \in \tau^*$, suppose that $i_s(\overrightarrow{P})_{H_i} \geq i_s(\overrightarrow{P})_{G_i}$ for every $w - y$ *di*-path $\overrightarrow{P}$ in $\overrightarrow{H_i}$. Then, $DICONN^*_{\overrightarrow{H_i}}(w, y) \leq DICONN^*_{\overrightarrow{G_i}}(w, y)$.

Proof Let $w, y \in \tau^*$ and $i_s(\overrightarrow{P})_{H_i} \geq i_s(\overrightarrow{P})_{G_i}$ for every $w - y$ *di*-path $\overrightarrow{P}$ in $\overrightarrow{H_i}$. In order to show that $DICONN^*_{\overrightarrow{H_i}}(w, y) \leq DICONN^*_{\overrightarrow{G_i}}(w, y)$, first we will show that $ii_s(\overrightarrow{P})_{H_i} \leq ii_s(\overrightarrow{P})_{G_i}$ for any $w - y$ *di*-path $\overrightarrow{P}$ in $\overrightarrow{H_i}$.

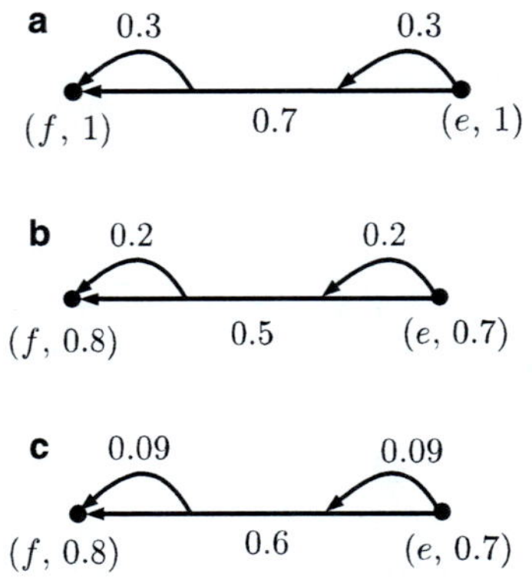

Fig. 6.7 **a** The DFIG $\overrightarrow{G_i}(\sigma, \mu, \psi)$ in Example 6.1.18. **b** The directed incidence subgraph $\overrightarrow{H_i}$ of $\overrightarrow{G_i}$. **c** The directed incidence subgraph $\overrightarrow{K_i}$ of $\overrightarrow{G_i}$

Let $\overrightarrow{Q}$ be a $w - y$ *di*-path in $\overrightarrow{H_i}$. By assumption, $i_s(\overrightarrow{Q})_{H_i} \geq i_s(\overrightarrow{Q})_{G_i}$. Then for each $w_i w_j \in \nu^*$ belonging to $\overrightarrow{Q}$, we have $\nu(w_i w_j) \leq \mu(w_i w_j)$ and for each $(s, t) \in \chi^*$ belonging to $\overrightarrow{Q}$, we have $\chi(s, t) \leq \psi(s, t)$. Hence $\wedge\{\nu(w_i w_j) : w_i w_j \in \overrightarrow{Q}\} \leq \wedge\{\mu(w_i w_j) : w_i w_j \in \overrightarrow{Q}\}$. Then

$$
\begin{aligned}
ii_s(\overrightarrow{Q})_{H_i} &= \wedge\{\nu(w_i w_j) : w_i w_j \in \overrightarrow{Q}\} - i_s(\overrightarrow{Q})_{H_i} \\
&\leq \wedge\{\mu(w_i w_j) : w_i w_j \in \overrightarrow{Q}\} - i_s(\overrightarrow{Q})_{G_i} \\
&= ii_s(\overrightarrow{Q})_{G_i}.
\end{aligned}
$$

Thus $\vee\{ii_s(\overrightarrow{P})_{H_i}\} \leq \vee\{ii_s(\overrightarrow{P})_{G_i}\}$ for every $w - y$ *di*-path $\overrightarrow{P}$ in $\overrightarrow{H_i}$ and hence $DICONN^*_{\overrightarrow{H_i}}(w, y) \leq DICONN^*_{\overrightarrow{G_i}}(w, y)$. □

If we ignore the extra condition put on the legal flow of each *di*-path in $\overrightarrow{H_i}$ and $\overrightarrow{G_i}$ in Proposition 6.1.19, the result may fail. Let us look at an example.

Example 6.1.20 Let e, f and g be the nodes of the DFIG $\overrightarrow{G_i}(\sigma, \mu, \psi)$ shown in Fig. 6.8. Each node is assigned the σ value 1. Let $\mu(ef) = 0.79$, $\mu(fg) = 0.49$, $\psi(e, ef) = 0.39$, $\psi(ef, f) = 0.19$, $\psi(f, fg) = 0.09$ and $\psi(fg, g) = 0.09$. Then there exists a unique $k - n$ *di*-path $\overrightarrow{Q}$: $e, (ef, f), ef, (ef, f), f, (f, fg), fg, (fg, g), g$ with $i_s(\overrightarrow{Q})_{G_i} = \wedge\{0.79 - 0.39, 0.49 - 0.09\} = 0.4$. Hence we get, $ii_s(\overrightarrow{Q})_{G_i} = DICONN^*_{\overrightarrow{G_i}}(e, g) = 0.49 - 0.4 = 0.09$.

Consider a directed incidence subgraph $\overrightarrow{H_i}(\tau, \nu, \chi)$ of $\overrightarrow{G_i}$ shown in Fig. 6.8a with $\tau(e) = 0.7$, $\tau(f) = 0.8$ and $\tau(g) = 0.9$. Define $\nu(ef) = 0.69$, $\nu(fg) = 0.49$, $\chi(e, ef) = 0.39$, $\chi(ef, f) = 0.19$, $\chi(f, fg) = 0.09$ and $\chi(fg, g) = 0.09$. Then $i_s(\overrightarrow{Q})_{H_i} = \wedge\{0.69 - 0.39, 0.49 - 0.09\} = 0.3$. That is, $i_s(\overrightarrow{Q})_{H_i} < i_s(\overrightarrow{Q})_{G_i}$. So, $ii_s(\overrightarrow{Q})_{H_i} = DICONN^*_{\overrightarrow{H_i}}(e, g) = 0.49 - 0.3 = 0.19$. Hence $DICONN^*_{\overrightarrow{H_i}}(e, g) > DICONN^*_{\overrightarrow{G_i}}(e, g)$.

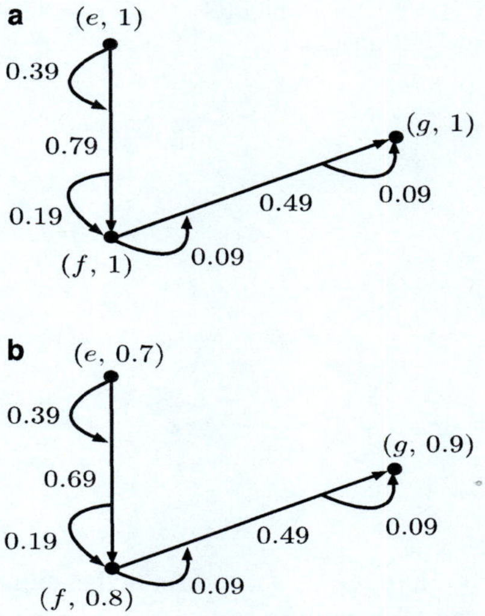

Fig. 6.8 **a** The DFIG $\overrightarrow{G_i}(\sigma, \mu, \psi)$ in Example 6.1.20. **b** The directed incidence subgraph $\overrightarrow{H_i}$ of $\overrightarrow{G_i}$

The *di*-incidence matrix of a DFIG can be defined as follows.

Definition 6.1.21 The ***di*-incidence matrix** $D = D(\overrightarrow{G_i})$ of a DFIG $\overrightarrow{G_i}(\sigma, \mu, \psi)$ is a $p \times p$ matrix where $\sigma^* = \{w_1, w_2, \ldots, w_p\}$ with

$d_{ij} = (DICONN_{\overrightarrow{G_i}}(w_i, w_j), DICONN^*_{\overrightarrow{G_i}}(w_i, w_j))$ if $i \neq j$ and $(\sigma(w_i), \sigma(w_j))$ if $i = j$.

The *di*-incidence matrix of the DFIG in Example 6.1.14 is given below. We take $\sigma^* = \{e, f, g, h\}$.

$$D(\overrightarrow{G_i}) = \begin{bmatrix} (1.0, 1.0) & (0.29, 0.29) & (0.0, 0.49) & (0.29, 0.49) \\ (0.0, 0.0) & (1.0, 1.0) & (0.0, 0.0) & (0.6, 0.19) \\ (0.0, 0.0) & (0.0, 0.0) & (1.0, 1.0) & (0.6, 0.09) \\ (0.0, 0.0) & (0.0, 0.0) & (0.0, 0.0) & (1.0, 1.0) \end{bmatrix}$$

Both legal and illegal flows between pairs of elements in σ^* can be easily obtained from this matrix.

We see that removing certain nodes or arcs decreases the illegal flow between some pairs of nodes, while removing some others reduces the legal flow between some nodes. As a result, we have the following definitions of legal flow reduction and illegal flow reduction nodes and arcs.

Definition 6.1.22 Let $\overrightarrow{G_i}(\sigma, \mu, \psi)$ be a DFIG. A node $w \in \sigma^*$ is said to be a **legal flow reduction node (LFR-node)** or an **illegal flow reduction node (IFR-node)** if $DICONN_{\overrightarrow{G_i}-w}(a, b) < DICONN_{\overrightarrow{G_i}}(a, b)$ or $DICONN^*_{\overrightarrow{G_i}-w}(a, b) < DICONN^*_{\overrightarrow{G_i}}(a, b)$, respectively for some nodes $a, b \in \sigma^*$, both distinct from w.

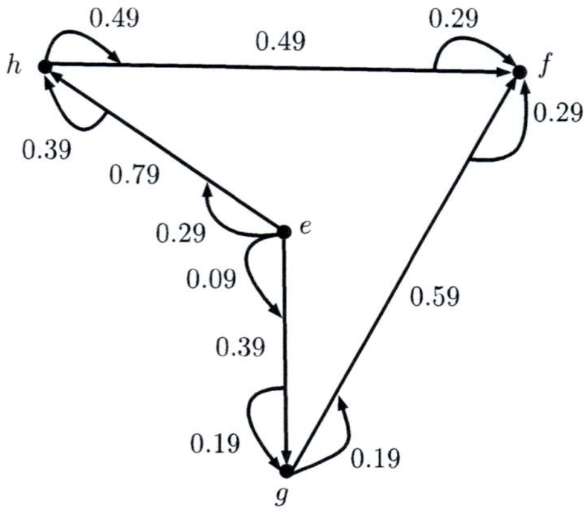

Fig. 6.9 DFIG with LFR and IFR-nodes

Here, $\overrightarrow{G_i} - w$ represents the subgraph of $\overrightarrow{G_i}$ obtained by removing the node w and all incoming and outgoing arcs at w including the d-pairs.

Example 6.1.23 Consider the DFIG given in Fig. 6.9 with four nodes e, f, g and h. Assign weight 1 to each node. The μ values and ψ values of arcs and d-pairs are defined by $\mu(eh) = 0.79$, $\mu(hf) = 0.49$, $\mu(eg) = 0.39$, $\mu(gf) = 0.59$, $\psi(e, eh) = \psi(hf, f) = \psi(gf, f) = 0.29$, $\psi(eh, h) = 0.39$, $\psi(h, hf) = 0.49$, $\psi(g, gf) = \psi(e.g., g) = 0.19$ and $\psi(e, eg) = 0.09$.

Here, g is an LFR-node. Because, there exist two nodes e and f such that $0 = DICONN_{\overrightarrow{G_i}-g}(e, f) < DICONN_{\overrightarrow{G_i}}(e, f) = 0.2$, by considering the two $e - f$ di-paths given by $\overrightarrow{P_1} : e, (eh, h), eh, (eh, h), h, (h, hf), hf, (hf, f), f$ and $\overrightarrow{P_2}$: $e, (e, eg), eg, (e.g., g), g, (g, gf), gf, (gf, f), f$. The node h is an IFR-node because $0.19 = DICONN^*_{\overrightarrow{G_i}-h}(e, f) < DICONN^*_{\overrightarrow{G_i}}(e, f) = 0.49$.

It is also worth noting that the nodes e and f are neither LFR nor IFR nodes. Because their removal has no effect on the legal or illegal flow between any other pair of nodes.

We may now have an equivalent condition of LFR and IFR-nodes.

Proposition 6.1.24 A node $w \in \sigma^*$ of a DFIG $\overrightarrow{G_i}(\sigma, \mu, \psi)$ is an LFR-node if and only if there exist $a, b\ (\neq w)$ in σ^* such that every widest $a - b$ legal di-path pass through w. Also, w is an IFR-node if and only if there exist $a, b\ (\neq w)$ such that every widest $a - b$ illegal di-path pass through w.

Proof Take $w \in \sigma^*$ as an LFR-node in $\overrightarrow{G_i}$. Then $DICONN_{\overrightarrow{G_i}-w}(a, b) < DICONN_{\overrightarrow{G_i}}(a, b)$ for some $a, b \in \sigma^* - \{w\}$. That is, $\vee\{i_s(\overrightarrow{Q}) : \overrightarrow{Q}$ is a di-path between a and b in $\overrightarrow{G_i} - \{w\}\} < \vee\{i_s(\overrightarrow{Q}) : \overrightarrow{Q}$ is a di-path between a and b in $\overrightarrow{G_i}\}$. That is, the maximum of the width of all legal $a - b$ di-paths in $\overrightarrow{G_i} - \{w\}$

is reduced by the removal of w from $\overrightarrow{G_i}$. That is, every widest $a-b$ legal *di*-paths pass through w. On converse, suppose there exist $a, b\ (\neq w)$ in σ^* such that every widest $a-b$ legal *di*-path pass through w. Then clearly w is an LFR-node because the removal of w from $\overrightarrow{G_i}$ removes all $a-b$ widest legal *di*-paths and hence $DICONN_{\overrightarrow{G_i}-w}(a,b) < DICONN_{\overrightarrow{G_i}}(a,b)$. □

The case of IFR-node is similar to that of LFR-node.

We may be able to limit illegal migration to some extent by examining some network paths or banning traffic through specific links. To that reason, we define the legal and illegal flow reduction link in a DFIG as follows.

Definition 6.1.25 Let $\overrightarrow{G_i}(\sigma, \mu, \psi)$ be a DFIG. An arc $wy \in \mu^*$ is said to be a **legal flow reduction link (LFR-link)** or an **illegal flow reduction link (IFR-link)** if $DICONN_{\overrightarrow{G_i}-wy}(a,b) < DICONN_{\overrightarrow{G_i}}(a,b)$ or $DICONN^*_{\overrightarrow{G_i}-wy}(a,b) < DICONN^*_{\overrightarrow{G_i}}(a,b)$, respectively for some nodes $a, b \in \sigma^*$.

Here, $\overrightarrow{G_i} - wy$ represents the subgraph of $\overrightarrow{G_i}$ obtained by removing the arc wy and all d-pairs (if any) incident on it from $\overrightarrow{G_i}$.

Example 6.1.26 Consider the DFIG given in Fig. 6.9 of Example 6.1.23. The arc eg is an LFR-link and the arc hf is an IFR-link. This may also be validated by applying the same nodes e and f. That is,

$$0 = DICONN_{\overrightarrow{G_i}-eg}(e,f) < DICONN_{\overrightarrow{G_i}}(e,f) = 0.2$$
$$0.19 = DICONN^*_{\overrightarrow{G_i}-hf}(e,f) < DICONN^*_{\overrightarrow{G_i}}(e,f) = 0.49.$$

Now, we provide an equivalent condition for LFR and IFR-links. The proof of the result is ignored since it is as easy to produce as the proof of Proposition 6.1.24 by simply substituting the node w with the arc zw.

Proposition 6.1.27 An arc $zw \in \mu^*$ of a DFIG $\overrightarrow{G_i}(\sigma, \mu, \psi)$ is an LFR-link (IFR-link) if and only if there exist a, b in σ^* such that every widest $a-b$ legal *di*-path (widest $a-b$ illegal *di*-path) contain zw.

It is noticeable that removing a d-pair can also minimize illegal flow. Such pairs are significant because they reflect the illegal component that contributes illegal flow into the network. There are also certain pairs whose removal improves the legal flow in the network. The recognition of such pairs will aid in network efficiency.

Definition 6.1.28 Let $\overrightarrow{G_i}(\sigma, \mu, \psi)$ be a DFIG. A d-pair $(w, wy) \in \psi^*$ is said to be an **illegal flow reduction pair (IFR-pair)** if $DICONN^*_{\overrightarrow{G_i}-(w,wy)}(a,b) < DICONN^*_{\overrightarrow{G_i}}(a,b)$ for some nodes $a, b \in \sigma^*$. The d-pair (w, wy) is said to be a **legal flow enhancement pair (LFE-pair)** if $DICONN_{\overrightarrow{G_i}-(w,wy)}(a,b) > DICONN_{\overrightarrow{G_i}}(a,b)$ for some nodes $a, b \in \sigma^*$.

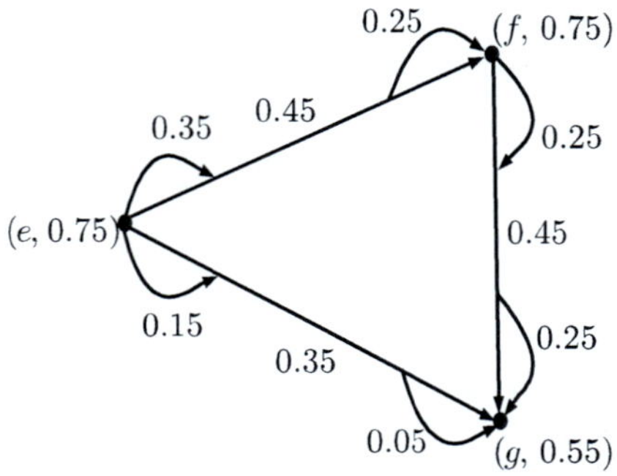

Fig. 6.10 DFIG with IFR and LFE-pairs

Here, $\overrightarrow{G_i} - (w, wy)$ represents the subgraph of $\overrightarrow{G_i}$ obtained by removing the pair (w, wy) from $\overrightarrow{G_i}$ and reassigning the flow through the corresponding arc wy as $\mu(wy) - \psi(wy, y)$.

Example 6.1.29 Figure 6.10 shows an example of a DFIG with LFE and IFR-pairs. It has three nodes e with $\sigma(e) = 0.75$, f with $\sigma(f) = 0.75$ and g with $\sigma(n) = 0.55$. Define the μ values and ψ values as in the figure. Then the d-pair (e, eg) is an LFE-Pair. Because, we can find two nodes $e, g \in \sigma^*$ such that $0.3 = DICONN_{\overrightarrow{G_i}-(e,eg)}(e, g) > DICONN_{\overrightarrow{G_i}}(e, g) = 0.2$,

where there are two $e - g$ *di*-paths given by $\overrightarrow{P_1} : e, (e, eg), eg, (eg, g), g$ with $i_s(\overrightarrow{P_1}) = 0.2$ and $\overrightarrow{P_2} : e, (e, ef), ef, (ef, f), f, (f, fg), fg, (fg, g), g$ with $i_s(\overrightarrow{P_2}) = 0.1$.
Moreover, the d-pair (e, ef) is an IFR-pair, because the nodes e and g satisfy

$$0.25 = DICONN^*_{\overrightarrow{G_i}-(e,ef)}(e, g) < DICONN^*_{\overrightarrow{G_i}}(e, g) = 0.35.$$

Proposition 6.1.30 In a DFIG $\overrightarrow{G_i}(\sigma, \mu, \psi)$, if $(w, wy) \in \psi^*$ is an IFR-pair, then there exist $a, b \in \sigma^*$ with the property that every widest $a - b$ illegal *di*-path contains (w, wy).

Proof Let $(w, wy) \in \psi^*$ be an IFR-pair. Then σ^* contains two nodes a, b with $DICONN^*_{\overrightarrow{G_i}-(w,wy)}(a, b) < DICONN^*_{\overrightarrow{G_i}}(a, b)$.

If possible, suppose there exists a widest $a - b$ illegal *di*-path $\overrightarrow{Q}$ in $\overrightarrow{G_i}$ that does not contain (w, wy). Then $\overrightarrow{Q}$ will also an $a - b$ illegal *di*-path in $\overrightarrow{G_i} - (w, wy)$ with $ii_s(\overrightarrow{Q}) = DICONN^*_{\overrightarrow{G_i}}(a, b)$. This is not possible because, every $a - b$ illegal *di*-path of $\overrightarrow{G_i} - (w, wy)$ has illegal incidence strength strictly less than $DICONN^*_{\overrightarrow{G_i}}(a, b)$. □

We cannot always trust the converse of Proposition 6.1.30. That is, if we have a d-pair $(w, wy) \in \psi^*$ belonging to each widest $a - b$ illegal *di*-path for some $a, b \in \sigma^*$,

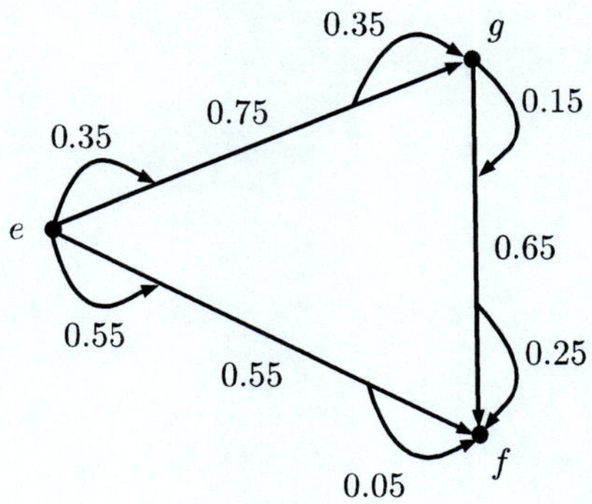

Fig. 6.11 DFIG given in Example 6.1.31

then (w, wy) need not be an IFR-pair. This idea may be shown using the example below.

Example 6.1.31 Let e, f and g be the nodes of the DFIG $\overrightarrow{G_i}(\sigma, \mu, \psi)$ given in Fig. 6.11 Each node is assigned with a weight 1. Let $\mu(eg) = 0.75$, $\mu(gf) = 0.65$, $\mu(ef) = 0.55$, $\psi(e, eg) = 0.35$, $\psi(e.g., g) = 0.35$, $\psi(e, ef) = 0.55$, $\psi(ef, f) = 0.05$, $\psi(g, gf) = 0.15$ and $\psi(gf, f) = 0.25$.

Choose the directed incidence pair $(ef, f) \in \psi^*$ and two nodes $e, f \in \sigma^*$. There are two $e - f$ *di*-paths $\overrightarrow{P}$: $e, (e, ef), ef, (ef, f), f$ and $\overrightarrow{Q}$: $e, (e, eg), eg, (e.g., g), g, (g, gf), gf, (gf, f), f$ with

$$\begin{aligned}
i_s(\overrightarrow{P}) &= \mu(ef) - \vee\{\psi(e, ef), \psi(ef, f)\} = 0.55 - \vee\{0.55, 0.05\} = 0 \\
ii_s(\overrightarrow{P}) &= \mu(ef) - i_s(\overrightarrow{P}) = 0.55 - 0 = 0.55 \\
i_s(\overrightarrow{Q}) &= \wedge\{\mu(eg) - \vee\{\psi(e, eg), \psi(e.g., g)\}, \mu(gf) - \vee\{\psi(g, gf), \psi(gf, f)\}\} \\
&= \wedge\{0.75 - \vee\{0.35, 0.35\}, 0.65 - \vee\{0.15, 0.25\}\} = \wedge\{0.4, 0.4\} = 0.4 \\
ii_s(\overrightarrow{Q}) &= \wedge\{\mu(eg), \mu(gf)\} - i_s(\overrightarrow{Q}) = 0.65 - 0.4 = 0.25.
\end{aligned}$$

Thus, $\overrightarrow{G_i}$ has a unique widest $e - f$ illegal *di*-path $\overrightarrow{P}$ having $(ef, f) \in \overrightarrow{P}$. But, (ef, f) is not an IFR-pair since removing it has no effect on the directed illegal incidence connectivity between any two nodes of $\overrightarrow{G_i}$.

Now, we present the idea of cycles in DFIG. Because the main focus is on legal and illegal incidence connectivity, and the weights of both arcs and d-pairs play an equal part in defining them, we describe fuzzy incidence cycles in terms of legal and illegal paths, with equal priority given to both arcs and d-pairs. A closed legal path is a legal cycle, which is analogous to cycles in graph theory.

Definition 6.1.32 A DFIG $\overrightarrow{G_i}(\sigma, \mu, \psi)$ is known to be a **legal cycle** when $\overrightarrow{G_i}$ itself forms a closed legal di-path.

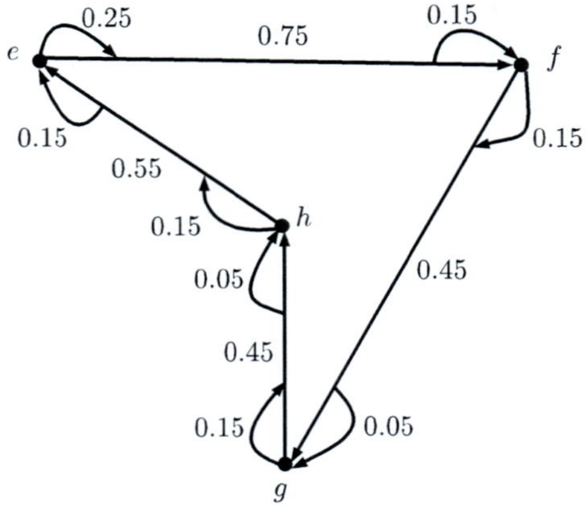

Fig. 6.12 Legal fuzzy incidence cycle

Definition 6.1.33 A **legal fuzzy cycle** is a legal cycle $\overrightarrow{G_i}(\sigma, \mu, \psi)$ such that $\overrightarrow{G_i}$ has no unique $zw \in \mu^*$ with $\mu(zw) = \wedge\{\mu(st) : st \in \mu^*\}$.

Definition 6.1.34 A **legal fuzzy incidence cycle** is a legal fuzzy cycle $\overrightarrow{G_i}(\sigma, \mu, \psi)$ such that it has no unique $zw \in \mu^*$ with $\mu(zw) - \vee\{\psi(z, zw), \psi(zw, w)\} = \wedge\{\mu(st) - \vee\{\psi(s, st), \psi(st, t)\} : st \in \mu^*\}$.

The example below illustrates a DFIG that is also a legal fuzzy incidence cycle.

Example 6.1.35 Figure 6.12 illustrates the DFIG $\overrightarrow{G_i}(\sigma, \mu, \psi)$ with $V = \{e, f, g, h\}$. Put σ value 1 for each node in σ^*. Define $\mu(ef) = 0.75$, $\mu(fg) = 0.45$, $\mu(gh) = 0.45$, $\mu(he) = 0.55$, $\psi(e, ef) = 0.25$, $\psi(ef, f) = 0.15$, $\psi(f, fg)=0.15$, $\psi(fg, g) = 0.05$, $\psi(g, gh) = 0.15$, $\psi(gh, h) = 0.05$, $\psi(h, he) = 0.15$ and $\psi(he, e) = 0.15$.

Based on the definitions, it is evident that this DFIG meets the requirements of a legal fuzzy cycle. Also, we are able to find arcs fg and gh such that

$$\begin{aligned}\mu(fg) - \vee\{\psi(f, fg), \psi(fg, g)\} &= \mu(gh) - \vee\{\psi(g, gh), \psi(gh, h)\}\\ &= \wedge\{\mu(st) - \vee\{\psi(s, st), \psi(st, t)\} : st \in \mu^*\}\\ &= 0.3.\end{aligned}$$

Hence, $\overrightarrow{G_i}$ is also an example of a legal fuzzy incidence cycle.

When working with directed fuzzy incidence graphs, connectivity is the most important structural attribute to consider. Next, the connectivity aspects of legal fuzzy incidence cycles, LFR and IFR-nodes, LFR and IFR-links, and LFE and IFR-pairs are addressed.

According to the following theorem, if we consider a legal fuzzy incidence cycle, each node and arc becomes an LFR-node and an LFR-link, respectively.

Theorem 6.1.36 If $\overrightarrow{G_i}(\sigma, \mu, \psi)$ is a legal fuzzy incidence cycle, then each node $x \in \sigma^*$ is an LFR-node and each $xy \in \mu^*$ is an LFR-link.

Proof Being a legal fuzzy incidence cycle, $\overrightarrow{G_i}(\sigma, \mu, \psi)$ contains exactly one directed incidence path joining any pair of nodes in σ^* and that *di*-path will be the widest legal path between those two nodes. Let $x \in \sigma^*$. Then there exist two nodes $y, z \in \sigma^*$, both different from x, such that x is contained in the unique widest $y - z$ legal *di*-path. Hence, x is an LFR-node by Proposition 6.1.24.

Now, for $xy \in \mu^*$, we can find two nodes $c, d \in \sigma^*$ so that xy belongs to the unique widest $c - d$ legal *di*-path. Hence, xy is an LFR-link by Proposition 6.1.27. □

We constantly try to enhance the legal flow through a network by assessing and regulating some illegal components influencing the flow. As a result, the analysis of LFE-pairs in a DFIG is critical since their removal promotes legal flow in the network. The following results show some of the properties of LFE-pairs. In Theorem 6.1.37, we find a relationship between LFE-pairs and incidence loss of arcs.

Theorem 6.1.37 Suppose $\overrightarrow{G_i}(\sigma, \mu, \psi)$ is a DFIG with $zw \in \mu^*$. If either (z, zw) or (zw, w) is an LFE-pair, then $L_i(zw) \neq 0$.

Proof Without loss of generality, take (z, zw) as an LFE-pair. Then
$DICONN_{\overrightarrow{G_i}-(z,zw)}(e, f) > DICONN_{\overrightarrow{G_i}}(e, f)$ for some $e, f \in \sigma^*$.

This means, the directed incidence connectivity between e and f in $\overrightarrow{G_i} - (z, zw)$ is strictly greater than that in $\overrightarrow{G_i}$. So, we can find a *di*-path $\overrightarrow{P}$ in $\overrightarrow{G_i} - (z, zw)$ such that $i_s(\overrightarrow{P}) = DICONN_{\overrightarrow{G_i}-(z,zw)}(e, f)$. Clearly, the *di*-path $\overrightarrow{P}$ should contain the arc zw. Otherwise, $\overrightarrow{P}$ will be an $e - f$ *di*-path in $\overrightarrow{G_i}$ with same $i_s(\overrightarrow{P})$ which is not possible because $i_s(\overrightarrow{P}) > DICONN_{\overrightarrow{G_i}}(e, f)$. Now, $\overrightarrow{P}$ together with (z, zw) forms an $e - f$ *di*-path $\overrightarrow{P'}$ in $\overrightarrow{G_i}$ which should have legal incidence strength $i_s(\overrightarrow{P'}) < i_s(\overrightarrow{P})$. Since there is no change in the legal flow through the arcs different from zw, we must have

$$\mu(zw) - \vee\{\psi(z, zw), \psi(zw, w)\} < \mu(zw) - \vee\{0, \psi(zw, w)\} = \mu(zw) - \psi(zw, w).$$

That is, $\vee\{\psi(z, zw), \psi(zw, w)\} > \psi(zw, w)$, which implies $\vee\{\psi(z, zw), \psi(zw, w)\} = \psi(z, zw)$ and $\psi(z, zw) > \psi(zw, w)$. Hence, $L_i(zw) = \psi(z, zw) - \psi(zw, w) > 0$. □

Corollary 6.1.38 Let $\overrightarrow{G_i}(\sigma, \mu, \psi)$ be a DFIG. If $(z, zw) \in \psi^*$ is an LFE-pair, then $\psi(z, zw) > \psi(zw, w)$.

The converse of Theorem 6.1.37 is not always true. That is, in a DFIG, an arc xy with nonzero incidence loss need not imply that either of the directed incidence pairs (x, xy) or (xy, y) is an LFE-pair. The example below demonstrates this.

Example 6.1.39 Let $V = \{e, f, g\}$ be the node set of the DFIG $\overrightarrow{G_i}(\sigma, \mu, \psi)$ given in Fig. 6.13 Let σ weight be 1 for each node in V. Define $\mu(eg) = 0.35$, $\mu(fg) = 0.65$, $\mu(ef) = 0.95$, $\psi(e, eg) = 0.35$, $\psi(e.g., g) = 0.05$, $\psi(e, ef) = 0.45$, $\psi(ef, f) = 0.35$, $\psi(f, fg) = 0.05$ and $\psi(fg, g) = 0.05$.

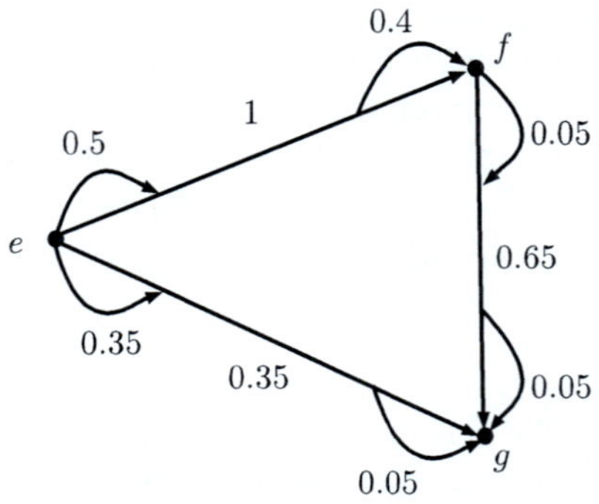

Fig. 6.13 DFIG violating the converse of Theorem 6.1.37

Here, the arc eg has an incidence loss, $L_i(eg) = 0.35 - 0.05 = 0.3 > 0$. However, neither (e, eg) nor (eg, g) are LFE-pairs. Because removing them has no effect on the directed incidence connectivity between any pair of nodes in $\overrightarrow{G_i}$.

Theorem 6.1.40 For an arc xy in a DFIG $\overrightarrow{G_i}(\sigma, \mu, \psi)$, at most one of (x, xy) or (xy, y) can be an LFE-pair.

Proof On contrary, suppose that both (x, xy) and (xy, y) are LFE-pairs. Since (x, xy) is an LFE-pair, from Corollary 6.1.38, we must have $\psi(x, xy) > \psi(xy, y)$. Also since (xy, y) is an LFE-pair, we must have $\psi(xy, y) > \psi(x, xy)$. Both the inequalities cannot be satisfied together. Hence we got a contradiction. □

It is worth noting that the converse of Corollary 6.1.38 is always true for a legal fuzzy incidence cycle. As a result, we establish the following theorem.

Theorem 6.1.41 If $\overrightarrow{G_i}(\sigma, \mu, \psi)$ is a legal fuzzy incidence cycle, then $(z, zw) \in \psi^*$ is an LFE-pair if and only if $\psi(z, zw) > \psi(zw, w)$.

Proof Assume $(z, zw) \in \psi^*$ is an LFE-pair. From Corollary 6.1.38, (z, zw) is an LFE-pair implies $\psi(z, zw) > \psi(zw, w)$.

Conversely suppose that $\psi(z, zw) > \psi(zw, w)$ for directed incidence pairs $(z, zw), (zw, w) \in \psi^*$. Since $zw \in \mu^*$ and $\overrightarrow{G_i}$ being a legal fuzzy incidence cycle, $\overrightarrow{P} : z, (z, zw), zw, (zw, w), w$ is the unique $z - w$ *di*-path with

$$
\begin{aligned}
DICONN_{\overrightarrow{G_i}}(z, w) &= i_s(\overrightarrow{P}) \\
&= \mu(zw) - \vee\{\psi(z, zw), \psi(zw, w)\} \\
&= \mu(zw) - \psi(z, zw).
\end{aligned}
$$

Also in $\overrightarrow{G_i} - (z, zw)$, the *di*-path $\overrightarrow{Q} : z, zw, (zw, w), w$ will be the unique $z - w$ *di*-path with

$$\begin{aligned} DICONN_{\vec{G}_i-(z,zw)}(z,w) &= i_s(\vec{Q}) \\ &= \mu(zw) - \psi(zw, w) \\ &> \mu(zw) - \psi(z, zw) = DICONN_{\vec{G}_i}(z, w). \end{aligned}$$

Hence, we have $z, w \in \sigma^*$ with $DICONN_{\vec{G}_i-(z,zw)}(z, w) > DICONN_{\vec{G}_i}(z, w)$. Thus (z, zw) becomes an LFE-pair. □

Assume we have a DFIG $\vec{G}_i(\sigma, \mu, \psi)$ and a d-pair (z, zw) in it. To become an LFE-pair, (z, zw) should meet $\psi(z, zw) > \psi(zw, w)$. By considering the example of (e, eg) in Example 6.1.39, it is evident that even while (z, zw) satisfies $\psi(z, zw) > \psi(zw, w)$, it may not become an LFE-pair. So we have the following proposition, which states that if (z, zw) has an extra property in addition to the above, then (z, zw) can be an LFE-pair.

Proposition 6.1.42 In a DFIG $\vec{G}_i(\sigma, \mu, \psi)$, let the d-pair $(z, zw) \in \psi^*$ satisfies $\psi(z, zw) > \psi(zw, w)$. If $DICONN_{\vec{G}_i}(z, w) = \mu(zw) - \psi(z, zw)$, then (z, zw) is an LFE-pair. □

Proof Since $DICONN_{\vec{G}_i}(z, w) = \mu(zw) - \psi(z, zw)$, the *di*-path $\vec{P} : z, (z, zw), zw, (zw, w), w$ will be a widest $z - w$ legal *di*-path. Define the *di*-path in $\vec{G}_i - (z, zw)$ by $\vec{P'} : z, zw, (zw, w), w$. All other $z - w$ *di*-paths of $\vec{G}_i - (z, zw)$ are also *di*-paths in $\vec{G}_i$ and have legal incidence strength at most $DICONN_{\vec{G}_i}(z, w)$. Since $\psi(z, zw) > \psi(zw, w)$, we have

$$\begin{aligned} DICONN_{\vec{G}_i}(z, w) &= \mu(zw) - \psi(z, zw) \\ &< \mu(zw) - \psi(zw, w) \\ &= i_s(\vec{P'}) = DICONN_{\vec{G}_i-(z,zw)}(z, w). \end{aligned}$$

Hence, we have two nodes $z, w \in \sigma^*$ with $DICONN_{\vec{G}_i-(z,zw)}(z, w) > DICONN_{\vec{G}_i}(z, w)$ implies that (z, zw) is an LFE-pair.

In general, the converse of Proposition 6.1.42 may fail. That is, if the directed incidence pair (z, zw) with $\psi(z, zw) > \psi(zw, w)$ is an LFE-pair, then $DICONN_{\vec{G}_i}(z, w)$ need not be equal to $\mu(zw) - \psi(z, zw)$. The following example demonstrates this.

Example 6.1.43 Figure 6.14 illustrates the DFIG $\vec{G}_i(\sigma, \mu, \psi)$ with $V = \{e, f, g\}$ such that each node has σ weight 1. Define $\mu(eg) = 0.45$, $\mu(fg) = 0.75$, $\mu(ef) = 0.75$, $\psi(e, eg) = 0.45$, $\psi(e.g., g) = 0.15$, $\psi(e, ef) = 0.55$, $\psi(ef, f) = 0.05$, $\psi(f, fg) = 0.15$ and $\psi(fg, g) = 0.15$.

Consider the directed incidence pair $(e, eg) \in \psi^*$ and nodes $e, g \in \sigma^*$. There are two *di*-paths between e and g, given by $\vec{Q} : e, (e, ef), ef, (ef, f), f, (f, fg), fg, (fg, g), g$ and $\vec{P} : e, (e, eg), eg, (e.g., g), g$. Then the legal incidence strengths of $\vec{P}$ and $\vec{Q}$ are given by

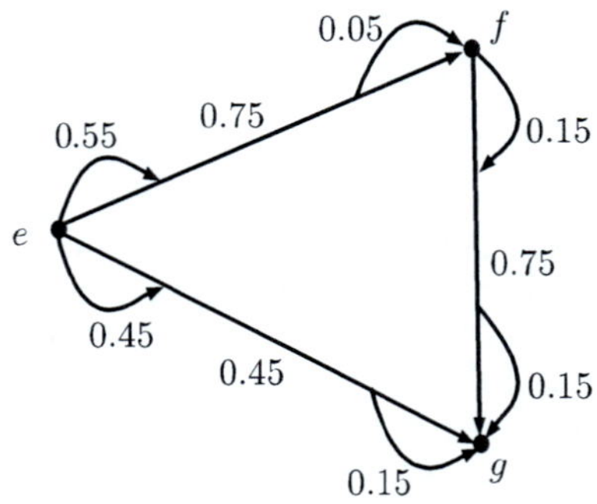

Fig. 6.14 DFIG of Example 6.1.43

$$i_s(\overrightarrow{P}) = \mu(eg) - \vee\{\psi(e, eg), \psi(e.g., g)\}$$
$$= 0.45 - \vee\{0.45, 0.15\} = 0.$$
$$i_s(\overrightarrow{Q}) = \wedge\{\mu(ef) - \vee\{\psi(e, ef), \psi(ef, f)\}, \mu(fg) - \vee\{\psi(f, fg), \psi(fg, g)\}\}$$
$$= \wedge\{0.75 - \vee\{0.55, 0.05\}, 0.75 - \vee\{0.15, 0.15\}\} = \wedge\{0.2, 0.6\} = 0.2.$$

Therefore, $DICONN_{\overrightarrow{G_i}}(x, z) = \vee\{0, 0.2\} = 0.2$.

After deleting (e, eg), the *di*-path $\overrightarrow{P'}$: $e, eg, (eg, g), g$ has legal incidence strength,

$$i_s(\overrightarrow{P'}) = \mu(eg) - \psi(e.g., g) = 0.45 - 0.15 = 0.3.$$

Thus we get $DICONN_{\overrightarrow{G_i}-(e,eg)}(e, g) = \vee\{0.3, 0.2\} = 0.3 > DICONN_{\overrightarrow{G_i}}(e, g) = 0.2$.

Hence, the directed incidence pair (e, eg) is an LFE-pair with $\psi(e, eg) > \psi(eg, g)$. But, $\mu(eg) - \psi(e, eg) = 0.45 - 0.45 = 0 \neq DICONN_{\overrightarrow{G_i}}(e, g) = 0.2$.

Corollary 6.1.44 A legal fuzzy incidence cycle $\overrightarrow{G_i}(\sigma, \mu, \psi)$ has at most $|\sigma|$ LFE-pairs. □

Proof Let $\overrightarrow{C}$: $u_1, (u_1, u_1u_2), u_1u_2, (u_1u_2, u_2), u_2, \ldots, (u_{i-1}u_i, u_i), u_i, (u_i, u_iu_1), u_iu_1, (u_iu_1, u_1), u_1$ be the closed legal *di*-path in $\overrightarrow{G_i}$. For each pair of consecutive nodes $u_k, u_{k+1} \in \sigma^*$, $\overrightarrow{P_k}$: $u_k, (u_k, u_ku_{k+1}), u_ku_{k+1}, (u_ku_{k+1}, u_{k+1}), u_{k+1}$ is the unique $u_k - u_{k+1}$ *di*-path. So, $DICONN_{\overrightarrow{G_i}}(u_k, u_{k+1}) = \mu(u_ku_{k+1}) - \vee\{\psi(u_k, u_ku_{k+1}), \psi(u_ku_{k+1}, u_{k+1})\}$. Now, there arises two cases,

Case 1. If $\psi(u_k, u_ku_{k+1}) > \psi(u_ku_{k+1}, u_{k+1})$, then by Proposition 6.1.42, we get (u_k, u_ku_{k+1}) is an LFE-pair. □

Case 2. If $\psi(u_k, u_ku_{k+1}) \leq \psi(u_ku_{k+1}, u_{k+1})$, then the removal of (u_k, u_ku_{k+1}) does not affect the legal flow between any two nodes in $\overrightarrow{G_i}$. Thus (u_k, u_ku_{k+1}) cannot be an LFE-pair.

That is, if we consider any two consecutive nodes $u_k, u_{k+1} \in \sigma^*$, then at most one of the d-pairs $(u_k, u_k u_{k+1}), (u_k u_{k+1}, u_{k+1})$ can be an LFE-pair. Hence, $\overrightarrow{G_i}$ can have at most $|\sigma|$ LFE-pairs.

In Proposition 6.1.42, we used the directed incidence connectivity between z and w to demonstrate (z, zw) is an LFE-pair. In general, the legal incidence connectivity between any pair of nodes c, d may be used to determine whether (z, zw) is an LFE-pair. However, we must apply some additional criteria to every widest $c - d$ legal path. The following result gives us a clear picture of them.

Theorem 6.1.45 A d-pair (z, zw) in a DFIG $\overrightarrow{G_i}(\sigma, \mu, \psi)$ is an LFE-pair if there exist two nodes $c, d \in \sigma^*$ satisfying

(i) (z, zw) belongs to every widest $c - d$ legal *di*-path in $\overrightarrow{G_i}$.
(ii) $\psi(z, zw) > \psi(zw, w)$.
(iii) zw is the unique arc in every widest $c - d$ legal *di*-path $\overrightarrow{P}$ of $\overrightarrow{G_i}$ such that

$$i_s(\overrightarrow{P}) = \mu(zw) - \vee\{\psi(z, zw), \psi(zw, w)\}.$$

Proof Let some nodes $c, d \in \sigma^*$ satisfies conditions (i), (ii) and (iii) for a d-pair $(z, zw) \in \psi^*$. We will prove (z, zw) is an LFE-pair by proving that $DICONN_{\overrightarrow{G_i}-(z,zw)}(c, d) > DICONN_{\overrightarrow{G_i}}(c, d)$.

Choose a widest $c - d$ legal *di*-path $\overrightarrow{Q}$ belonging to $\overrightarrow{G_i}$. Then by conditions (i), (ii) and (iii), $(z, zw) \in \overrightarrow{Q}$ and zw is the unique arc in $\overrightarrow{Q}$ such that

$$DICONN_{\overrightarrow{G_i}}(c, d) = i_s(\overrightarrow{Q}) = \mu(zw) - \vee\{\psi(z, zw), \psi(zw, w)\} = \mu(zw) - \psi(z, zw).$$

Choose $\overrightarrow{Q}'$ as the $c - d$ *di*-path in $\overrightarrow{G_i} - (z, zw)$ obtained by removing (z, zw) from $\overrightarrow{Q}$. Since $\mu(zw) - \psi(z, zw) < \mu(zw) - \psi(zw, w)$, we get $i_s(\overrightarrow{Q}) \leq i_s(\overrightarrow{Q}')$. If $i_s(\overrightarrow{Q}) = i_s(\overrightarrow{Q}')$, then there exists an arc $st \in \overrightarrow{Q}'$ such that $i_s(\overrightarrow{Q}') = \mu(st) - \vee\{\psi(s, st), \psi(st, t)\} = i_s(\overrightarrow{Q})$. Also note that, the arc st is different from the arc zw because, $\mu(zw) - \psi(z, zw) < \mu(zw) - \psi(zw, w)$. This is not possible by the uniqueness property of zw mentioned in condition (iii). Thus we get $i_s(\overrightarrow{Q}) < i_s(\overrightarrow{Q}')$ and $DICONN_{\overrightarrow{G_i}}(c, d) = i_s(\overrightarrow{Q}) < i_s(\overrightarrow{Q}') \leq DICONN_{\overrightarrow{G_i}-(z,zw)}(c, d)$. Hence, (z, zw) is an LFE-pair.

We may understand the necessity of each of the criteria mentioned in the theorem statement if we closely study the proof of Theorem 6.1.45. The need of condition (i) is clear from the definition of an LFE-pair.

If criterion (ii) is omitted, even if a d-pair meets the other two conditions, it need not be an LFE-pair. This is clear from the DFIG shown in Example 6.1.39. In that example, consider the d-pair (ef, f) with $\psi(ef, f) = 0.4 < 0.5 = \psi(e, ef)$. We can see that (ef, f) is not an LFE-pair, even though it belongs to the unique $e - g$ legal *di*-path, $\overrightarrow{P} : e, (e, ef), ef, (ef, f), f, (f, fg), fg, (fg, g), g$ and ef is the unique arc in $\overrightarrow{P}$ with $0.5 = i_s(\overrightarrow{P}) = \mu(ef) - \vee\{\psi(e, ef), \psi(ef, f)\} = 1 - 0.5 = 0.5$.

Further more, in the proof of Theorem 6.1.45, we used to prove (z, zw) is an LFE-pair by proving the increase in directed incidence connectivity between c and d after the removal of (z, zw). In that sense, the "uniqueness" of zw in condition (iii) of c, d is mandatory. But, we cannot say that the theorem fails if we omit "uniqueness" from condition (iii). Because, in such cases, if c, d satisfies $(i), (ii)$ and (iii) without the "uniqueness" property of zw, then also we can prove (z, zw) is an LFE-pair by proving the increase in directed incidence connectivity between z and w itself. This can be understood by the next theorem.

Theorem 6.1.46 A d-pair (z, zw) in a DFIG $\overrightarrow{G_i}(\sigma, \mu, \psi)$ is an LFE-pair if there exist two nodes $c, d \in \sigma^*$ satisfying

(i) (z, zw) belongs to every widest $c - d$ legal *di*-path in $\overrightarrow{G_i}$.
(ii) $\psi(z, zw) > \psi(zw, w)$.
(iii) $i_s(\overrightarrow{Q}) = \mu(zw) - \vee\{\psi(z, zw), \psi(zw, w)\}$, where $\overrightarrow{Q}$ is a widest $c - d$ legal *di*-path in $\overrightarrow{G_i}$.

Proof Suppose the nodes $c, d \in \sigma^*$ satisfies conditions $(i), (ii)$ and (iii) for a d-pair $(z, zw) \in \psi^*$. If zw is the unique arc in each widest $c - d$ legal *di*-path $\overrightarrow{Q}$ of $\overrightarrow{G_i}$ such that $i_s(\overrightarrow{Q}) = \mu(zw) - \vee\{\psi(z, zw), \psi(zw, w)\}$, then using Theorem 6.1.45, we can prove (z, zw) is an LFE-pair. Now, we assume $\overrightarrow{G_i}$ has a widest $c - d$ legal *di*-path, $\overrightarrow{Q}$ such that

$$i_s(\overrightarrow{Q}) = \mu(zw) - \vee\{\psi(z, zw), \psi(zw, w)\} = \mu(st) - \vee\{\psi(s, st), \psi(st, t)\},$$

for some arc $st\ (\neq zw) \in \overrightarrow{Q}$.

Now, suppose that (z, zw) is not an LFE-pair. Then, the removal of (z, zw) causes no change in the legal flow between any pair of nodes in $\overrightarrow{G_i}$. In particular, if we consider z and w, we must able to find an $z - w$ *di*-path, say $\overrightarrow{P}$ without (z, zw) satisfying

$$\begin{aligned} DICONN^*_{\overrightarrow{G_i}-(z,zw)}(z, w) &= DICONN^*_{\overrightarrow{G_i}}(z, w) \\ &= i_s(\overrightarrow{P}) = \mu(zw) - \vee\{\psi(z, zw), \psi(zw, w)\}. \end{aligned}$$

Now, we can form a new $c - d$ *di*-path, say $\overrightarrow{R}$ in $\overrightarrow{G_i}$ by removing zw from $\overrightarrow{Q}$ and joining $\overrightarrow{P}$ to $\overrightarrow{Q}$. Since $i_s(\overrightarrow{P}) = \mu(zw) - \vee\{\psi(z, zw), \psi(zw, w)\}$, we get $i_s(\overrightarrow{R}) = i_s(\overrightarrow{Q})$. Also, since $st \in \overrightarrow{R}$ and $i_s(\overrightarrow{Q}) = \mu(st) - \vee\{\psi(s, st), \psi(st, t)\}$, we get $i_s(\overrightarrow{R}) = \mu(st) - \vee\{\psi(s, st), \psi(st, t)\}$. Thus, we get a widest $c - d$ legal *di*-path of $\overrightarrow{G_i}$ not containing (z, zw). This is a contradiction to condition (i) and hence (z, zw) must be an LFE-pair. □

Another kind of d-pair, known as IFR-pair, significantly impacts the illegal flow through a network since their removal lowers the illegal flow between pairs of nodes. The following theorem is more significant since it states that IFR-pairs and LFE-pairs coincide for a legal fuzzy incidence cycle.

Theorem 6.1.47 A d-pair (z, zw) in a legal fuzzy incidence cycle $\overrightarrow{G_i}(\sigma, \mu, \psi)$ is an IFR-pair if and only if it is an LFE-pair.

Proof Let $(z, zw) \in \psi^*$ be an IFR-pair. Then σ^* has two nodes c, d with $DICONN^*_{\overrightarrow{G_i}-(z,zw)}(c, d) < DICONN^*_{\overrightarrow{G_i}}(c, d)$.
So, $\overrightarrow{G_i}$ contains a $c - d$ *di*-path $\overrightarrow{P}$ and being a legal incidence cycle, $\overrightarrow{P}$ becomes the unique $c - d$ *di*-path. Now, we will prove that $\psi(z, zw) > \psi(zw, w)$. Suppose not. That is, $\psi(z, zw) \leq \psi(zw, w)$. Then,

$$\begin{aligned}
i_s(\overrightarrow{P}) &= \wedge\{\mu(st) - \vee\{\psi(s, st), \psi(st, t)\} : st \in \overrightarrow{P}\} \\
&= \mu(zw) - \vee\{\psi(z, zw), \psi(zw, w)\} \wedge \{\mu(st) - \vee\{\psi(s, st), \psi(st, t)\} : \\
&\quad st(\neq zw) \in \overrightarrow{P}\} \\
&= \mu(zw) - \psi(zw, w) \wedge \{\mu(st) - \vee\{\psi(s, st), \psi(st, t)\} : st(\neq zw) \in \overrightarrow{P}\} \\
&= i_s(\overrightarrow{P'}),
\end{aligned}$$

where $\overrightarrow{P'}$ is taken as the unique $c - d$ *di*-path belonging to $\overrightarrow{G_i} - (z, zw)$, obtained by removing (z, zw) from $\overrightarrow{P}$. Thus,

$$\begin{aligned}
DICONN^*_{\overrightarrow{G_i}-(z,zw)}(c, d) &= ii_s(\overrightarrow{P'}) \\
&= \wedge\{\mu(st) : st \in \overrightarrow{P'}\} - i_s(\overrightarrow{P'}) \\
&= \wedge\{\mu(st) : st \in \overrightarrow{P}\} - i_s(\overrightarrow{P}) \\
&= ii_s(\overrightarrow{P}) = DICONN^*_{\overrightarrow{G_i}}(c, d).
\end{aligned}$$

This contradicts that (z, zw) is an IFR-pair. Hence, $\psi(z, zw) > \psi(zw, w)$ and therefore we get (z, zw) is an LFE-pair from Theorem 6.1.41.

On converse, let $(z, zw) \in \psi^*$ be an LFE-pair. Then from Theorem 6.1.41, we get $\psi(z, zw) > \psi(zw, w)$. Since (z, zw) becomes an LFE-pair, σ^* contains c, d with $DICONN_{\overrightarrow{G_i}-(z,zw)}(c, d) > DICONN_{\overrightarrow{G_i}}(c, d)$.
Thus, $\overrightarrow{G_i} - (z, zw)$ contains a widest $c - d$ *di*-path $\overrightarrow{Q}$ including zw such that $i_s(\overrightarrow{Q}) = DICONN_{\overrightarrow{G_i}-(z,zw)}(c, d) > DICONN_{\overrightarrow{G_i}}(c, d)$. Take $\overrightarrow{Q'}$ as the $c - d$ *di*-path in $\overrightarrow{G_i}$ obtained by adding the d-pair (z, zw) to $\overrightarrow{Q}$. Since $\overrightarrow{G_i}$ being a legal fuzzy incidence cycle, $\overrightarrow{Q'}$ will be the unique $c - d$ *di*-path in $\overrightarrow{G_i}$. So, $i_s(\overrightarrow{Q'}) = DICONN_{\overrightarrow{G_i}}(c, d)$. Also we get, $ii_s(\overrightarrow{Q}) = DICONN^*_{\overrightarrow{G_i}-(z,zw)}(c, d)$ and $ii_s(\overrightarrow{Q'}) = DICONN^*_{\overrightarrow{G_i}}(c, d)$. Now, $i_s(\overrightarrow{Q'}) < i_s(\overrightarrow{Q})$ gives,

$$\begin{aligned}DICONN^*_{\overrightarrow{G_i}}(c,d) &= ii_s(\overrightarrow{Q'})\\ &= \wedge\{\mu(st) : st \in \overrightarrow{Q'}\} - i_s(\overrightarrow{Q'})\\ &> \wedge\{\mu(st) : st \in \overrightarrow{Q}\} - i_s(\overrightarrow{Q})\\ &= ii_s(\overrightarrow{Q}) = DICONN^*_{\overrightarrow{G_i}-(z,zw)}(c,d)\end{aligned}$$

Hence, we get $c, d \in \sigma^*$ satisfying $DICONN^*_{\overrightarrow{G_i}-(z,zw)}(c,d) < DICONN^*_{\overrightarrow{G_i}}(c,d)$ implies (z, zw) forms an IFR-pair. □

Using the following theorem, we relate an IFR-pair and the directed illegal incidence connectivity between its incident nodes.

Theorem 6.1.48 Let the d-pair (z, zw) in a DFIG $\overrightarrow{G_i}(\sigma, \mu, \psi)$ satisfies $\psi(z, zw) > \psi(zw, w)$. Then (z, zw) is an IFR-pair if $DICONN^*_{\overrightarrow{G_i}-(z,zw)}(z, w) < \psi(z, zw)$.

Proof Let $(z, zw) \in \psi^*$ has $\psi(z, zw) > \psi(zw, w)$. We will prove (z, zw) is an IFR-pair by showing that $DICONN^*_{\overrightarrow{G_i}-(z,zw)}(z, w) < DICONN^*_{\overrightarrow{G_i}}(z, w)$. Suppose not. Then, $DICONN^*_{\overrightarrow{G_i}-(z,zw)}(z, w) \geq DICONN^*_{\overrightarrow{G_i}}(z, w)$. So,
$\psi(z, zw) > DICONN^*_{\overrightarrow{G_i}-(z,zw)}(z, w) \geq DICONN^*_{\overrightarrow{G_i}}(z, w)$.

This implies, $\psi(z, zw) > DICONN^*_{\overrightarrow{G_i}}(z, w)$. This is not possible, because $\overrightarrow{P}$: $z, (z, zw), zw, (zw, w), w$ is an $z - w$ *di*-path with $i_s(\overrightarrow{P}) = \mu(zw) - \psi(z, zw)$ and $ii_s(\overrightarrow{P}) = \mu(zw) - \{\mu(zw) - \psi(z, zw)\} = \psi(z, zw)$ implying that $\psi(z, zw) \leq DICONN^*_{\overrightarrow{G_i}}(z, w)$. □

The converse of Theorem 6.1.48 cannot be generalized. That is, we may able to find an IFR-pair (z, zw) in $\overrightarrow{G_i}(\sigma, \mu, \psi)$ with $DICONN^*_{\overrightarrow{G_i}-(z,zw)}(z, w) \not< \psi(z, zw)$. Let's look at an example.

Example 6.1.49 Let Fig. 6.15 illustrates the DFIG $\overrightarrow{G_i}$ (σ, μ, ψ) with $V=\{e, f, g, h\}$ such that each node of V is assigned a weight 1. Define $\mu(ef) = 0.7$, $\mu(fg) = 0.7$, $\mu(gh) = 0.5$, $\mu(eh) = 0.8$, $\mu(hf) = 0.8$, $\psi(e, ef) = 0.3$, $\psi(ef, f) = 0.1$, $\psi(f, fg) = 0.1$, $\psi(fg, g) = 0.1$, $\psi(g, gh) = 0.4$, $\psi(gh, h)=0.3$, $\psi(e, eh) = 0.3$, $\psi(eh, h) = 0.1$, $\psi(h, hf) = 0.1$ and $\psi(hf, f) = 0.3$.

Now, consider the d-pair (e, ef) with $\psi(e, ef) > \psi(ef, f)$. Then (e, ef) is an IFR-pair. For, we can consider two nodes $e, g \in \sigma^*$ with two $e - g$ *di*-paths in $\overrightarrow{G_i}$ given by $\overrightarrow{P}$: $e, (e, ef), ef, (ef, f), f, (f, fg), fg, (fg, g), g$ and $\overrightarrow{Q}$: $e, (e, eh), eh, (eh, h), h, (h, hf), hf, (hf, f), f, (f, fg), fg, (fg, g), g$. Then we have

$$\begin{aligned}i_s(\overrightarrow{P}) &= \wedge\{0.7 - \vee\{0.3, 0.1\}, 0.7 - \vee\{0.1, 0.1\}\} = \wedge\{0.4, 0.6\} = 0.4\\ ii_s(\overrightarrow{P}) &= 0.7 - 0.4 = 0.3\\ i_s(\overrightarrow{Q}) &= \wedge\{0.8 - \vee\{0.3, 0.1\}, 0.8 - \vee\{0.1, 0.3\}, 0.7 - \vee\{0.1, 0.1\}\} = 0.5\\ ii_s(\overrightarrow{Q}) &= 0.7 - 0.5 = 0.2.\end{aligned}$$

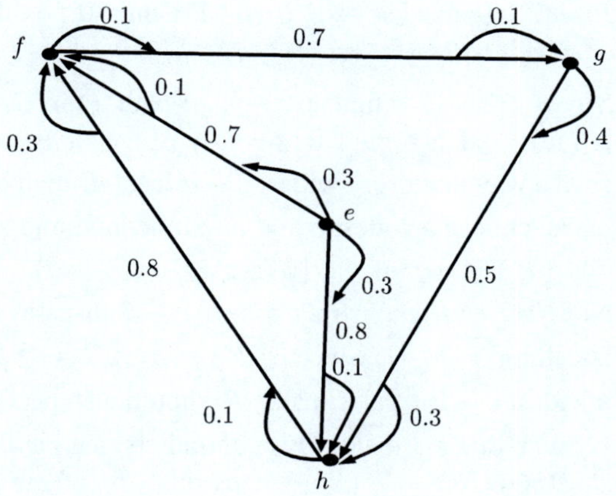

Fig. 6.15 DFIG given in Example 6.1.49

Therefore, $DICONN^*_{\overrightarrow{G_i}}(e, g) = 0.3$. Now, the removal of (e, ef) from $\overrightarrow{G_i}$ gives two $e - g$ *di*-paths in $\overrightarrow{G_i} - (e, ef)$ given by $\overrightarrow{Q} : e, (e, eh), eh, (eh, h), h, (h, hf), hf, (hf, f), f, (f, fg), fg, (fg, g), g$ and $\overrightarrow{P'} : e, ef, (ef, f), f, (f, fg), fg, (fg, g), g$. We can find the legal incidence strength of $\overrightarrow{P'}$ as similar as that of $\overrightarrow{P}$. Thus, we get $i_s(\overrightarrow{P'}) = 0.6, ii_s(\overrightarrow{P}) = 0.1$ and $i_s(\overrightarrow{Q}) = 0.5, ii_s(\overrightarrow{Q}) = 0.2$. Hence,
$0.2 = DICONN^*_{\overrightarrow{G_i}-(e,ef)}(e, g) < DICONN^*_{\overrightarrow{G_i}}(e, g) = 0.3$.
Next, we find $DICONN^*_{\overrightarrow{G_i}-(e,ef)}(e, f)$. For that, consider the $e - f$ *di*-paths in $\overrightarrow{G_i} - (e, ef)$ given by $\overrightarrow{R} : e, (e, eh), eh, (eh, h), h, (h, hf), hf, (hf, f), f$ and $\overrightarrow{S} : e, ef, (ef, f), f$. Then, we get

$$i_s(\overrightarrow{R}) = \wedge\{0.8 - \vee\{0.3, 0.1\}, 0.8 - \vee\{0.1, 0.3\}\} = \wedge\{0.5, 0.5\} = 0.5$$
$$ii_s(\overrightarrow{R}) = 0.8 - 0.5 = 0.3$$
$$i_s(\overrightarrow{S}) = \{0.7 - \vee\{0, 0.1\}\} = 0.6$$
$$ii_s(\overrightarrow{Q}) = 0.7 - 0.6 = 0.1.$$

Thus, $DICONN^*_{\overrightarrow{G_i}-(e,ef)}(e, f) = 0.3$. Hence (e, ef) violates the converse of Theorem 6.1.48 as $0.3 = DICONN^*_{\overrightarrow{G_i}-(e,ef)}(e, f) \nless \psi(e, ef) = 0.3$.

Next, we establish a condition that is necessary and sufficient to make an arc as LFR-link.

Theorem 6.1.50 An arc $zw \in \mu^*$ of a DFIG $\overrightarrow{G_i}(\sigma, \mu, \psi)$ is an LFR-link if and only if $DICONN_{\overrightarrow{G_i}-zw}(z, w) < \mu(zw) - \vee\{\psi(z, zw), \psi(zw, w)\}$.

Proof Suppose $zw \in \mu^*$ is an LFR-link. If possible, suppose that
$DICONN_{\overrightarrow{G_i}-zw}(z, w) \geq \mu(zw) - \vee\{\psi(z, zw), \psi(zw, w)\}$.
So, we are able to find an $z - w$ *di*-path $\overrightarrow{P}$ in $\overrightarrow{G_i} - zw$ with $zw \notin \overrightarrow{P}$ and $i_s(\overrightarrow{P}) \geq \mu(zw) - \vee\{\psi(z, zw), \psi(zw, w)\}$. Since zw is an LFR-link, there exist $c, d \in \sigma^*$ in such a way that every widest $c - d$ legal *di*-path of $\overrightarrow{G_i}$ includes zw. Among all such paths, choose a widest $c - d$ *di*-path containing zw, say $\overrightarrow{Q}$. Then,
$i_s(\overrightarrow{Q}) \leq \mu(zw) - \vee\{\psi(z, zw), \psi(zw, w)\}$.
Now, we can reconstruct a new $c - d$ *di*-path $\overrightarrow{R}$ by removing zw from $\overrightarrow{Q}$ and attaching $\overrightarrow{P}$ to $\overrightarrow{Q}$. Since $i_s(\overrightarrow{P}) \geq \mu(zw) - \vee\{\psi(z, zw), \psi(zw, w)\}$ and $\overrightarrow{Q}$ being a widest $c - d$ legal *di* path, $\overrightarrow{R}$ should also become a widest $c - d$ legal *di*-path in $\overrightarrow{G_i}$ such that $zw \notin \overrightarrow{R}$. This contradicts that zw is an LFR-link. Thus,
$DICONN_{\overrightarrow{G_i}-zw}(z, w) < \mu(zw) - \vee\{\psi(z, zw), \psi(zw, w)\}$.

Conversely, let $DICONN_{\overrightarrow{G_i}-zw}(z, w) < \mu(zw) - \vee\{\psi(z, zw), \psi(zw, w)\}$. Since $zw \in \mu^*$, we have $DICONN_{\overrightarrow{G_i}}(z, w) \geq \mu(zw) - \vee\{\psi(z, zw), \psi(zw, w)\}$. If zw is not an LFR-link, then between any two nodes, there exists a widest legal *di*-path does not containing zw. In particular, if we consider z and w, we get

$$DICONN_{\overrightarrow{G_i}-zw}(z, w) = DICONN_{\overrightarrow{G_i}}(z, w) \geq \mu(zw) - \vee\{\psi(z, zw), \psi(zw, w)\},$$

which is a contradiction to the assumption. Hence, zw must be an LFR-link. □

Proposition 6.1.51 If an arc $zw \in \mu^*$ of a DFIG $\overrightarrow{G_i}(\sigma, \mu, \psi)$ is an LFR-link, then $DICONN_{\overrightarrow{G_i}}(z, w) = \mu(zw) - \vee\{\psi(z, zw), \psi(zw, w)\}$.

Proof Suppose that zw is an LFR-link. Then σ^* contains two nodes c, d with zw belonging to every widest $c - d$ legal *di*-path of $\overrightarrow{G_i}$. Let $\overrightarrow{P}$ be one among such widest $c - d$ legal *di*-path. If possible, suppose that $DICONN_{\overrightarrow{G_i}}(z, w) > \mu(zw) - \vee\{\psi(z, zw), \psi(zw, w)\}$. Then, $\overrightarrow{G_i}$ has a widest $z - w$ legal *di*-path $\overrightarrow{Q}$ with $i_s(\overrightarrow{Q}) > \mu(zw) - \vee\{\psi(z, zw), \psi(zw, w)\}$. Now, consider a new $c - d$ *di*-path, say $\overrightarrow{R}$ in $\overrightarrow{G_i}$ obtained by removing zw from $\overrightarrow{P}$ and joining the *di*-path $\overrightarrow{Q}$ to $\overrightarrow{P}$. Since $i_s(\overrightarrow{Q}) > \mu(zw) - \vee\{\psi(z, zw), \psi(zw, w)\}$, we get $i_s(\overrightarrow{R}) \geq i_s(\overrightarrow{P})$. Thus, we get a widest $c - d$ legal *di*-path belonging to $\overrightarrow{G_i}$ without zw, which is a contradiction to the fact that zw is an LFR-link. Hence, we must have $DICONN_{\overrightarrow{G_i}}(z, w) = \mu(zw) - \vee\{\psi(z, zw), \psi(zw, w)\}$. □

However, the converse of the above proposition may fail in general cases. That is, even though an arc zw in a DFIG, $\overrightarrow{G_i}(\sigma, \mu, \psi)$ satisfies $DICONN_{\overrightarrow{G_i}}(z, w) = \mu(zw) - \vee\{\psi(z, zw), \psi(zw, w)\}$, that arc need not be an LFR-link. We can see this though the next example.

Example 6.1.52 Let $V = \{e, f, g\}$ be the node set of the the DFIG $\overrightarrow{G_i}(\sigma, \mu, \psi)$ given in Fig. 6.16 Each node of V is assigned σ value 1. Define $\mu(ef) = \mu(gf) = \mu(eg) = 0.7$, $\psi(e, ef) = \psi(g, gf) = \psi(gf, f) = \psi(e, eg) = 0.3$ and $\psi(e.g., g) = \psi(ef, f) = 0.1$.

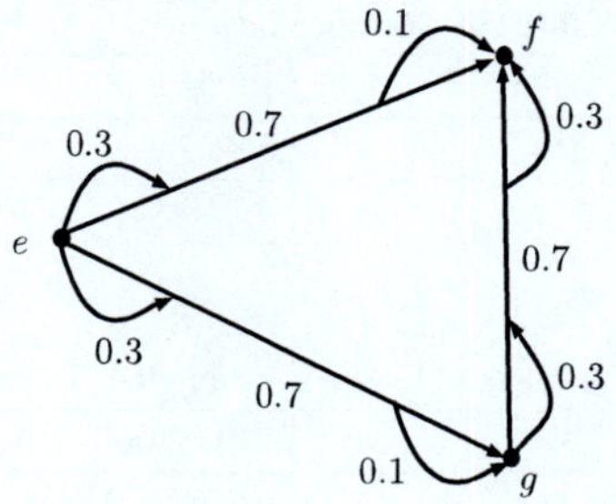

Fig. 6.16 DFIG violating the converse of Proposition 6.1.51

For $ef \in \mu^*$, the $e - f$ *di*-paths $\overrightarrow{Q} : e, (e, eg), eg, (e.g., g), g, (g, gf), gf, (gf, f), f$ and $\overrightarrow{P} : e, (e, ef), ef, (ef, f), f$ have

$$i_s(\overrightarrow{P}) = \mu(ef) - \vee\{\psi(e, ef), \psi(ef, f)\} = 0.7 - \vee\{0.3, 0.1\} = 0.4$$
$$i_s(\overrightarrow{Q}) = \wedge\{\mu(eg) - \vee\{\psi(e, eg), \psi(e.g., g)\}, \mu(gf) - \vee\{\psi(g, gf), \psi(gf, f)\}\}$$
$$= \wedge\{0.7 - \vee\{0.3, 0.1\}, 0.7 - \vee\{0.3, 0.3\}\} = \wedge\{0.4, 0.4\} = 0.4.$$

Thus, $DICONN_{\overrightarrow{G_i}}(e, f) = 0.4 = \mu(ef) - \vee\{\psi(e, ef), \psi(ef, f)\}$. But, ef is not an LFR-link, because its removal never cause a reduction in the directed incidence connectivity between any two nodes of $\overrightarrow{G_i}$.

Now, we arrive at an interesting result that connects an LFR-link and an LFR-pair. If we have an LFR-link, then the corresponding d-pair incident with that link and having maximum weight will becomes an LFR-pair. Formally, we can say,

Corollary 6.1.53 In a DFIG $\overrightarrow{G_i}(\sigma, \mu, \psi)$, suppose $zw \in \mu^*$ and $(z, zw) \in \psi^*$ satisfies $\psi(z, zw) > \psi(zw, w)$. If zw is an LFR-link, then (z, zw) is an LFE-pair.

Proof From Proposition 6.1.51, if zw is an LFR-link, then
$DICONN_{\overrightarrow{G_i}}(z, w) = \mu(zw) - \vee\{\psi(z, zw), \psi(zw, w)\}$.
Since $\psi(z, zw) > \psi(zw, w)$, we get
$DICONN_{\overrightarrow{G_i}}(z, w) = \mu(zw) - \psi(z, zw)$.
Then by Proposition 6.1.42, since $DICONN_{\overrightarrow{G_i}}(z, w) = \mu(zw) - \psi(z, zw)$, we get (z, zw) is an LFE-pair. □

6.2 Application of DFIG in the Migration of Refugees

According to [2], the United States is carrying out a major deportation of migrants or refugees. Similarly, during peak years of immigration, a huge number of Mexicans were deported. The Central Americans likewise crossed the border in comparable numbers without being subjected to widespread deportation, despite Mexico's agreement to welcome them from the United States under authority on pandemics. Mexico does not accept expelled Haitians or persons from countries other than Mexico,

Table 6.3 Refugee flow

	1995–2000	2005–2010	2009–2014	2013–2018
$\psi(x, xy)$	0.81	0.50	0.47	0.55
$\psi(xy, y)$	0.19	0.50	0.53	0.45

Table 6.4 Incidence strength

	1995–2000	2005–2010	2009–2014	2013–2018
$i_s(xy)$	$\mu(xy) - 0.81$	$\mu(xy) - 0.50$	$\mu(xy) - 0.53$	$\mu(xy) - 0.55$
$ii_s(xy)$	0.81	0.50	0.53	0.55

Guatemala, Honduras, and El Salvador. Mexico has announced that it would begin deporting Haitians to their home country.

Through Mexico, there are various immigration routes from Central America to the United States. The tools given in this research can be used to investigate immigration via various channels. The return of Haitians to Haiti from the United States is massive. At the time of this writing, data on refugee expulsions is continuously changing, making it difficult to analyze whole pathways. However, we will demonstrate how our method may be employed.

We analyze immigration from Mexico to the United States as well as refugee returns from the United States to Mexico. Let x represents Mexico and y represents the United States. The directed edge from x to y is then denoted by xy. In [3], the number of refugees n from Mexico to the United States and the number of refugees m returned from the United States to Mexico was given for certain time periods. We let $\psi(x, xy) = n/(n+m)$ and $\psi(xy, y) = m/(n+m)$. These percentages are shown in Table 6.3.

In general, we have $i_s(xy) = \mu(xy) - \psi(x, xy) \vee \psi(xy, y)$ and $ii_s(xy) = \mu(xy) - i_s(xy) = \psi(x, xy) \vee \psi(xy, y)$. Also, $i_s(xy) + ii_s(xy) = \mu(xy)$. We thus have (Table 6.4).

Country x has a restriction on the amount of refugees it permits to flee to y, while country y has a restriction on the amount of refugees it lets to enter. Let $\sigma(x)$ and $\sigma(y)$ denote measures of these limits. Also, xy may have additional limitations due to difficult terrain and/or kidnapping. Hence $\psi(x, xy) \vee \psi(xy, y) \leq \sigma(x) \wedge \mu(xy) \wedge \sigma(y) = \mu(xy)$ since $\mu(xy) \leq \sigma(x) \wedge \sigma(y)$.

Now $i_s(\overrightarrow{Q})$ is the smallest $\mu(z_i z_{i+1}) - \psi(z_i, z_i z_{i+1}) \vee \psi(z_i z_{i+1}, z_{i+1})$, say $xy = z_i z_{i+1}$. Then $i_s(\overrightarrow{Q}) = i_s(xy)$ and $ii_s(\overrightarrow{Q}) = ii_s(xy)$.

References

1. Gayathri, G., Mathew, S., Mordeson, J.N.: Directed fuzzy incidence: a model for illicit flow networks. Inf. Sci. **608**, 1375–1400 (2022)
2. Lozano, J.A., Gay, E., Spaget, E., Sanon, E.: US launches mass expulsion of Haitian migrants from Texas. AP News, Sept 19 (2021)
3. Gonzalez-Barrera, A.: Before COVID-19, More Mexicans Came to the U. S. Than Left for Mexico for the First Time in Years. Pew Research Center, July 9 (2021)

Chapter 7
Max-flow Min-cut Theorem for Directed Fuzzy Incidence Networks

The internet and transportation networks may be most successfully evaluated when regarded as directed fuzzy incidence graphs with certain extra properties. This chapter presents directed fuzzy incidence network (DFIN), a network model ideal for studying the dynamism and stability of many unpredictable networks. DFINs are both edge and node capacitated connected networks. Such fuzzified models are really useful for including the massive size of various modern networks. This chapter focuses on legal flow, saturated and unsaturated arc, arc cut, and legal flow enhancing path in a DFIN. This study also explores the relationship between the value of the legal flow and the capacity of the arc cut. Our main purpose is to produce a DFIN-analog of the max-flow min-cut theorem in graph theory. Furthermore, the study suggests and shows an algorithm for calculating maximum legal flow in a DFIN. The contents of this chapter are from [1].

7.1 Directed Fuzzy Incidence Networks and Legal Flows

This section develops the fundamental definition of directed fuzzy incidence network (DFIN) and the idea of legal flow in a DFIN. We go further into these ideas using examples and illustrations.

Definition 7.1.1 A **directed fuzzy incidence network (DFIN)** is a directed fuzzy incidence graph $\overrightarrow{G}_i(\sigma, \mu, \psi)$ with the following properties.

(i) $\overrightarrow{G}_i$ has only one node s, termed the source, that has no incoming arcs.
(ii) $\overrightarrow{G}_i$ has only one node t, termed the sink, that has no outgoing arcs.
(iii) The underlying fuzzy graph of $\overrightarrow{G}_i$ is connected.

J. N. Mordeson et al., *Fuzzy Graph Theory*, Studies in Fuzziness and Soft Computing 424, https://doi.org/10.1007/978-3-031-23108-7_7

It is denoted by $\overrightarrow{N}(\sigma, \mu, \psi)$. Here, σ^* is the node set of $\overrightarrow{N}$ and μ^* is the arc set of $\overrightarrow{N}$. The members of ψ^* are known as directed incidence pairs or d-pairs.

It is worth noting that a DFIN is both node and edge capacitated. In $\overrightarrow{N}(\sigma, \mu, \psi)$, $\sigma(m)$ is the capacity of the node m and $\mu(mn)$ is the capacity of the arc mn. Also, there are some nodes in $\overrightarrow{N}$ that are neither a source nor a sink. Such nodes are called the intermediate nodes of $\overrightarrow{N}$.

As mentioned in Chap. 6, the value of legal flow along a directed incidence path $\overrightarrow{P}$ in a DFIG $\overrightarrow{G_i}(\sigma, \mu, \psi)$ is given by $i_s(\overrightarrow{P}) = \wedge\{\mu(mn) - \vee\{\psi(m, mn), \psi(mn, n)\} : mn \in \overrightarrow{P}\}$. Thus the maximum legal flow possible through an arc mn in $\overrightarrow{G_i}$ can be obtained as $i_s(mn) = \mu(mn) - \vee\{\psi(m, mn), \psi(mn, n)\}$. The following example depicts *di*-paths and their legal flow values in a directed fuzzy incidence graph.

Example 7.1.2 Figure 7.1 depicts the DFIG $\overrightarrow{G_i}(\sigma, \mu, \psi)$, which contains 5 nodes e, f, d, g and h, each with σ-value 1. The μ and ψ values are shown in the picture.

Consider the nodes $f, g \in \sigma^*$. Then the $f - g$ *di*-path $\overrightarrow{Q} : f, (f, fh), fh, (fh, h), h, hg, (hg, g), g$ has legal flow, $i_s(\overrightarrow{Q}) = \wedge\{\mu(fh) - \vee\{\psi(f, fh), \psi(fh, h)\}, \mu(hg) - \psi(hg, g)\} = \wedge\{0.4, 0.5\} = 0.4$.

Similarly, we find the legal flow of each *di*-path in $\overrightarrow{G_i}$ and tabulate them in Table 7.1.

Now, we introduce the concept of legal flow in a DFIN $\overrightarrow{N}(\sigma, \mu, \psi)$ as a function defined on the arc set μ^*.

Definition 7.1.3 Let $\overrightarrow{N}(\sigma, \mu, \psi)$ be a DFIN with source s and sink t. A **legal flow** on $\overrightarrow{N}$ is a function $\eta : \mu^* \to [0, 1]$ such that

(i) $\eta(a) \leq i_s(a) \, \forall a \in \mu^*$.

(ii) $\sum\limits_{m \in \sigma^*} \eta(sm) \leq \sigma(s)$ and $\sum\limits_{m \in \sigma^*} \eta(mn) \leq \sigma(n) \, \forall n \in \sigma^* \backslash \{s\}$.

(iii) $\sum\limits_{m \in \sigma^*} \eta(mn) = \sum\limits_{l \in \sigma^*} \eta(nl) \, \forall n \in \sigma^* \backslash \{s, t\}$.

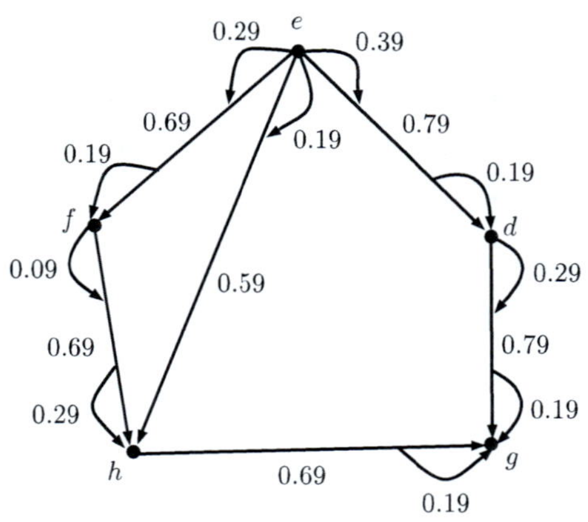

Fig. 7.1 Directed fuzzy incidence graph $\overrightarrow{G_i}(\sigma, \mu, \psi)$

Table 7.1 Legal flow of *di*-paths in DFIG given in Fig. 7.1

Ordered pair of nodes	*Di*-path	Legal flow (i_s)
(e, f)	$e, (e, ef), ef, (ef, f), f$	0.4
(e, g)	$e, (e, ef), ef, (ef, f), f, (f, fh), fh, (fh, h), h, hg, (hg, g), g$	0.4
	$e, (e, eh), eh, h, hg, (hg, g), g$	0.4
	$e, (e, ed), ed, (ed, d), d, (d, dg), dg, (dg, g), g$	0.4
(e, h)	$e, (e, eh), eh, h$	0.4
	$e, (e, ef), ef, (ef, f), f, (f, fh), fh, (fh, h), h$	0.4
(e, d)	$e, (e, ed), ed, (ed, d), d$	0.4
(f, h)	$f, (f, fh), fh, (fh, h), h$	0.4
(f, g)	$f, (f, fh), fh, (fh, h), h, hg, (hg, g), g$	0.4
(h, g)	$h, hg, (hg, g), g$	0.5
(d, g)	$d, (d, dg), dg, (dg, g), g$	0.5

Conditions (ii) and (iii) are referred to as the node capacity constraint and flow conservation constraint respectively. Since $i_s(a) \leq \mu(a)\, \forall\, a \in \mu^*$, it follows from condition (i) that the arc capacity constraint $\eta(a) \leq \mu(a)$, $\forall\, a \in \mu^*$ also holds.

For a subset U of the node set V of $\overrightarrow{N}(\sigma, \mu, \psi)$, let $\overline{U}$ denotes $V \backslash U$. For a legal flow η on μ^* and a subset E of μ^*, let $\eta(E)$ denotes $\sum_{a \in E} \eta(a)$. If $(U, \overline{U}) \subseteq \mu^*$ indicates the set of all arcs in $\overrightarrow{N}$ with tail in U and head in $\overline{U}$, then let $\eta^+(U)$ denotes $\eta(U, \overline{U})$ and $\eta^-(U)$ denotes $\eta(\overline{U}, U)$.

Then condition (ii) of Definition 7.1.3 can be rewritten as

$$\eta^+(\{s\}) \leq \sigma(s) \text{ and } \eta^-(\{n\}) \leq \sigma(n)\ \forall n \in \sigma^* \backslash \{s\}.$$

Also condition (iii) of Definition 7.1.3 can be modified as

$$\eta^-(\{n\}) = \eta^+(\{n\})\ \forall n \in \sigma^* \backslash \{s, t\}.$$

It is always possible to define a 'zero legal flow' in any DFIN $\overrightarrow{N}(\sigma, \mu, \psi)$ by $\eta(a) = 0\ \forall a \in \mu^*$. It satisfies all the requirements of a legal flow and hence there exists at least one legal flow in any DFIN.

Let us look at an example of a DFIN with a legal flow defined on it.

Example 7.1.4 Example of a DFIN is shown in Fig. 7.2. $\overrightarrow{N}(\sigma, \mu, \psi)$ has 3 nodes s, e, t with capacities $\sigma(s) = 0.7$, $\sigma(m) = 0.4$ and $\sigma(t) = 0.6$. The capacities of the arcs are defined by $\mu(sm) = 0.59$, $\mu(st) = 0.29$ and $\mu(mt) = 0.39$. The $\psi-$values are given by $\psi(s, sm) = 0.39$, $\psi(sm, m) = 0.29$, $\psi(s, st) = 0.09$ and $\psi(st, t) = \psi(m, mt) = \psi(mt, t) = 0.19$. Here, s is the source, t is the sink, and e is the single intermediate node in $\overrightarrow{N}$.

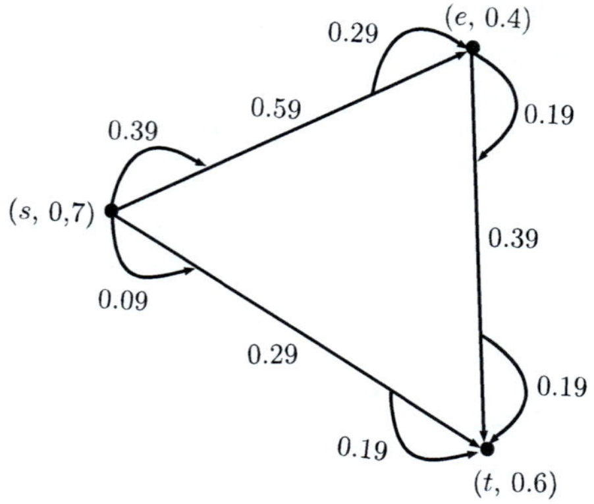

Fig. 7.2 Example of a DFIN

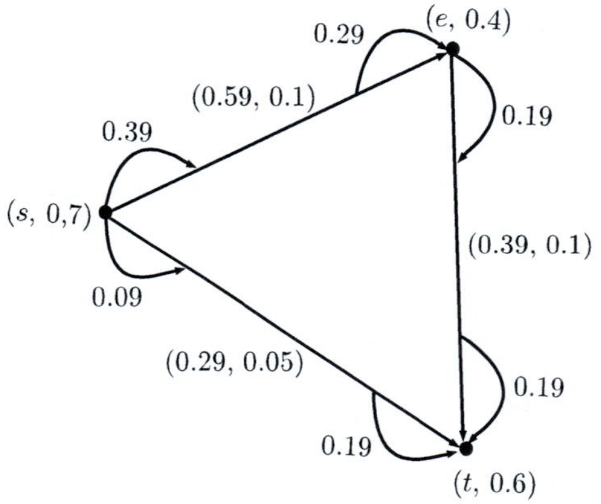

Fig. 7.3 Legal flow given in the DFIN in Fig. 7.2

Now, define a function $\eta : \mu^* \to [0, 1]$ by $\eta(a) = \frac{i_s(a)}{2}$. Then $\eta(se) = \frac{0.59 - \vee\{0.39, 0.29\}}{2} = 0.1$, $\eta(st) = \frac{0.29 - \vee\{0.09, 0.19\}}{2} = 0.05$ and $\eta(et) = \frac{0.39 - \vee\{0.19, 0.19\}}{2} = 0.1$. Then

(i) $\eta(a) \leq i_s(a) \, \forall \, a \in \{se, st, et\}$.
(ii) $\eta^+(\{s\}) = \eta(se) + \eta(st) = 0.1 + 0.05 = 0.15 < 0.7 = \sigma(s)$
$\eta^-(\{e\}) = \eta(se) = 0.1 < 0.4 = \sigma(e)$ and
$\eta^-(\{t\}) = \eta(et) + \eta(st) = 0.1 + 0.05 = 0.15 < 0.6 = \sigma(t)$.
(iii) The unique intermediate node e satisfies $\eta^-(\{e\}) = \eta(se) = 0.1 = \eta(et) = \eta^+(\{e\})$.

Hence, η meets all of the requirements for a legal flow in $\overrightarrow{N}$. Figure 7.3 depicts the DFIN $\overrightarrow{N}(\sigma, \mu, \psi)$ with the legal flow η. We use an ordered pair to express the capacity of an arc followed by the legal flow along that arc.

Next, we present the concept of resultant legal flow into and out of a subset of the arc set of a DFIN. Since the legal flow function is bounded by 0 and 1, the resultant legal flow into and out of a subset is considered only if it is non-negative.

Definition 7.1.5 Let η be a legal flow defined on a DFIN $\overrightarrow{N}(\sigma, \mu, \psi)$ with arc set $A(\overrightarrow{N})$. Then for a subset U of σ^*, **the resultant legal flow into** U is defined as

$$I_\eta(U) = \eta^-(U) - \eta^+(U).$$

There is a resultant legal flow into U if $I_\eta(U) > 0$.

Also, for a subset U of σ^*, **the resultant legal flow out of** U is defined as

$$O_\eta(U) = \eta^+(U) - \eta^-(U).$$

There is a resultant legal flow out of U if $O_\eta(U) > 0$.

The following proposition states a key characteristic of all legal flows in a DFIN.

Proposition 7.1.6 *With respect to any legal flow η defined on a DFIN $\overrightarrow{N}(\sigma, \mu, \psi)$, the resultant legal flow out of the source s equals the resultant legal flow into the sink t. That is,*

$$O_\eta(\{s\}) = I_\eta(\{t\}).$$

Proof Let η be the legal flow defined on $\overrightarrow{N}(\sigma, \mu, \psi)$. Then for any $a \in \mu^*$, $\eta(a)$ is contributed to $\eta^+(\{m\})$ of exactly one node $m \in \sigma^*$ and $\eta^-(\{n\})$ of exactly one node $n \in \sigma^*$ such that $n \neq m$. Hence

$$\sum_{k\in\sigma^*} O_\eta(\{k\}) = \sum_{k\in\sigma^*} I_\eta(\{k\}).$$

But by condition (iii) of Definition 7.1.3,

$$\eta^-(\{m\}) = \eta^+(\{m\}) \, \forall m \in \sigma^* \backslash \{s, t\}.$$

This implies, $O_\eta(\{m\}) = I_\eta(\{m\}) = 0 \forall m \in \sigma^* \backslash \{s, t\}$.

Thus $O_\eta(\{s\}) + O_\eta(\{t\}) = I_\eta(\{s\}) + I_\eta(\{t\})$.

Since $O_\eta(\{t\}) = I_\eta(\{s\}) = 0$, the above equation becomes $O_\eta(\{s\}) = I_\eta(\{t\})$. ■

According to Proposition 7.1.6, the resulting legal flow out of the source s equals the resultant legal flow into the sink t for any legal flow η. Hence, we define this quantity as the value of the legal flow η.

Definition 7.1.7 For a legal flow η defined on a DFIN $\overrightarrow{N}(\sigma, \mu, \psi)$, the **value of** η is defined by $Val\, \eta = O_\eta(\{s\}) = \sum_{m\in\sigma^*} \eta(sm)$. Or equivalently, $Val\, \eta = I_\eta(\{t\}) = \sum_{m\in\sigma^*} \eta(mt)$.

Arcs in a DFIN with a legal flow η may be classified into two categories: η - zero or η - positive arcs in the first class, and η - saturated or η - unsaturated arcs in the second class.

Definition 7.1.8 Let η be a legal flow defined on $\overrightarrow{N}(\sigma, \mu, \psi)$.

1. An arc $a \in \mu^*$ is said to be η **- zero** is $\eta(a) = 0$ and η **- positive** if $\eta(a) > 0$.
2. An arc $a \in \mu^*$ is said to be η **- saturated** is $\eta(a) = i_s(a)$ and η **- unsaturated** if $\eta(a) < i_s(a)$.

Next, we introduce the notion of maximum legal flow in a DFIN. Since each DFIN has at least one legal flow, the concept of maximum legal flow is valid.

Definition 7.1.9 A legal flow η on $\overrightarrow{N}(\sigma, \mu, \psi)$ is said to be a **maximum legal flow** if there exists no legal flow η' in $\overrightarrow{N}$ such that $Val\,\eta' > Val\,\eta$.

Consider the following example, in which we divide the arcs of a DFIN into η - zero, η - positive, η - saturated, and η - unsaturated arcs.

Example 7.1.10 Let Fig. 7.4 depicts a DFIN $\overrightarrow{N}(\sigma, \mu, \psi)$ with a legal flow η. The source s, sink t and four intermediate nodes e, f, g and h have σ value 1. The μ, ψ and legal flow values of the arcs in $\overrightarrow{N}$ are as in the figure. Then the value of the legal flow η in $\overrightarrow{N}$ is

$$Val\,\eta = O_\eta(\{s\}) = 0.3 + 0.2 = 0.5.$$

We have $\eta(a) > 0 \forall a \in \mu^*$. Hence, all arcs in $\overrightarrow{N}$ are η - positive. Also,

$$\eta(sf) = 0.2 = i_s(sf) \text{ and } \eta(fg) = 0.2 = i_s(fg).$$

That is, the arcs sf and fg are η - saturated whereas all other arcs of $\overrightarrow{N}$ are η - unsaturated.

Note that, a DFIN $\overrightarrow{N}(\sigma, \mu, \psi)$ is **connected** if for any pair of nodes m and n in $\overrightarrow{N}$, there is at least one directed incidence path from m to n or n to m.

Now we look at an interesting concept in DFIN called 'arc cut'. Their determination is critical in analyzing legal flows in the network.

Definition 7.1.11 Let $\overrightarrow{N}(\sigma, \mu, \psi)$ be a DFIN. An **arc cut K** in $\overrightarrow{N}$ is a collection of arcs $(U, \overline{U})$ with $s \in U$ and $t \in \overline{U}$ such that the removal of $\mathbf{K}$ from $\overrightarrow{N}$ produce a disconnection of $\overrightarrow{N}$ where s and t belong to distinct directed components.

The capacity of an arc cut $\mathbf{K}$ in $\overrightarrow{N}$ is given by $C(\mathbf{K}) = \sum_{a \in \mathbf{K}} i_s(a)$.

An arc cut $\mathbf{K}$ in $\overrightarrow{N}$ is said to be a **minimum arc cut** if there exists no arc cut $\mathbf{K}'$ in $\overrightarrow{N}$ such that $C(\mathbf{K}') < C(\mathbf{K})$.

The following example demonstrates the notion of arc cut in a DFIN.

Example 7.1.12 Take the same DFIN $\overrightarrow{N}(\sigma, \mu, \psi)$ with the same legal flow η of Fig. 7.4. Let $U = \{s, f\}$. Then $\overline{U} = \{e, h, g, t\}$. The collection of arcs $\mathbf{K} = (U, \overline{U}) = \{se, fg\}$ forms an arc cut in $\overrightarrow{N}$. Because, $s \in U$, $t \in \overline{U}$ and $\overrightarrow{N} - \mathbf{K}$ has two directed components with s and t belonging to distinct directed components. $\overrightarrow{N}$ with the arc cut $\mathbf{K}$ is given in Fig. 7.5, where the arcs of $\mathbf{K}$ are thickened. Moreover, Fig. 7.5(a) shows the directed components of $\overrightarrow{N} - \mathbf{K}$.

In this case, the arc cut $\mathbf{K}$ has capacity $C(\mathbf{K}) = i_s(se) + i_s(fg) = 0.4 + 0.2 = 0.6$. Also, any other arc cut in $\overrightarrow{N}$ have capacity at least 0.6. In other words, $\mathbf{K}$ is a minimum arc cut in $\overrightarrow{N}$.

We can observe from Definition 7.1.7 that, $Val\,\eta = O_\eta(\{s\})$ for a legal flow η on $\overrightarrow{N}(\sigma, \mu, \psi)$. However, in the following proposition, we can see that the value of a legal flow may be computed using any arc cut in the DFIN.

Proposition 7.1.13 *Let η be any legal flow defined on $\overrightarrow{N}(\sigma, \mu, \psi)$. Then for any arc cut $\boldsymbol{K} = (U, \overline{U})$ in $\overrightarrow{N}$,*

$$Val\,\eta = \eta^+(U) - \eta^-(U).$$

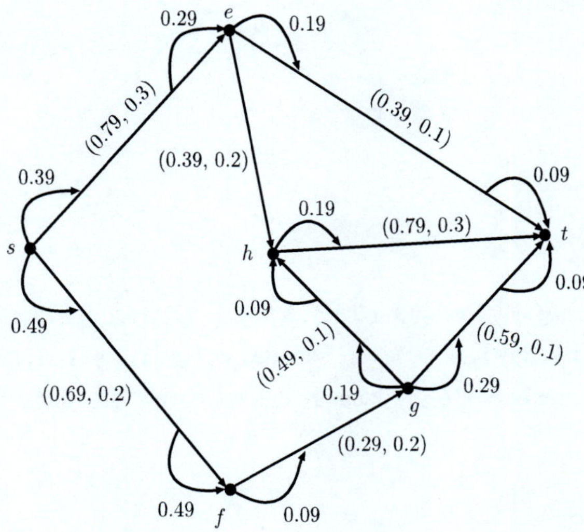

Fig. 7.4 DFIN with η—saturated and η—unsaturated arcs

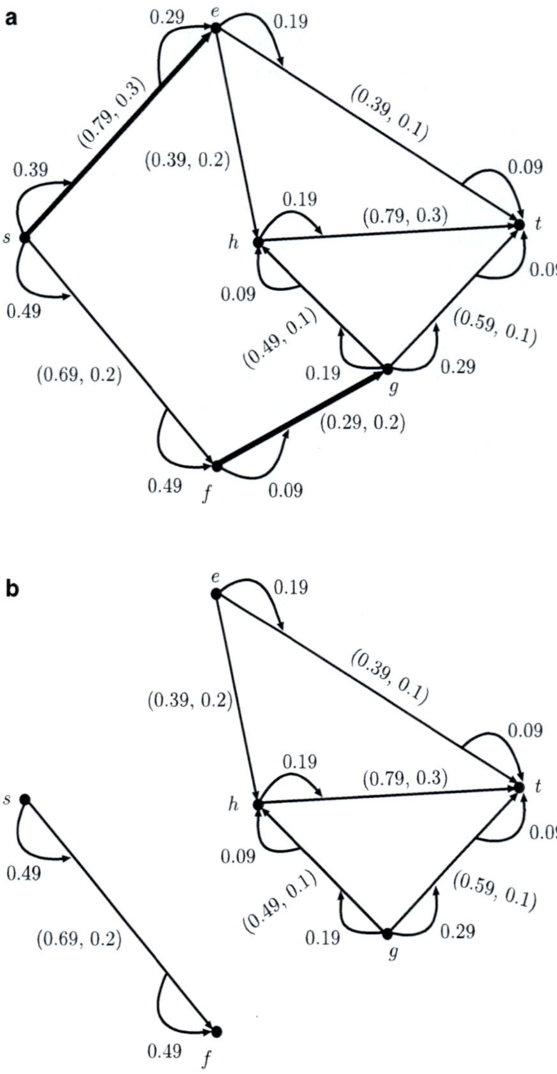

Fig. 7.5 The DFIN $\overrightarrow{N}(\sigma, \mu, \psi)$ with arc cut **K**. **a**. The directed components of $\overrightarrow{N}$ - **K**

Proof By Definition 7.1.7, $Val\,\eta = O_\eta(\{s\}) = \eta^+(\{s\}) - \eta^-(\{s\})$.
Also, any $m \in \sigma^*\backslash\{s, t\}$ satisfies $\eta^+(\{m\}) - \eta^-(\{m\}) = 0$.
Since $\mathbf{K} = (U, \overline{U})$ is an arc cut, we have $s \in U$ and $t \notin U$.

Hence,

$$\begin{aligned}\eta^+(U) - \eta^-(U) &= \sum_{u \in U} \eta^+(\{u\}) - \sum_{u \in U} \eta^-(\{u\}) \\ &= \sum_{u \in U} \{\eta^+(\{u\}) - \eta^-(\{u\})\} \\ &= \eta^+(\{s\}) - \eta^-(\{s\}) + \sum_{u \in U \setminus \{s\}} \{\eta^+(\{u\}) - \eta^-(\{u\})\} \\ &= \eta^+(\{s\}) - \eta^-(\{s\}) \\ &= Val\, \eta.\end{aligned}$$

■

There is always a relationship between the value of η and the capacity of $\mathbf{K}$ in a DFIN for any legal flow η and arc cut $\mathbf{K}$. The following theorem proves this. Furthermore, the theorem specifies a necessary and sufficient condition for the equality of the two quantities mentioned above.

Theorem 7.1.14 *Let η be any legal flow and $\mathbf{K} = (U, \overline{U})$ be any arc cut in $\overrightarrow{N}(\sigma, \mu, \psi)$. Then $Val\, \eta \leq C(\mathbf{K})$. Moreover, $Val\, \eta = C(\mathbf{K})$ if and only if the arcs of $(U, \overline{U})$ are η - saturated and the arcs of $(\overline{U}, U)$ are η - zero.*

Proof From Definition 7.1.3, we have $\eta(a) \leq i_s(a) \forall a \in (U, \overline{U})$. So,

$$\eta^+(U) = \sum_{a \in (U, \overline{U})} \eta(a) \leq \sum_{a \in (U, \overline{U})} i_s(a) = C(\mathbf{K}).$$

Since $\eta(a) \in [0, 1] \forall a \in (\overline{U}, U)$, we have

$$\eta^-(U) = \sum_{a \in (\overline{U}, U)} \eta(a) \geq 0.$$

Hence,

$$Val\, \eta = \eta^+(U) - \eta^-(U) \leq C(\mathbf{K}).$$

Now, $Val\, \eta = C(\mathbf{K}) \iff \eta^+(U) - \eta^-(U) = C(\mathbf{K})$.

$$\iff \sum_{a \in (U, \overline{U})} \eta(a) = \sum_{a \in (U, \overline{U})} i_s(a) \text{ and } \eta^-(U) = 0.$$

$$\iff \eta(a) = i_s(a) \forall a \in (U, \overline{U}) \text{ and } \eta(a) = 0 \forall a \in (\overline{U}, U).$$

$$\iff a \text{ is } \eta \text{ - saturated } \forall a \in (U, \overline{U}) \text{ and } a \text{ is } \eta \text{ - zero } \forall a \in (\overline{U}, U).$$ ■

Using the preceding result and the definitions of maximum legal flow and minimum arc cut, we can draw some conclusions about the legal flow η and arc cut $\mathbf{K}$ fulfilling $Val\, \eta = C(\mathbf{K})$. This can be seen in the following proposition.

Proposition 7.1.15 *If a DFIN $\overrightarrow{N}(\sigma, \mu, \psi)$ has a legal flow η and an arc cut $\mathbf{K} = (U, \overline{U})$ such that $Val\, \eta = C(\mathbf{K})$, then η is a maximum legal flow and $\mathbf{K}$ is a minimum arc cut in $\overrightarrow{N}$.*

Proof Suppose η' is a maximum legal flow and $\mathbf{K}'$ is a minimum arc cut in $\overrightarrow{N}$. Then we have

$Val\,\eta \leq Val\,\eta'$ and $C(\mathbf{K}') \leq C(\mathbf{K})$.

Also from Theorem 7.1.14, we get $Val\,\eta' \leq C(\mathbf{K}')$.

Hence,

$Val\,\eta \leq Val\,\eta' \leq C(\mathbf{K}') \leq C(\mathbf{K})$.

But by assumption, $Val\,\eta = C(\mathbf{K})$.

Thus we get $Val\,\eta = Val\,\eta'$ and $C(\mathbf{K}') = C(\mathbf{K})$. That is, η is a maximum legal flow and $\mathbf{K}$ is a minimum arc cut in $\overrightarrow{N}$. ■

We want to enhance the legal flow in a DFIN $\overrightarrow{N}(\sigma, \mu, \psi)$ to improve system efficiency. In every DFIN, a maximum legal flow may always be found. However, the challenge is to determine if a particular legal flow is maximum or not. For this purpose, we operate on the underlying fuzzy incidence network $\tilde{G}(\sigma, \mu, \Psi)$ of $\overrightarrow{N}$. $\tilde{G}$ is obtained by replacing the directed arcs and d-pairs of $\overrightarrow{N}$ by edges and incidence pairs with no direction. Let $A(\overrightarrow{N})$ be the arc set of $\overrightarrow{N}$.

Definition 7.1.16 Let P be a fuzzy incidence path in $\tilde{G}(\sigma, \mu, \Psi)$ with $E(P) = \{mn \in \mu^* : mn \text{ is an edge in } P\}$. For each edge $mn \in E(P)$, define the **enhancement factor**, $e(mn)$ by

$$e(mn) = \begin{cases} i_s(mn) - \eta(mn) & \text{if } mn \in A(\overrightarrow{N}) \\ \eta(nm) & \text{if } nm \in A(\overrightarrow{N}) \end{cases}$$

Then, the **enhancement factor of** P is defined by

$e(P) = \wedge\{e(mn) : mn \in E(P)\}$.

Clearly, $e(P) \geq 0$. The quantity $e(P)$ gives the maximum value by which the legal flow η through P can be enhanced under the definition of η.

Definition 7.1.17 Let P be a fuzzy incidence path in $\tilde{G}(\sigma, \mu, \Psi)$. Then P is

1. η - **saturated** if $e(P) = 0$.
2. η - **unsaturated** if $e(P) > 0$.

Definition 7.1.18 A path P in $\tilde{G}(\sigma, \mu, \Psi)$ is said to be an η - **enhancing path** if P is an η - unsaturated path from the source s to sink t.

The following example gives an idea of the terminologies defined above.

Example 7.1.19 Figure 7.6 depicts the underlying fuzzy incidence network $\tilde{G}(\sigma, \mu, \Psi)$ of the DFIN $\overrightarrow{N}(\sigma, \mu, \psi)$ given in Fig. 7.4 of Example 7.1.10.

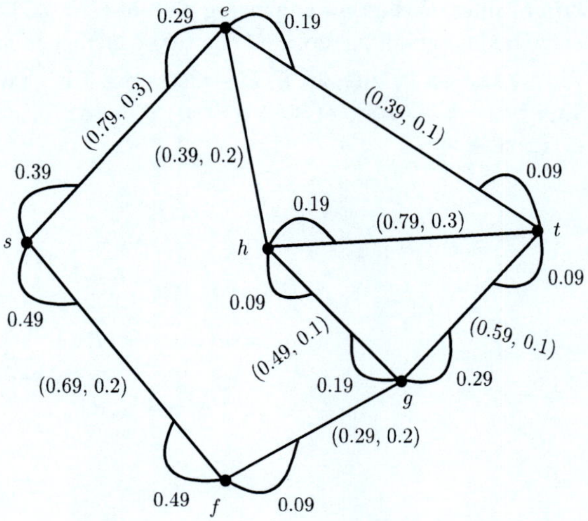

Fig. 7.6 Underlying fuzzy incidence network $\tilde{G}(\sigma, \mu, \Psi)$ of $\overrightarrow{N}(\sigma, \mu, \psi)$

In $\tilde{G}(\sigma, \mu, \Psi)$, choose the fuzzy incidence path $R : s, (s, se), se, (e, se), e, (e, et), et, (t, et), t$. Then

$e(se) = i_s(se) - \eta(se) = 0.4 - 0.3 = 0.1$ and

$e(et) = i_s(mt) - \eta(mt) = 0.2 - 0.1 = 0.1.$

Hence, $e(R) = \wedge\{0.1, 0.1\} = 0.1$, which implies that R is an η - unsaturated path. Since R is an η - unsaturated path from s to t, it is an η - enhancing path in $\tilde{G}$.

The existence of an η - enhancing path in a DFIN $\overrightarrow{N}(\sigma, \mu, \psi)$ implies that the legal flow η is not maximum. Using the enhancement factor $e(P)$ for an η - enhancing path P in $\overrightarrow{N}$, it is possible to construct a new legal flow $\bar{\eta}$ in $\overrightarrow{N}$ such that $Val\,\bar{\eta} > Val\,\eta$. This new legal flow $\bar{\eta}$ is as follows.

Definition 7.1.20 Let $\overrightarrow{N}(\sigma, \mu, \psi)$ be a DFIN and P be an η - enhancing path in $\overrightarrow{N}$. Then the **enhanced legal flow based on** P, denoted by $\bar{\eta}$, is defined by

$$\bar{\eta}(mn) = \begin{cases} \eta(mn) + e(P) & \text{if } mn \in E(P) \text{such that mn } \in A(\overrightarrow{N}) \\ \eta(nm) - e(P) & \text{if } mn \in E(P) \text{such that nm } \in A(\overrightarrow{N}) \\ \eta(mn) & \text{if } mn \notin E(P) \end{cases}$$

Proposition 7.1.21 *Let $\overrightarrow{N}(\sigma, \mu, \psi)$ be a DFIN and P be an η - enhancing path in $\tilde{G}(\sigma, \mu, \Psi)$. Let $\bar{\eta}$, the enhanced legal flow based on P, be defined as Definition 7.1.20. Then $Val\,\bar{\eta} = Val\,\eta + e(P)$.*

Proof Since P is an η - enhancing path in $\tilde{G}(\sigma, \mu, \Psi)$, the source s belongs to P. Since nodes cannot be repeated in a fuzzy incidence path, exactly one edge incident with s, say sm belongs to $E(P)$. Also since s has no incoming arcs, $sm \in A(\overrightarrow{N})$. So, $\bar{\eta}(sm) = \eta(sm) + e(P)$ and $\bar{\eta}(sn) = \eta(sn)$ for all other outgoing arcs sn from s. Hence,

$$\begin{aligned} Val\,\bar{\eta} &= O_{\bar{\eta}}(\{s\}) \\ &= \bar{\eta}(sm) + \sum_{n \neq m} \bar{\eta}(sn) \\ &= \eta(sm) + e(P) + \sum_{n \neq m} \eta(sn) \\ &= Val\,\eta + e(P). \end{aligned}$$

■

Consider the following example, which shows how to develop an enhanced legal flow.

Example 7.1.22 In the DFIN $\overrightarrow{N}(\sigma, \mu, \psi)$ of Example 7.1.19, we found an η - enhancing path $R : s, (s, se), se, (e, se), e, (e, et), et, (t, et), t$ in with $e(R) = 0.1$. Now using R, we define a new legal flow $\bar{\eta}$ in $\overrightarrow{N}$ as

$\bar{\eta}(se) = \eta(se) + e(R) = 0.3 + 0.1 = 0.4$

$\bar{\eta}(et) = \eta(et) + e(R) = 0.1 + 0.1 = 0.2$

$\bar{\eta}(a) = \eta(a)$ for all other arcs $a \in \mu^*$.

Hence, $\bar{\eta}$ is an enhanced legal flow based on R with $Val\,\bar{\eta} = Val\,\eta + e(P) = 0.5 + 0.1 = 0.6$.

The DFIN $\overrightarrow{N}(\sigma, \mu, \psi)$ with enhanced legal flow $\bar{\eta}$ based on R is shown in Fig. 7.7.

The following theorem is crucial as it addresses a necessary and sufficient condition for a legal flow to become maximum.

Theorem 7.1.23 *Let η be a legal flow defined in a DFIN $\overrightarrow{N}(\sigma, \mu, \psi)$. Then η is a maximum legal flow if and only if $\overrightarrow{N}$ has no η - enhancing paths.*

Proof Suppose η is a maximum legal flow in $\overrightarrow{N}(\sigma, \mu, \psi)$. If possible, suppose that $\overrightarrow{N}$ has an η - enhancing path P. Then, it is possible to construct an enhanced legal flow $\bar{\eta}$ based on P with $Val\,\bar{\eta} > Val\,\eta$. This contradicts the fact that η is a maximum legal flow. Hence $\overrightarrow{N}$ cannot have an η - enhancing path.

Conversely, suppose that $\overrightarrow{N}$ has no η - enhancing path. Define a subset U of σ^* by

$U = \{$either $m = s$ or $\overrightarrow{N}$ has a $s - m$ unsaturated path$\}$.

Since $\overrightarrow{N}$ has no η - enhancing path, $\overrightarrow{N}$ has no $s - t$ unsaturated path. Thus $t \notin U$.

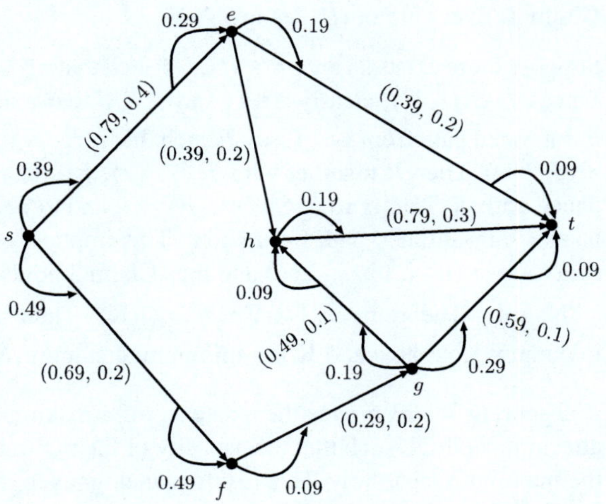

Fig. 7.7 $\overrightarrow{N}(\sigma, \mu, \psi)$ with enhanced legal flow $\bar{\eta}$ based on R

Claim 1. $\mathbf{K} = (U, \overline{U})$ is an arc cut in $\overrightarrow{N}$.

Clearly, $s \in U$ and $t \notin \overline{U}$. We need to prove that the removal of $\mathbf{K}$ from $\overrightarrow{N}$ disconnects $\overrightarrow{N}$ such that s and t belong to distinct components. Suppose not. Then s and t are connected in $\overrightarrow{N}$ - $\mathbf{K}$ by a $s - t$ directed incidence path, say P. Note that, since s is the source and t is the sink, the directed incidence path from t to s is not possible. Let w_j be the node just prior to t in P. Since $w_j t$ is an arc belonging to $\overrightarrow{N} - \mathbf{K}$ and $t \in \overline{U}$, we must have $w_j \notin U$ by the definition of arc cut. Now since $w_j \in \overline{U}$ and $w_{j-1}w_j$ belongs to $\overrightarrow{N} - \mathbf{K}$, we must have $w_{j-1} \notin U$. Similarly preceding, since $w_2 \in \overline{U}$ and $w_1 w_2$ belongs to $\overrightarrow{N} - \mathbf{K}$, we must have $w_1 \notin U$. Now since $w_1 \in \overline{U}$ and sw_1 belongs to $\overrightarrow{N} - \mathbf{K}$, we must have $s \notin U$, a contradiction. Hence s and t must be disconnected in $\overrightarrow{N} - \mathbf{K}$. Thus Claim 1 holds.

Claim 2. Every arc in $(U, \overline{U})$ is η - saturated.

Suppose there exists an arc $mn \in (U, \overline{U})$ such that mn is η - unsaturated. Since $mn \in (U, \overline{U})$, clearly $m \in U$ and $n \in \overline{U}$. Since $m \in U$, there exists an η - unsaturated path from s to m, say Q with $e(Q) > 0$. Now, mn is η - unsaturated implies $\eta(mn) < i_s(mn)$. Also, $mn \in A(\overrightarrow{N})$. Hence $e(mn) > 0$. So, the fuzzy incidence path Q together with m, (m, mn), mn, (n, mn), n forms a new fuzzy incidence path Q' from s to n such that $e(Q) > 0$. So, Q' is an η - unsaturated path from s to n. This implies, $n \in U$, a contradiction. Hence, every arc in $(U, \overline{U})$ is η - saturated and thus Claim 2 holds.

Claim 3. Every arc in $(\overline{U}, U)$ is η - zero.

Suppose there exists an arc $ef \in (U, \overline{U})$ such that ef is η - positive. Then $\eta(ef) > 0$. Since $ef \in (\overline{U}, U)$, clearly $e \in \overline{U}$ and $f \in U$. Now since $f \in U$, there exists an η - unsaturated path from s to f, say R such that $e(R) > 0$. Since $ef \in A(\overrightarrow{N})$, $e(fe) = \eta(ef) > 0$. Then R together with f, (f, fe), fe, (e, fe), e forms a new fuzzy incidence path R' from s to e. Since $e(R) > 0$ and $e(fe) > 0$, $e(R') > 0$. That is, R' is an η - unsaturated path from s to e. This implies, $e \in U$, a contradiction. Hence, every arc in $(\overline{U}, U)$ is η - zero and thus Claim 3 holds.

Now by Theorem 7.1.14, $Val\, \eta = C(\mathbf{K})$. Then by Proposition 7.1.15, η is a maximum legal flow and $\mathbf{K}$ is a minimum arc cut in $\overrightarrow{N}$. ■

Theorem 7.1.23 proves the existence of a maximum legal flow and a minimum arc cut in a DFIN, such that the capacity of the minimum arc cut equals the value of the maximum legal flow. This result is analogous to the max-flow min-cut theorem in graph theory.

Theorem 7.1.24 *If $\overrightarrow{N}(\sigma, \mu, \psi)$ is any DFIN, then the value of a maximum legal flow equals to the capacity of a minimum arc cut in $\overrightarrow{N}$.*

Proof The proof follows directly from the proof of Theorem 7.1.23. ■

Hence using these results, we can conclude that the legal flow $\bar{\eta}$ obtained in Example 7.1.22 with $Val\, \bar{\eta} = 0.6$ is maximum. Because, we have obtained a minimum arc cut $\mathbf{K}$ for the same DFIN with capacity 0.6 in Example 7.1.12.

Now, the remaining question is how to determine a maximum legal flow in any DFIN without finding a minimum arc cut. For this reason, we create an algorithm that is inspired by the Ford-Fulkerson method [2].

7.2 Algorithm to Find a Maximum Legal Flow in a DFIN

Let $\overrightarrow{N}(\sigma, \mu, \psi)$ be a DFIN with source s and sink t. To find a maximum legal flow, start with a known legal flow (otherwise, consider the zero legal flow) in $\overrightarrow{N}$. In each stage, we identify whether an enhanced path based on the existing legal flow is present or not. If an enhancing path is present, then the current legal flow is enhanced based on that enhancing path. If the enhancing path is absent, then the current legal flow is maximum. That is, in each stage, we recursively build up an enhanced legal flow based on the enhancing path existing in the DFIN. The process gets continued up to the stage where there is no enhancing path exists. Then that legal flow will be the maximum legal flow possible in the given DFIN. So, the main procedure involved is the identification of η - enhancing path based on the available legal flow η in each stage. We apply the following procedure to examine the presence of an η - enhancing path based on the legal flow η in each of the stages.

For this purpose, we need to define a new sub DFIG $\overrightarrow{H}(\tau, \nu, \chi)$ of $\overrightarrow{N}$ having the legal flow η as follows.

1. $s \in \tau^*$.
2. For each $m \in \tau^*$, there exists a unique $s - m$ fuzzy incidence path in the underlying fuzzy incidence graph $\tilde{H}(\tau, \nu, \chi)$ of $\overrightarrow{H}$.
3. For each $m \in \tau^*$, the unique $s - m$ fuzzy incidence path in $\tilde{H}$ is η - unsaturated.

Seeking an enhanced path in each stage involves the enhancement of $\overrightarrow{H}(\tau, \nu, \chi)$ in $\overrightarrow{N}(\sigma, \mu, \psi)$. At first, let s be the only node available in τ^*. The enlargement of $\overrightarrow{H}(\tau, \nu, \chi)$ is possible in the following ways.

Let $U = V(\overrightarrow{H})$, the node set of $\overrightarrow{H}$.

(a) If uv is an arc in $(U, \overline{U})$ such that uv is η - unsaturated, then uv and v are adjoined to $\overrightarrow{H}$.
(b) If uv is an arc in $(\overline{U}, U)$ such that uv is η - positive, then uv and u are adjoined to $\overrightarrow{H}$.

Clearly, the above two possibilities cause the enlargement of $\overrightarrow{H}$. Eventually performing the steps (a) and (b), there are two possibilities for $\overrightarrow{H}$. Either,

(i) $t \in V(\overrightarrow{H})$ or
(ii) $\overrightarrow{H}$ stops enlarging before arriving at t.

In case (i), there will be a unique $s - t$ unsaturated path in $\overrightarrow{H}$, which is the required η - enhancing path. Then enhance the legal flow η using this path according to Definition 7.1.20. Again continue the construction and enlargement of $\overrightarrow{H}$ for the new enhanced legal flow.

In case (ii), since $\overrightarrow{H}$ did not reach at t, there will not be an $s - t$ unsaturated path. Hence by Theorem 7.1.23, η will be the maximum legal flow.

Illustration of Algorithm 7.2 Consider the DFIN $\overrightarrow{N}(\sigma, \mu, \psi)$ with initial legal flow η_0 given in Fig. 7.8. It has source s, sink t, and four intermediate nodes e, f, g, h. For convenience, we take the capacity of all nodes as 1. The capacity and legal flow values of the arcs as well as the weight of d-pairs are defined as in the figure. Then

$$Val\, \eta_0 = O_{\eta_0}(\{s\}) = 0.35 + 0.15 + 0.35 = 0.85.$$

Stage 1. To examine the presence of an η_0 enhancing path, we define $\overrightarrow{H}(\tau, \nu, \chi)$ as in Algorithm 7.2 and carry the enlargement process of $\overrightarrow{H}$. This process has 4 stages and each stage is illustrated in Fig. 7.8(a). The enlargement of $\overrightarrow{H}$ is represented with thicker arcs.

As a result, we obtain a situation as in Case (i). That is, $t \in V(\overrightarrow{H})$ and there is a unique $s - t$ unsaturated path in $\overrightarrow{H}$, which is an η_0 - enhancing path in $\overrightarrow{N}$. So, we get the required η_0 - enhancing path $P_1 : s, (s, sg), sg, (g, sg), g, (g, ge), ge, (e, ge), e, gh, (h, gh), h, (h, ht), ht, (t, ht), t$ with $e(P_1) = \wedge\{(0.6 - 0.2) - 0.15, (0.7 -$

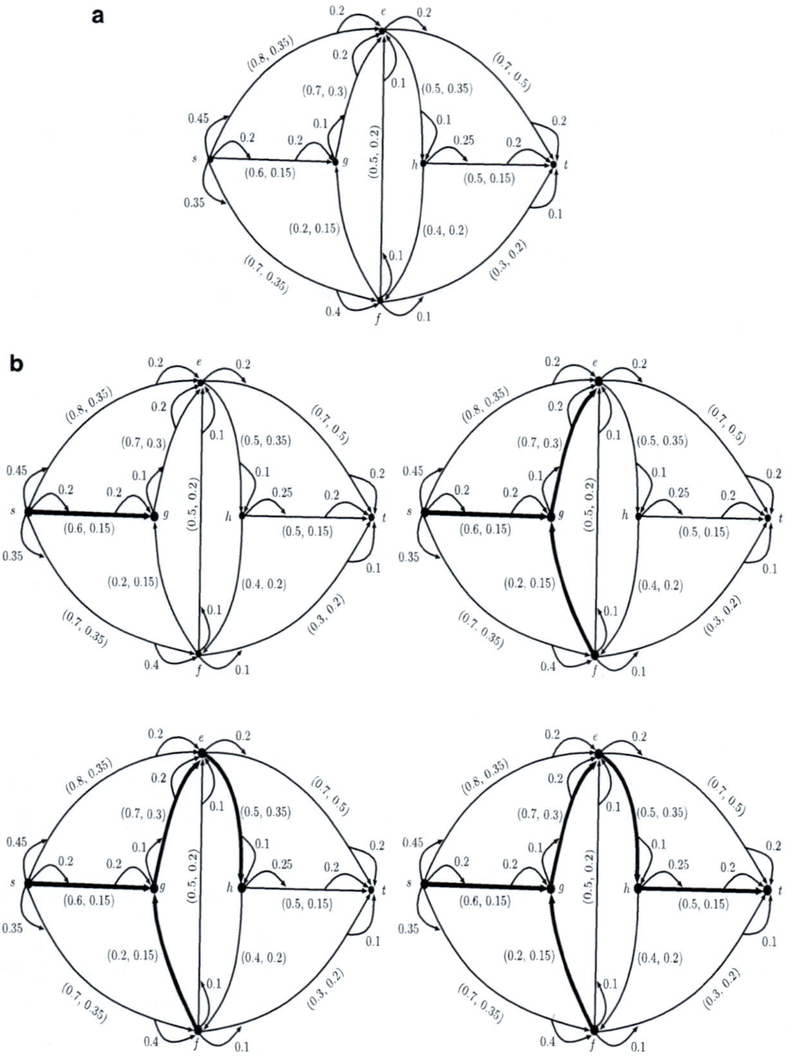

Fig. 7.8 $\overrightarrow{N}(\sigma, \mu, \psi)$ with initial legal flow η_0. **a**. Four stages in the enlargement of $\overrightarrow{H}$ of Stage 1

$0.2) - 0.3, (0.5 - 0.1) - 0.35, (0.5 - 0.25) - 0.15\} = \wedge\{0.25, 0.2, 0.05, 0.1\} = 0.05$.

Hence by Definition 7.1.20, an enhanced legal flow $\bar{\eta_1}$ based on P_1 can be defined as follows.

$$\bar{\eta_1}(sg) = \eta_0(sg) + e(P_1) = 0.15 + 0.05 = 0.2$$

$$\bar{\eta_1}(ge) = \eta_0(ge) + e(P_1) = 0.3 + 0.05 = 0.35$$

$$\bar{\eta_1}(eh) = \eta_0(eh) + e(P_1) = 0.35 + 0.05 = 0.4$$

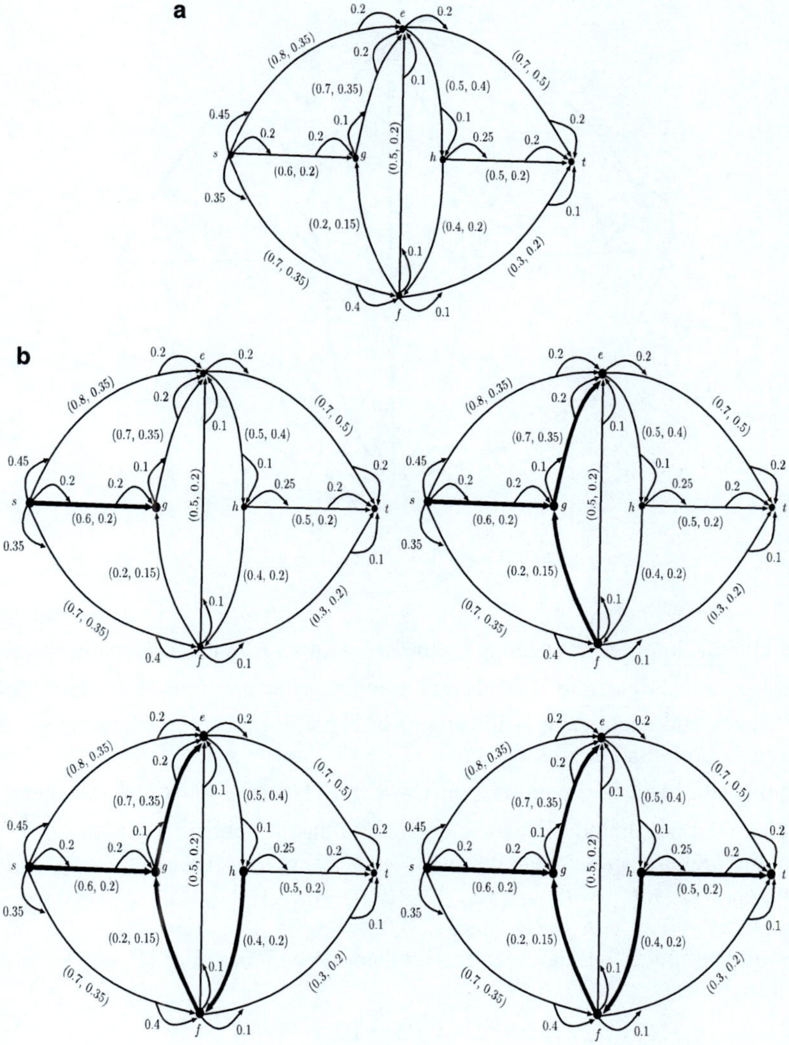

Fig. 7.9 $\overrightarrow{N}(\sigma, \mu, \psi)$ with enhanced legal flow $\bar{\eta_1}$ based on P_1. (a).Four stages in the enlargement of $\overrightarrow{H}$ of Stage 2

$$\bar{\eta_1}(ht) = \eta_0(ht) + e(P_1) = 0.15 + 0.05 = 0.2 \text{ and}$$

$$\bar{\eta_1}(a) = \eta_0(a) \text{ for all other arcs } a \in \mu^*.$$

Then $Val\,\bar{\eta_1} = Val\,\eta_0 + e(P_1) = 0.85 + 0.05 = 0.9$.

The DFIN $\overrightarrow{N}(\sigma, \mu, \psi)$ with enhanced legal flow $\bar{\eta_1}$ based on P_1 is given in Fig. 7.9.

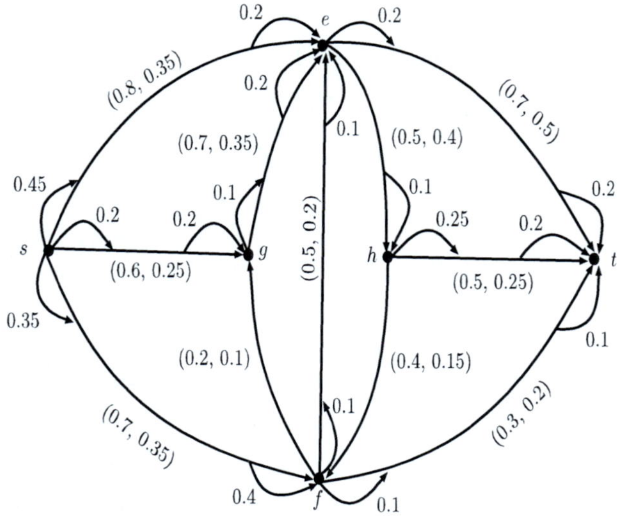

Fig. 7.10 $\overrightarrow{N}(\sigma, \mu, \psi)$ with enhanced legal flow $\bar{\eta_2}$ based on P_2

Stage 2. Now, in order to examine the presence of an $\bar{\eta_1}$ enhancing path, we define $\overrightarrow{H}(\tau, \nu, \chi)$ as in Algorithm 7.2 and carry the enlargement process of $\overrightarrow{H}$. This process has 4 stages and each stage is illustrated in Fig. 7.9(a). The enlargement of $\overrightarrow{H}$ is represented with thicker arcs.

Again we obtain a situation as in Case (i). That is, $t \in V(\overrightarrow{H})$ and there is a unique $s - t$ unsaturated path in $\overrightarrow{H}$, which is an $\bar{\eta_1}$ - enhancing path in $\overrightarrow{N}$. So, we get the required $\bar{\eta_1}$ - enhancing path $P_2 : s, (s, sg), sg, (g, sg), g, gf, f, fh, h, (h, ht), ht, (t, ht), t$ with $e(P_2) = \wedge\{(0.6 - 0.2) - 0.2, 0.2 - 0.15, 0.4 - 0.2, (0.5 - 0.25) - 0.2\} = \wedge\{0.2, 0.05, 0.2, 0.05\} = 0.05$.
Hence by Definition 7.1.20, an enhanced legal flow $\bar{\eta_2}$ based on P_2 can be defined as follows.

$$\bar{\eta_2}(sg) = \bar{\eta_1}(sg) + e(P_2) = 0.2 + 0.05 = 0.25$$
$$\bar{\eta_2}(gf) = \bar{\eta_1}(gf) - e(P_2) = 0.15 - 0.05 = 0.1$$
$$\bar{\eta_2}(fh) = \bar{\eta_1}(fh) - e(P_1) = 0.2 - 0.05 = 0.15$$
$$\bar{\eta_2}(ht) = \bar{\eta_1}(ht) + e(P_1) = 0.2 + 0.05 = 0.25 \quad \text{and}$$
$$\bar{\eta_2}(a) = \bar{\eta_1}(a) \quad \text{for all other arcs } a \in \mu^*.$$

Then $Val\,\bar{\eta_2} = Val\,\eta_1 + e(P_2) = 0.9 + 0.05 = 0.95$.

The DFIN $\overrightarrow{N}(\sigma, \mu, \psi)$ with enhanced legal flow $\bar{\eta_1}$ based on P_1 is given in Fig. 7.10.

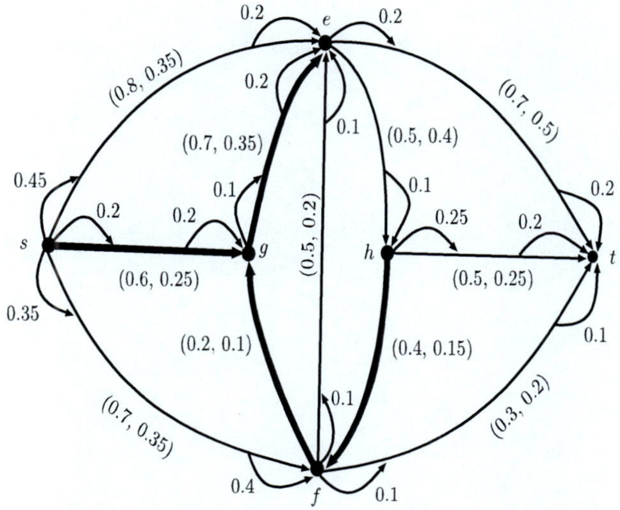

Fig. 7.11 Final stage of the enlargement of $\overrightarrow{H}$

But in the next stage of construction of $\overrightarrow{H}$, Case (ii) of the algorithm happens. That is, $\overrightarrow{H}$ stops enlarging before arriving at t. This is illustrated below in Fig. 7.11. Since $\overrightarrow{H}$ did not reach at t, there will not be an $s - t$ unsaturated path. Hence by Theorem 7.1.23, $\bar{\eta_2}$ is the maximum legal flow in $\overrightarrow{N}$.

References

1. Gayathri, G., Mathew, S., Mordeson, J. N.: Max-flow min-cut theorem for directed fuzzy incidence networks. Fuzzy Sets and Systems (Under Review)
2. Ford, L.R., Fulkerson, D.R.: Maximal flow through a network. Can. J. Math. **8**, 399–404 (1956)

Chapter 8
Cycle Connectivity of Fuzzy Graphs with Applications

Cyclic reachability is a novel concept connected to the dynamics of networks. Cyclic connectivity determines cyclic reachability, in terms of strong cycles available in the network. Different aspects of cyclic connectivity are discussed in this chapter. Structures like cycles, Cartesian products of graphs and blocks are examined for this purpose. Boost edges and fair fuzzy graphs are also discussed in detail. Also, an application of these concepts related with migration chains is proposed towards the end.

8.1 Cycle Connectivity of Fuzzy Graphs

Cycle connectivity is a measure of connectedness of a fuzzy graph. Cycle connectivity between a pair of vertices u and v is always bounded by the strength of connectedness between u and v. Most of the results discussed in this section are related to the definition given by Mathew and Sunitha in [1]. The following basic results from [1] are relevant for further development.

Theorem 8.1.1 *A fuzzy graph G is a fuzzy tree if and only if $CC(G) = 0$.*

Theorem 8.1.2 *Let G be a complete fuzzy graph with vertices $v_1, v_2, \ldots, v_n$ such that $\sigma(v_i) = t_i$ and $t_1 \leq t_2 \leq \cdots \leq t_{n-2} \leq t_{n-1} \leq t_n$. Then $CC(G) = t_{n-2}$.*

Using Theorem 8.1.4, we arrived at the following result.

Corollary 8.1.3 *For every $t \in (0, 1]$, there exists a complete fuzzy graph G with $CC(G) = t$.*

Theorem 8.1.4 *Cycle connectivity of a weak fuzzy cycle G is either 0 or the minimum weight of strong edges in G.*

Proof For a weak fuzzy cycle G, there exists a non zero real number $t \in (0, h(\mu)]$ such that G^t is a cycle, which implies that any $e \in G \setminus G^t$ is a δ- edge and for any

J. N. Mordeson et al., *Fuzzy Graph Theory*, Studies in Fuzziness and Soft Computing 424, https://doi.org/10.1007/978-3-031-23108-7_8

arbitrary number of vertices of G, there is a unique cycle G^t passing through them with $s(G^t) > t$. Thus,

$$CC(G) = \begin{cases} \wedge\{\text{Weight of strong edges in G}\} & \text{if } G^t \text{ is a strong cycle} \\ 0 & \text{if } G^t \text{ is a fuzzy tree.} \end{cases}$$

■

Next we have a proposition whose proof is obvious.

Proposition 8.1.5 *$F = (\sigma, \mu)$ is a fuzzy tree if and only if $CC(F - uv) = CC(F) = 0$ for any $uv \in \mu^*$.*

Theorem 8.1.6 *If $G = (\sigma, \mu)$ is a fuzzy graph such that for any pair $x, z \in \sigma^*$, both x and z lie on a common strongest strong cycle, then $CC(G) = \vee\{CONN_G(x, z) : x, z \in \sigma^*\}$.*

Proof Assume that any $x, z \in \sigma^*$ lie on a strongest strong cycle. We have $\theta(x, z) = \{\alpha \in (0, 1] : \alpha$ is the strength of a strong cycle passing through x and $z\}$ and $C^G_{x,z} = \vee\{\alpha : \alpha \in \theta(x, z), x, z \in \sigma^*\}$. Thus, $C^G_{x,z}$ has a unique value, $C^G_{x,z} = CONN_G(x, z)$. Hence $CC(G) = \vee\{CONN_G(x, z) : x, z \in \sigma^*\}$. ■

Next we compare the cycle connectivity of two different threshold fuzzy graphs.

Theorem 8.1.7 *If $0 < t_1 \leq t_2 \leq 1$, then for any fuzzy graph G, $CC(G^{t_2}) \leq CC(G^{t_1})$.*

Proof For any fuzzy graph G, $G^{t_2} \subseteq G^{t_1}$, $0 < t_1 \leq t_2 \leq 1$. Thus, we have $CONN_{G^{t_2}}(a_1, a_2) \leq CONN_{G^{t_1}}(a_1, a_2)$. Cycle connectivity between any pair of vertices is always less than or equal to its strength of connectedness, $C^{G^{t_2}}_{a_1,a_2} \leq C^{G^{t_1}}_{a_1,a_2}$. Hence, $CC(G^{t_2}) \leq CC(G^{t_1})$. ■

Corollary 8.1.8 *If $0 < t_1 \leq t_2 \leq \cdots \leq t_k \leq 1$, then for any fuzzy graph G, $CC(G^{t_1}) \leq CC(G^{t_2}) \leq \cdots \leq CC(G^{t_k})$.*

In Theorem 8.1.9, we discuss the cycle connectivity of the Cartesian product of two fuzzy graphs. To avoid confusion, sometimes we use the notation $(\mu(x, y)$ instead of $\mu(xy)$ in the proof.

Theorem 8.1.9 *Let $G' = (\sigma', \mu')$ and $G'' = (\sigma'', \mu'')$ be two fuzzy graphs with $\sigma'(u) = \sigma''(t) = 1 \; \forall \; u \in \sigma'^*$ and $t \in \sigma''^*$ and have maximum weight of edges in G' and G'' are k and m, respectively with $k \leq m$. Then cycle connectivity of the Cartesian product $G' \times G'' = \vee\{CC(G'), CC(G''), k\}$.*

Proof Since G' and G'' are two fuzzy graphs having $|\sigma'^*| = n_1$ and $|\sigma''^*| = n_2$, we have Cartesian product $G' \times G'' = (\sigma'^* \times \sigma''^*, \mu' \times \mu'')$ with $(\sigma' \times \sigma'')(u) = 1 \, \forall \, u \in (\sigma' \times \sigma'')^*$, and $|(\sigma' \times \sigma'')^*| = n_1 n_2$. Let the vertex set of G' be $u'_1, u'_2, \ldots, u'_{n_1}$, and the vertex set of G'' be $u''_1, u''_2, \ldots, u''_{n_2}$. If there exists a cycle C'_1 in G' with $s(C'_1) = s_1$, then there corresponds exactly n_2 copies of cycles in $G' \times G''$

whose strengths equal to s_1. If possible, $C_1' : u_1', u_2', \ldots, u_k',\ k \le n_1$; then, $C_1 : u_l''u_1', u_l''u_2', \cdots, u_l''u_k',\ k \le n_1, l \le n_2$ is an arbitrary cycle in $G = G' \times G''$, and by the definition of cross products $s(C_1) = s(C_1') = s_1$. Since C_1 is a random cycle, we can find such n_2 copies of cycles in G. Similarly in the case for each cycle C_1'' with strength s_2 there are exactly n_1 copies of cycles in $G' \times G''$ with same strength as that of C_1''.

Now we claim that all such cycles corresponding to a strong cycle in G' is strong. Let $C_1' : u_1', u_2', \ldots, u_p',\ p \le n_1$ be a strong cycle in G'. Then we have to show the cycles corresponds to C_1' in G are also strong. Make C_1 to be one such cycle in G such that $C_1 : u_q''u_1', u_q''u_2', \ldots, u_q''u_p',\ p \le n_1, q \le n_2$. If possible, assume that C_1 is not strong, which implies there exists an edge in C_1 which is not strong. Let it be $e = (u_q''u_j', u_q''u_{j+1}'),\ j \le p$. That is, there exists a strongest path P between $u_q''u_j'$ and $u_q''u_{j+1}'$, such that $s(P) > (\mu' \times \mu'')(u_q''u_j', u_q''u_{j+1}')$. For reduction of complexity, P simply writes as $u_q''u_j', \ldots, u_q''u_{j+1}'$. If there is an edge in P of the form $(u_r''u_s', u_t''u_s')$ with $\mu''(u_r''u_t'') \le \mu'(u_j', u_{j+1}')$, then $s(P) \le (\mu' \times \mu'')(u_q''u_j', u_q''u_{j+1}')$, a contradiction.

Now removing the vertices of G' from the path P, we get a path P_1' in G' with $s(P_1') \le \mu'(u_j', u_{j+1}')$, which results in $s(C_1) \le s(P_1') \le (\mu' \times \mu'')(u_q''u_j, u_q''u_{j+1})$. Thus $e = (u_q''u_j', u_q''u_{j+1}'),\ j \le p$ is strong, implies that the cycle C_1 is strong. Thus there exists a cycle C_1 in G corresponding to a strong cycle C_k' in G' which contribute to the cyclic connectivity of G', such that $CC(G') = s(C_1)$. Similarly, we can see a cycle C_2 in G with $CC(G'') = s(C_2)$. Next we claim the existence of a strong cycle C in G with $s(C) = k$. For that, let $e' = (u_k'u_{k+1}')$ with maximum weight k in G' and $e'' = (u_m''u_{m+1}'')$ with maximum weight m in G'' having $k \le m$. Now using the definition of Cartesian product of two fuzzy graphs, we have a cycle $C : u_k'u_m', u_{k+1}'u_m', u_{k+1}'u_{m+1}', u_k'u_{m+1}', u_k'u_m'$ in G such that $s(C) = k$. Now we claim that the cycle C is strong. On the contrary assume C is not strong, which implies the edge $(u_k'u_m'', u_{k+1}'u_m'')$ or the edge $(u_{k+1}'u_{m+1}'', u_k'u_{m+1}'')$ is a δ-edge. If possible let $(u_k'u_m'', u_{k+1}'u_m'')$ be a δ-edge. Then there exists a strongest path between $u_k'u_m''$ and $u_{k+1}'u_m''$. But any path connecting these two vertices should pass through at least one edge of the form $(u_p'u_q'', u_r'u_q'')$ and by definition of the Cartesian product $(\mu' \times \mu'')(u_p'u_q'', u_r'u_q'') \le (\mu' \times \mu'')(u_k'u_m'', u_{k+1}'u_m'')$. Thus any strongest path P between $u_k'u_m''$ and $u_{k+1}'u_m''$ has $s(P) = \mu''(u_k'u_m'', u_{k+1}'u_m'')$. Hence our assumption is wrong, and C is strong, which completes the proof (Fig. 8.1). ∎

Theorem 8.1.9 is illustrated using the figures given below. Here, Fig. 8.2 is a product of two fuzzy graphs given in Fig. 8.1.

A fuzzy graph is said to be a block if it has no fuzzy cutvertices [2]. Figure 8.3 is an example of a fuzzy block.

Theorem 8.1.10 *Let $G = (\sigma, \mu)$ be a block fuzzy graph. Then* $CC(G) = \vee\{CONN_G(x, y) : xy$ *is not an* $\alpha -$ *strong edge in* $G\}$.

Proof In a fuzzy graph G, none of the α—strong edges contribute to its cycle connectivity. Using Theorem 1.2.59 for a block fuzzy graph, any two vertices x and y such that xy is not a fuzzy bridge are joined by two internally disjoint strongest paths say,

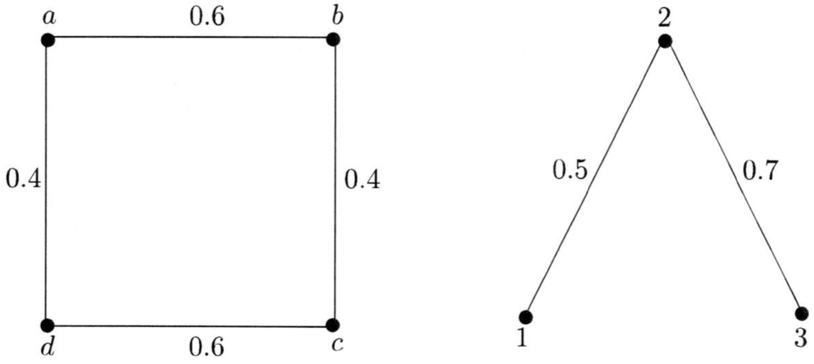

Fig. 8.1 Fuzzy graphs G' and G'' with $k = 0.6$ and $m = 0.7$

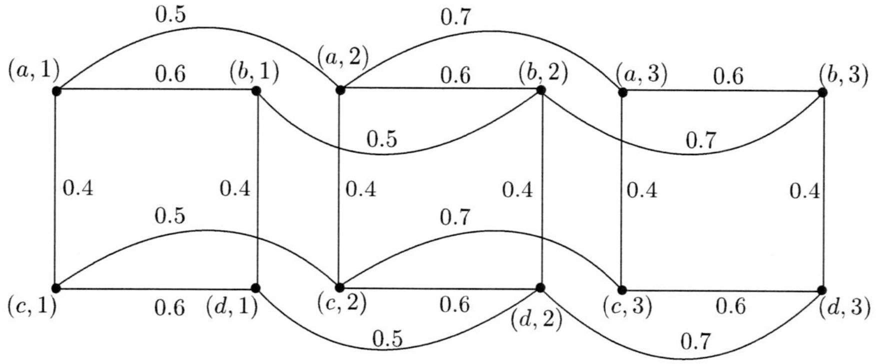

Fig. 8.2 Fuzzy graph $G = G \times G''$ with $CC(G) = 0.5$

Fig. 8.3 A block fuzzy graph

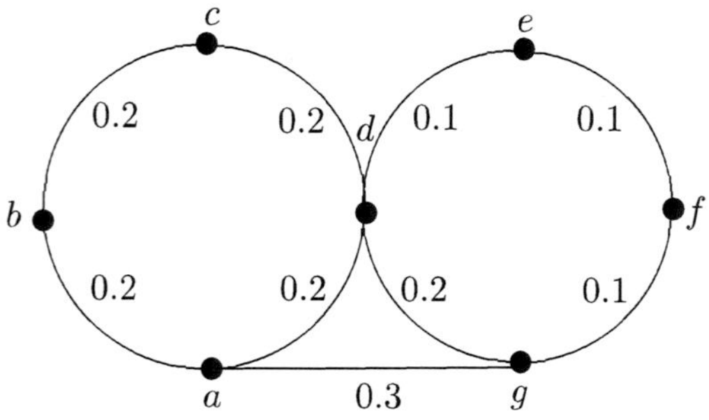

P_1 and P_2. Then $C = P_1 \cup P_2$ is a strong cycle in G with $s(C) = CONN_G(x, y)$. Since cycle connectivity is the maximum strength of all strong cycles, proof follows. ■

Theorem 8.1.11 *Isomorphic fuzzy graphs have the same cycle connectivity.*

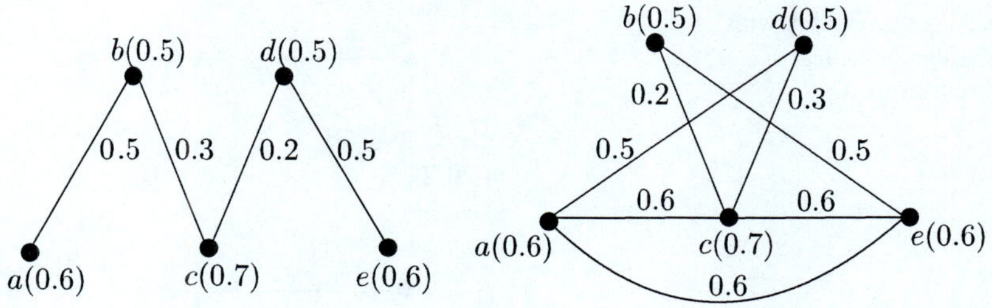

Fig. 8.4 A fuzzy graph G and its complement

Proof Let $G = (\sigma, \mu)$ and $G' = (\eta, \xi)$ be two isomorphic fuzzy graphs. Then by definition, there exists a bijective function from the vertex set of G to the vertex set of G' given by $f : \sigma^* \to \eta^*$ such that,

$$\sigma(w) = \eta(f(w)), \quad \forall w \in \sigma^* \quad \text{and}$$
$$\mu(xy) = \xi(f(x)f(y)), \quad \forall x, y \in \sigma^*.$$

Let $\sigma^* = \{w_1, w_2, \ldots, w_n\}$ and let $f(w_i) = x_i$ for $i = 1, 2, \ldots, n$. Then,

$$\mu\left(w_i w_j\right) = \xi\left(f\left(w_i\right) f\left(w_j\right)\right) = \xi\left(x_i x_j\right).$$

Let C be a strong fuzzy cycle of G with $s(C_1) = CC(G)$. Without loss of generality, assume $C_1 : w_1, w_2, \ldots, w_k,\ k \leq n$. Then $C_1' : x_1, x_2, \ldots, x_k,\ k \leq n$, is a fuzzy cycle in G'. Here C' is a strong cycle in G'. If not, there exists a non strong edge in C_1' and let it be $x_i x_{i+1}$. Then, there exists a strongest path P' between x_i and x_{i+1}, having $s(P') > \xi(x_i x_{i+1})$ and without loss of generality take P' as $x_j, x_{j+1}, \ldots, x_{j+p},\ j + p \leq n$. Since G and G' are isomorphic, there exists a one to one correspondence between $x_i x_j$ path in G' and $w_i w_j$ path in G. Thus P' is a strongest path in G' if and only if there exist $w_i - w_j$ strongest path P in G with $s(P) > \mu(w_i w_j)$, which is a contradiction. So the cycle C_1' is strong.

Now suppose C_2' is another strong cycle in G' such that $s(C_2') > s(C_1')$. Let $C_2' : x_1, x_2, \ldots, x_l,\ l \leq n, f(w_i) = x_i,\ i = 1, 2, \ldots, l$. Then there exists a strong cycle $C_2 : w_1, w_2, \ldots, w_l,\ l \leq n$. Thus, $s(C_2) > s(C_1) = CC(G)$, a contradiction. ■

The study of complement of a fuzzy graph was made in [3] (see Fig. 8.4). If a fuzzy graph G is isomorphic to its complement, then G is called a self-complementary fuzzy graph. Now using Theorem 8.1.11, we can obtain the cycle connectivity of self complementary fuzzy graphs.

Corollary 8.1.12 *If G is a self-complementary fuzzy graph then, $CC(G) = CC(G^c)$.*

Proposition 8.1.13 *Let $G = (\sigma, \mu)$ be a fuzzy graph. Then $CC(G) < CC(G')$, where G' is a CFG spanned by the vertex set of G.*

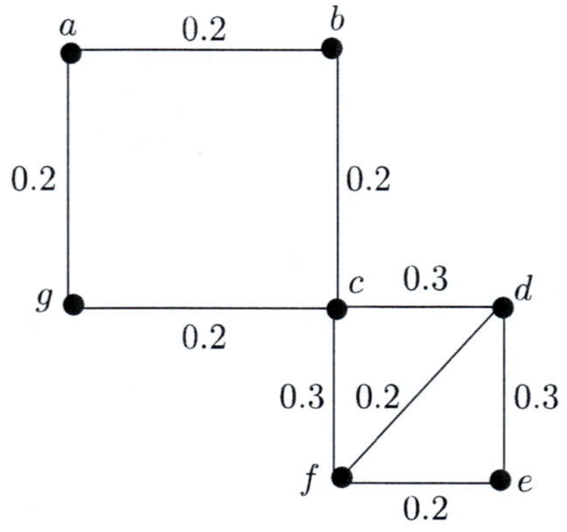

Fig. 8.5 A fuzzy graph with fuzzy bridge *bc* with a unique cutvertex *d*

Theorem 8.1.14 *Let G be an edge disjoint fuzzy graph. Then G has a unique strong cycle C with $s(C) = CC(G)$ if and only if it has more than one cyclic cutvertex.*

Proof Let G be an edge disjoint fuzzy graph with a unique strong cycle C and $s(C) = CC(G)$. Then the removal of any vertex from C reduces the $CC(G)$. That is, there are more than one cyclic cutvertices in C.

Conversely, assume that G has more than one cyclic cutvertices, and also assume that there is at least two strong cycles with maximum strength. Since the graph is edge disjoint, any two cycles of G have at most one common vertex. ■

Theorem 8.1.15 *If $CC(H) < CC(G)$ for a fuzzy graph H obtained by deleting an edge from G, then G has a unique strong cycle C with $s(C) = CC(G)$.*

Proof Let H be the fuzzy graph obtained by deleting an edge from G with $CC(H) < CC(G)$. If possible, assume that G has at least two strong cycles say C_1 and C_2 with $s(C_1) = s(C_2) = CC(G)$. Then there are two possibilities, C_1 and C_2 are vertex disjoint or have at least one common vertex. In both cases, the removal of any edge e from C_1 does not change the strength of C_2. Thus, $CC(G - e) = s(C_2) = CC(G)$, a contradiction. We can conclude that the graph has a unique strong cycle C with $s(C) = CC(G)$. ■

Proposition 1.2.53 states that, if an edge uv is a cyclic bridge in G, then both u and v are cyclic cutvertices. But it is not true in general. It is illustrated in Fig. 8.5 Here edge *bc* is a cyclic bridge but the vertex *c* is not a cyclic cutvertex. These type of vertices and edges will be discussed in the next section.

8.2 Cyclically Balanced and Cyclically Fair Fuzzy Graphs

Cut vertices and bridges affect the connectivity between certain pair of vertices in a fuzzy graph when they are removed. Also, cyclic cutvertices and cyclic bridges affect the cycle connectivity of a fuzzy graph on their removal. We know that the removal of a vertex or edge never enhances the connectivity between any pair of vertices. But it is not the case with cycle connectivity. There are vertices and edges whose removal increases the cycle connectivity between pairs of vertices and we name them as cyclic boost vertex and cyclic boost edge respectively.

A fuzzy graph is called **cyclically balanced** if it has no cyclic cutvertices or cyclic bridges. Now we provide some results on cyclically balanced fuzzy graphs, and prove the existence of certain types of cyclically balanced fuzzy graphs.

Theorem 8.2.1 *Let $G = (\sigma, \mu)$ be a CFG with $\sigma^* = \{b_1, b_2, \dots, b_m\}$ and $\sigma(b_i) = k_i$ for $i = 1, 2, \dots, m$ with $k_1 \le k_2 \le \cdots \le k_m$. Then G is cyclically balanced if and only if $k_{m-2} = k_{m-3}$.*

Proof Let $b_1, b_2, \dots, b_m \in \sigma^*$, $\sigma(b_i) = k_i$ for $i = 1, 2, \dots, m$ and $k_1 \le k_2 \le \cdots \le k_m$. Suppose that G is cyclically balanced. Then there is no $u \in \sigma^*$ or $e \in \mu^*$ whose removal reduces the cycle connectivity of G. Since G is CFG, all cycles are strong and the cycle $C = b_{m-2}b_{m-1}b_m b_{m-2}$ has $s(C) = k_{m-2}$. Now from Theorem 8.2.1, we have $CC(G) = k_{m-2}$. Hence the removal of vertex b_{m-2} or edge $e = b_{m-2}b_{m-1}$ changes the cycle connectivity of the graph to $k_{m-3} \le k_{m-2}$. The case $k_{m-3} < k_{m-2}$, contradicts to our assumption that G is cyclically balanced. Hence $k_{m-2} = k_{m-3}$.

Conversely, assume that $k_{m-2} = k_{m-3}$. Then there exist cycles $C_1 = b_m b_{m-1} b_{m-2} b_m$, $C_2 = b_m b_{m-1} b_{m-3} b_m$, $C_3 = b_{m-1} b_{m-2} b_{m-3} b_{m-1}$, and $C_4 = b_m b_{m-2} b_{m-3} b_m$, in G having the same strength. Thus the removal of any one of b_m, b_{m-1}, b_{m-2} or b_{m-3} will not reduce $CC(G)$. The edge case is similar. Hence the converse is proved. ■

Theorem 8.2.1 points out the existence of a non-trivial cyclically balanced fuzzy graph. So we have the following lemma.

Lemma 8.2.2 *For any $n \ge 4$, there is a connected cyclically balanced fuzzy graph with $|\sigma^*| = n$.*

Theorem 8.2.3 *Complement of a fuzzy cycle $G = (\sigma, \mu)$ for $|\sigma^*| \ge 6$ and $\sigma(u) = \gamma\ \forall u \in \sigma^*$ is cyclically balanced.*

Proof Let G be a fuzzy cycle with $n \ge 6$. Then for every set of 5 vertices in G^c, there exists a strong fuzzy cycle C passing through them having $s(C) = \gamma = CC(G^c)$. Hence, the removal of any vertex from G^c never reduces $CC(G)$. In a similar way we can prove the edge case. ■

Definition 8.2.4 A vertex w in a fuzzy graph is called a **cyclic boost vertex** if $CC(G - w) > CC(G)$ or there exist at least one pair of vertices u, v such that $C_{u,v}^{G-w} > C_{u,v}^{G}$ and $C_{u,v}^{G-w} = CC(G)$. An edge e of a fuzzy graph is called a **cyclic boost edge** if $CC(G - e) > CC(G)$, or $C_{u,v}^{G-e} > C_{u,v}^{G}$ and $C_{u,v}^{G-e} = CC(G)$, for some pair of vertices u and v.

Definition 8.2.5 A vertex w in a fuzzy graph is called a **cyclic boost vertex of** G if $CC(G - w) > CC(G)$ and an edge e of a fuzzy graph is called a **cyclic boost edge** of G if $CC(G - e) > CC(G)$.

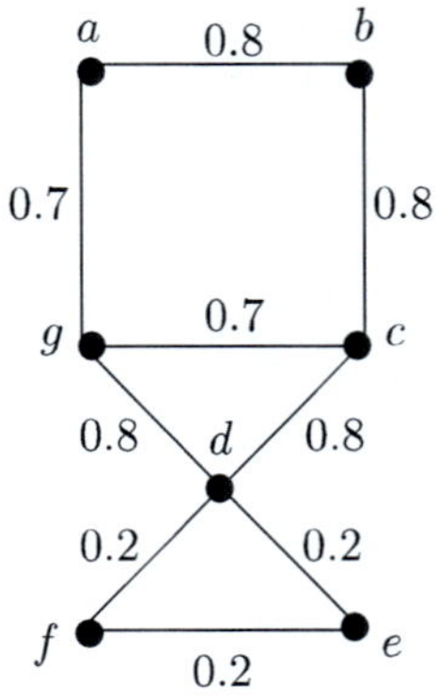

Fig. 8.6 Fuzzy graph with cyclic boost vertices and edges

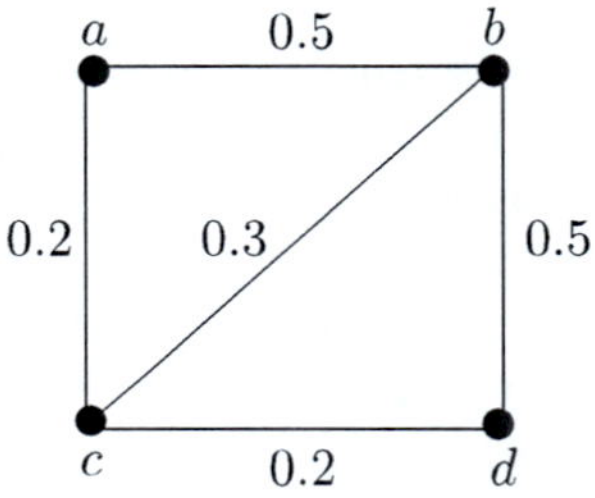

Fig. 8.7 A fuzzy graph with a boost edge bc but no boost vertex

Note that cyclic boost vertex of graph G is a special type of boost vertex in G.

Definition 8.2.6 A vertex w in a fuzzy graph is called a **local cyclic boost vertex** if there exist at least a pair of vertices u, v such that $C_{u,v}^{G-w} > C_{u,v}^{G}$ and an edge e of a fuzzy graph is called a **local cyclic boost edge** if $C_{u,v}^{G-e} > C_{u,v}^{G}$, for some pair of vertices u and v.

Example 8.2.7 Let $G = (\sigma, \mu)$ be the fuzzy graph with $\sigma^* = \{a, b, c, d, e, f, g\}$ and $\mu^* = \{ac, bc, cd, de, ef, fg, cg\}$. Let $\mu(ac) = 0.2$, $\mu(bc) = 0.5$, $\mu(cd) = 0.1$, $\sigma(x) = 1\ \forall x \in \sigma^*$ (Fig. 8.6). Then, $CC(G) = 0.2$ whereas $CC(G - de) = CC(G - fe) = CC(G - fd) = 0$ and $CC(G - cd) = CC(G - gd) = 0.7$. Thus de, fe and fd are cyclic bridges whereas cd and gd are cyclic boost edges. Also since $0.7 = C_{a,b}^{G-d} > C_{a,b}^{G} = 0$, d is a cyclic boost vertex.

Existence of cyclic boost vertices and cyclic boost edges are independent, we cannot say anything in term of the other. Figure 8.7 is an example of a fuzzy graph with no cyclic boost vertex, but has a cyclic boost edge.

Fig. 8.8 is a fuzzy graph without cyclic boost edges. But it has a cyclic boost vertex a.

Theorem 8.2.8 gives the relationship between weight of a boost edge and cycle connectivity of the subgraph obtained by deleting the boost edge.

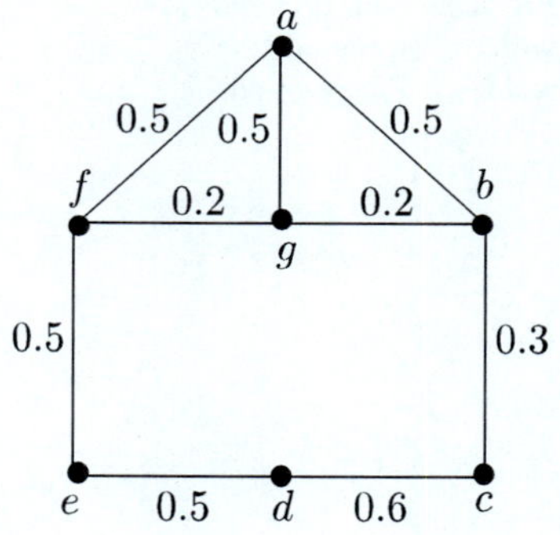

Fig. 8.8 A fuzzy graph G with no boost edges but have a boost vertex a

Theorem 8.2.8 *For a boost edge e in a fuzzy graph G,* $w(e) > CC(G - e)$.

Proof For a boost edge e in $G = (\sigma, \mu)$, $CC(G - e) > CC(G)$ or $C_{u,v}^{G-e} > C_{u,v}^{G}$ and $C_{u,v}^{G-e} = CC(G)$ for some $u, v \in \sigma^*$. Specifically, the removal of the boost edge e makes some cycle of $G - e$ strong. Let C_1 be one such strong cycle. Thus e is in the unique strongest path between the end vertices of some weakest edge e_1 in C_1, with $s(C_1) = \mu(e_1)$, which implies, $\mu(e) > \mu(e_1)$ and so, $s(C_1) < \mu(e)$. Let $C_1, C_2, \ldots, C_k$ be non strong cycles in G and strong in $G - e$, with $e_1, e_2, \ldots, e_k$, the set of weakest edges. Then by aforementioned argument, we have $\mu(e) > \mu(e_k)$, and $s(C_k) < \mu(e)$. Now using the definitions of cycle connectivity and a boost edge, we have $CC(G - e) = \vee\{C_{u,v}^{G-e} : u,\ v\ \in \sigma^*\} = \vee\{s(C) : C$ is a strong cycle in $G - e\}$, where C is not strong in $G = \vee\{\mu(e_i) : i = 1, 2, \ldots k\} < \mu(e)$. Hence the proof follows. ∎

Theorem 8.2.9 *Let* $G = (\sigma, \mu)$ *be a fuzzy graph. If* $CC(G - uv) = CC(G) \neq 0$ *for any* $uv \in \mu^*$, *then there exist at least two strong cycles with strength equals to the cycle connectivity of* G.

Proof Let $G = (\sigma, \mu)$ be a fuzzy graph with $CC(G - uv) = CC(G) \neq 0$ for any $uv \in \mu^*$. This means, the graph has at least one strong cycle and removal of any of its edge never reduce the cycle connectivity. If G has exactly one strong cycle C, then there is at least one edge e in C with $CC(G - e) < CC(G) \neq 0$. On the contrary, if for all edges e in C, $CC(G - e) \geq CC(G) \neq 0$, then e's are boost edges such that removal of e makes some δ-edge strong. Let E be the collection of such edges and $e = uv\ \in E$ be one such edge with minimum weight. Now, uv belongs to all strongest paths between x and y. Hence, $\mu(uv) \leq CONN_G(x, y) > \mu(xy)$ and $\mu(uv) < \mu(xy)$. Thus, any cycle C' containing the edge xy has $s(C') < s(C) = CC(G)$. Therefore no edge in C is a boost edge and so $CC(G - uv) < CC(G)$, which shows that the removal of an edge from a strong cycle with strength as cycle connectivity of G reduces its $CC(G)$. Therefore there exists at least two strong cycles with strength equals to the cycle connectivity of G. ∎

Converse of Theorem 8.2.9 need not be true. It is be shown in Fig. 8.9a. It has two strong cycles $C_1 = a - b - d - g - a$ and $C_2 = b - c - e - d - a$,

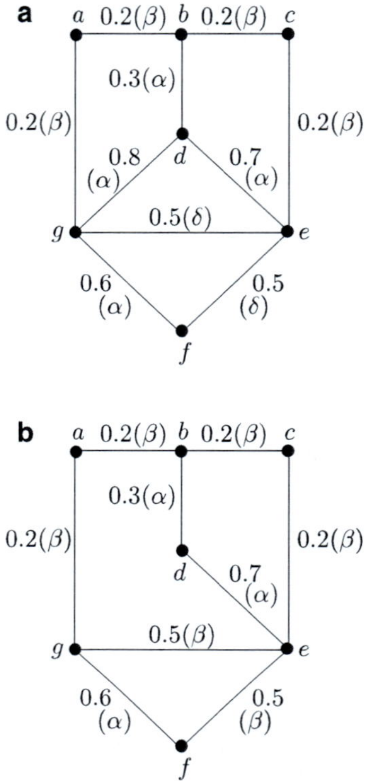

Fig. 8.9 **a** Fuzzy graph G with two strong cycles C_1 and C_2. **b** Fuzzy graph $G-dg$ with $CC(G-dg)=0.5$

where $s(C_1)=s(C_2)=CC(G)=0.2$. But $CC(G-dg)=0.5\neq CC(G)$ (Refer Fig. 8.9b).

Next theorem gives the relationship between a boost edge and a fuzzy bond.

Theorem 8.2.10 *Every local boost edge in a fuzzy graph G is a fuzzy bond.*

Proof Let $e=uv$ be a local boost edge in a fuzzy graph G. Then, $C^{G-e}_{u',v'} > C^{G}_{u',v'}$, for some pair of vertices u and v. This means that, the removal of e makes some cycle C containing u and v to a fuzzy cycle. It is equivalent to say that the weakest edge $e'=u'v'$ in C of G becomes strong in $G-e$. Implies, e is an edge of the unique strongest $u'-v'$ path in G. Thus, $CONN_{G-uv}(u',v') < CONN_G(u',v')$, for the pair of vertices u and v with at least one of them different from x and y. Hence, e is a fuzzy bond. ■

Theorem 8.2.10 shows that only a fuzzy bond can be a boost edge. But the converse is not true, which is clear from a fuzzy tree G, because no fuzzy bond of G is a local boost edge. Now we have the result; at least one of the end vertices of a fuzzy bond is a fuzzy cutvertex [4]. Applying this result in Theorem 8.2.10, we can derive the following corollary.

Corollary 8.2.11 *At least one end vertex of a local boost edge is a fuzzy cutvertex.*

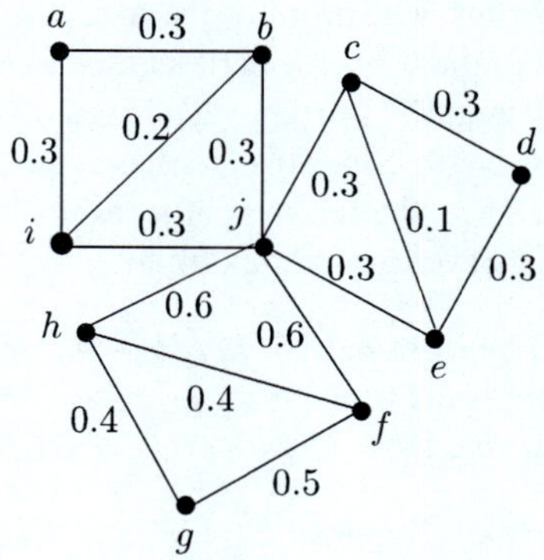

Fig. 8.10 A fuzzy graph G with cycle connectivity 0.3

Definition 8.2.12 A fuzzy graph G is said to be **cyclically fair** if it has no cyclic boost vertices and cyclic boost edges.

Theorem 8.2.13 *Let $G = (\sigma, \mu)$ be a cyclically fair fuzzy graph and uv be a cyclic bridge. Then both u and v are cyclic cutvertices.*

Proof Let $G = (\sigma, \mu)$ be a cyclically fair fuzzy graph and $e = uv$ be a cyclic bridge. Then the removal of uv reduces the cycle connectivity of G, which implies e is a common edge of every strong cycle C with $s(C) = CC(G)$. Since the fuzzy graph G is cyclically fair, no vertex of G is a boost vertex. Hence the removal of u or v do not increase the cycle connectivity of G. So, the removal of u may or may not reduce the cycle connectivity. But since the removal of u removes the edge $e = uv$, which is a cyclic bridge, u becomes a cyclic cutvertex. The case of v is similar. Hence follows the proof. ■

Theorem 8.2.14 proves the necessary condition for a vertex to be a cyclic cutvertex.

Theorem 8.2.14 *If a vertex u is a cyclic cutvertex of a fuzzy graph G, then u is a common vertex of every strong cycle passing through it.*

Proof Let G be a fuzzy graph. Let u be a cyclic cutvertex of G. Then $CC(G - u) < CC(G)$. That is, $\vee\{s(C') :$ where C' is a strong cycle in $G - u\} < \vee\{s(C) :$ where C is a strong cycle in $G\}$. Therefore, every strong cycle in G of maximum strength will be removed by the deletion of u. Hence, u is a common vertex of every strong cycle with maximum strength. Thus the proof of the theorem is completed. ■

Converse of Theorem 8.2.14 is not true in general. This is explained in Fig. 8.10 Cycle connectivity of G is 0.3, and the vertex j is a common vertex of all strong cycles, but j is not a cyclic cutvertex.

Theorem 8.2.15 *For a cyclically fair fuzzy graph G, a vertex is a cyclic cutvertex if and only if it is a common vertex of every strong cycle in G..*

Proof The proof of the first part follows from Theorem 8.2.14.
For the converse part, let u be a common vertex of every strong cycle with maximum strength. The removal of u results in the deletion of every strong cycle with maximum strength. Since the graph is cyclically fair, removal of u never makes any cycle strong. Hence, the removal of u results in the reduction of cycle connectivity of G. Thus, u is a cyclic cutvertex of G. ■

Theorem 8.2.16 *Let H be a fuzzy graph obtained by deleting an edge from a strong fuzzy cycle of a fuzzy graph G. Then $CC(H) < CC(G)$ if and only if G has a unique strong cycle C with $s(C) = CC(G)$.*

Proof Suppose $CC(H) < CC(G)$, for a fuzzy subgraph H obtained by deleting an edge e from a strong cycle say C in G. This implies e is a cyclic bridge and is a common edge of all strong cycles in G. Let $\mathbb{C}$ be the set of all strong cycles in G with $s(C) = CC(G)$. Now we need to show that $|\mathbb{C}| = 1$. On the contrary assume that $|\mathbb{C}| > 1$ and let $C_1,\ C_2\ \in \mathbb{C}$. Since $CC(H) < CC(G)$, e should be a common edge of both C_1 and C_2. Then $(C_1 \cup C_2) \setminus (C_1 \cap C_2)$ will be a strong fuzzy cycle with $S\left((C_1 \cup C_2) \setminus (C_1 \cap C_2)\right) = CC(G)$. Since $e \notin (C_1 \cup C_2) \setminus (C_1 \cap C_2)$ we have $(C_1 \cup C_2) \setminus (C_1 \cap C_2) \in H$, a strong cycle in H, a contradiction to our assumption. Thus, $|\mathbb{C}| = 1$.

Conversely, let C be the unique strong cycle in G, with $s(C) = CC(G)$. Removal of $e \in C$ results in $C^{G-e}_{u,v} < C^{G}_{u,v}\ \forall u, v \in C$. If $C^{G-e}_{u,v} = C^{G}_{u,v}$ for some $u, v \in C$, then there exists a cycle C_1 containing the vertices u, v such that $C \cap C_1 \neq \phi$ and $s(C_1) = CC(G)$, a contradiction to the assumption that C is a unique strong cycle. Thus, $C^{G-e}_{u,v} < C^{G}_{u,v}\ \forall u, v \in C$. Since, $C^{G-e}_{x,y} = C^{G}_{x,y} < C^{G}_{u,v}\ \forall x, y\ \in G \setminus C$. Hence $CC(G-e) = \vee\{C^{G-e}_{u,v} : u,\ v\ \in \sigma^*\} < \vee\{C^{G}_{u,v} : u,\ v\ \in \sigma^*\} = CC(G)$, which completes the proof. ■

Theorem 8.2.17 *Block fuzzy graphs are cyclically fair.*

Proof Let $G = (\sigma, \mu)$ a block fuzzy graph. Then G has no fuzzy cut vertices and bridges. In other words, the removal of any vertex or edge do not change the connectivity between any pair of vertices. Thus, $G - w$ is a subgraph with $CONN_{G-w}(x, y) = CONN_G(x, y)\ \forall x, y \in \sigma^* \setminus w$. Assume, w is a cyclic boost vertex and so, $CC(G - w) > CC(G)$ or $C^{G-w}_{x,y} > C^{G}_{x,y}$ and $C^{G-w}_{x,y} = CC(G)$ for some pair of vertices $x,\ y$. That is, $C^{G-w}_{x,y} > C^{G}_{x,y}$ for some pair of vertices $x, y \in \sigma^* \setminus w$, which implies that, a non strong cycle in G becomes a strong cycle in $G - w$. That is, some δ-edge $e = ab$ in G becomes a strong edge in $G - w$. It is same as saying $\mu(ab) = CONN_{G-w}a, b < CONN_G(a, b)$, which implies w is a fuzzy cut vertex, a contradiction to our supposition. Hence our assumption is wrong. So, no vertex in a block is a cyclic boost vertex. Similarly the edge case also. Thus a block fuzzy graph has no cyclic boost vertices or boost edges. ■

Corollary 8.2.18 *For a block fuzzy graph $G = (\sigma, \mu)$, $CC(G - w) \leq CC(G) \forall w \in \sigma^*$ and $CC(G - e) \leq CC(G)\ \forall e \in \mu^*$.*

Lemma 8.2.19 *Cycle connectivity of a fuzzy graph is the weight of a β-strong edge in G.*

Theorem 8.2.20 *If cycle connectivity of a fuzzy graph is the minimum weight of its strong edges, then the graph is a θ-fuzzy graph and hence all the strong cycles has the same strength.*

Proof Let $q_1 \leq q_2 \leq \cdots \leq q_p$ be the weights of strong edges in G. Then by condition of the theorem, $CC(G) = q_1$. It is required to prove that $|\theta(x, z)| \leq 1$. On the contrary assume that $|\theta(x, z)| \geq 2$, for some $x, z \in \sigma^*$. Now, let $q_i, q_j \in \theta(x, z), i \leq j$. Then, $C^G_{x,z} \geq q_j$, which entails $CC(G) \geq q_j$. So the only possible value of j is 1. Hence, $\theta(x, z) = q_1 \; \forall x, z \in \sigma^*$ and thus, $C^G_{x,z} = q_1 \; \forall x, z \in \sigma^*$. ■

Now we focus on cyclic vertex connectivity and cyclic edge connectivity of some special types of fuzzy graphs. In [5], it is proved that for a CFG and a fuzzy tree $\kappa_c(G) \leq \kappa'(G)$.

Theorem 8.2.21 *If all the cycles in a fuzzy graph $F = (\tau, \nu)$ are strong, then $\kappa_c(F) \leq \kappa'_c(F)$.*

Proof Let $Y = \{e_i = u_i v_i : \text{ for some } i = 1, 2, \cdots, n\}$ be a cyclic edge cut of F with strong weight $S'_c(Y)$ with X^p be the set of pairs of vertices (u, v) with $C^F_{u,v} = CC(F)$. Then, $C^{F \setminus Y}_{u,v} < CC(F) \; \forall (u, v) \in X^p$ and so $CC(F \setminus Y) < CC(F)$. Since all cycles in F are strong, removal of any set of vertices X never make a cycle in $F \setminus X$ strong. The edge case is Similar. Now we have to prove that, $\kappa_c(F) \leq \kappa'_c(F)$. We have the following cases.

Case 1. Every edge in Y has a common end vertex v.

In this case, Y is of the form $Y = \{e_i = u_i v : \text{ for some } i = 1, 2, \cdots, n\}$. Then clearly, $X = \{v\}$ is a cyclic vertex cut. Then $\wedge_{u_i \in \tau^*}\{\nu(u_i v)\} \leq S'_c(Y)$. Thus, we can find an X for any Y in the above mentioned form such that, $S_c(X) \leq S'_c(Y)$. Hence, $\kappa_c(F) \leq \kappa'_c(F)$.

Case 2. Any two edges in Y have a common vertex.

Let $X' = \{v_i : v_i \text{ is a common vertex of at least two edges in } Y\}$, for some $i \in N$ and $Y' = \{u_j v_j \in Y : u_j \text{ or } v_j \text{ is not an end vertex of any other edges in } Y\}$ for some i. Let X be the set of all end vertices in Y'. Then X can be partitioned in two disjoint vertex sets X_1 and X_2 such that $X_1 \cap X_2 = \phi$ and $X_1 \cup X_2 = X$. Now, $K = X' \cup X_1$ or $K = X' \cup X_2$ are cyclic vertex cuts because its removal from F deletes all edges in Y. Also, $S_c(K) \leq S'_c(Y)$. Thus, $\kappa_c(F) \leq S_c(K) \leq S'_c(Y) \leq \kappa'_c(F)$.

Case 3. No two edges have a common vertex.

Let $X_1 = \{u_1, u_2, \ldots, u_n\}$ and $X_2 = \{v_1, v_2, \cdots, v_n\}$. Then the removal of X_1 or X_2 from F deletes all edges in Y. Thus, X_1 and X_2 are cyclic vertex cuts with $S_c(X) \leq S'_c(Y)$ and so, $\kappa_c(F) \leq S_c(X) \leq S'_c(Y) \leq \kappa'_c(F)$. ■

Theorem 8.2.22 *In a block fuzzy graph G, $\kappa_c(G) \leq \kappa'_c(G)$.*

Theorem 8.2.23 *If any minimal fuzzy vertex cut of a fuzzy graph $G = (\sigma, \mu)$ contains more than one element, then*

(i) $C^G_{u,v} \neq 0 \quad \forall u, v \in \sigma^*$.
(ii) $CC(G) = \vee\{s(C) : C \text{ is a cycle in } G\}$.

Proof (i) Let G be a fuzzy graph with a minimal fuzzy vertex cut having more than one element. We claim that, any two vertices $x, y \in \sigma^*$ belong to a fuzzy cycle in G. Since any minimal fuzzy vertex cut has more than one element, by Theorem 1.2.43 there are at least two internally disjoint strongest $x - y$ paths in G, for two vertices $x, y \in \sigma^*$ such that xy is not strong. Let P' and P'' be two such paths. Then $P' \cup P''$ is a fuzzy cycle containing x and y. Suppose xy is a β-strong edge. Then evidently x and y belong to a fuzzy cycle. Thus, $C^G_{x,y} \neq 0$.

Now, if xy is an α-strong edge, there are three possibilities, case 1, if xy is an edge of fuzzy cycle, then we are done. Now in the case that xy belongs to a cycle C' which is not a fuzzy cycle, then C' has an edge, say $x'y'$ which is not strong. By our assumption from Theorem 1.2.43, there exist at least two internally disjoint strongest $x' - y'$ paths.

If $\mu(xy) = CONN_G(x', y')$, then xy is a β-strong edge, which is a contradiction. If $\mu(uv) < CONN_G(x'y')$, then xy is a δ-edge, which is again a contradiction. Hence, xy belongs to a fuzzy cycle. If xy is an α-strong edge and xy does not belong to a cycle, then either x or y is a fuzzy cutvertex. But any minimum fuzzy vertex cut of G has at least two elements, which is a contradiction. Thus our claim is true. Hence the proof.

(ii) As from the proof of (i) we obtained that for any two vertices $x, y \in \sigma^*$, there exists a fuzzy cycle C in G containing both x and y. Let C be any cycle G which is not strong. It is enough to prove that there exists a strong cycle C' in G such that $s(C) < s(C')$. Let $x'y'$ be the weakest edge in C. Since any minimal fuzzy edge cut of G has more than one element, there are at least two internally disjoint strongest $x' - y'$ paths in G. Let P' and P'' be two such paths. Then, $C' = P' \cup P''$ is a fuzzy cycle containing x and y. Hence, $s(C) < s(C')$. Thus, $CC(G) = \vee\{s(C) : C \text{ is a cycle in } G\} = \vee\{s(C) : C \text{ a strong cycle in } G\}$. ■

By Theorem 1.2.60 any two arbitrary vertices lie on a strong cycle. Thus the cycle connectivity between any pair of vertices always greater than zero.

Theorem 8.2.24 *Let $G = (\sigma, \mu)$ be a θ-fuzzy graph which is a block. Then the cycle connectivity of G is the minimum weight of strong edges in G.*

Proof Let G be a θ-fuzzy graph which is a block. For a θ-fuzzy graph, θ-evaluation of each pair of vertices is either empty or a singleton set. Furthermore, in a block fuzzy graph, any minimal fuzzy vertex cut of G has more than one element. Thus by Theorem 8.2.23, $C^G_{u,v} \neq 0 \ \forall u, v \in \sigma^*$. Therefore θ-evaluation of each pair of vertices is always a singleton set. Let $\alpha_1 \leq \alpha_2 \leq \cdots \leq \alpha_p$, be the strong edges in G and the cycle connectivity of G be α_k. Then there exist a pair of vertices g, h such that $\mu(gh) = \alpha_k$ and $C^G_{g,h} \leq \alpha_k$. Else if $C^G_{g,h} > \alpha_k$, as G is θ-fuzzy graph and

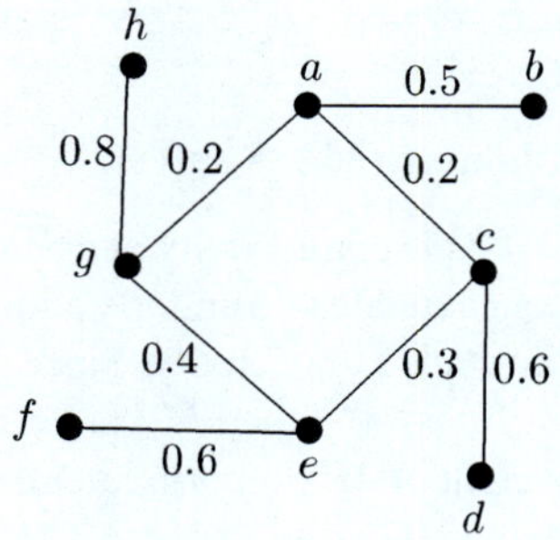

Fig. 8.11 Fuzzy graph with $CC(G) = 0.2$

$|\theta(g, h)| = 1$, then $C^G_{g,h} = \alpha_i$ for some $i > k$. This implies there exists a strongest strong $g - h$ path of strength α_i, a contradiction to the strong property of α_k. So it is enough to show that $\alpha_k = \alpha_1$. While combining strong property of α_1 and above mentioned arguments we have a pair of vertices c, d such that $\mu(cd) = C^G_{c,d} = \alpha_1$. By Theorem 1.2.60, any two strong edges of G lie on a common strong cycle. Hence, there exists a cycle C containing both gh and cd having $s(C) = \alpha_1$. Thus, α_1 is an element of $\theta(g, h)$. Thus, $\alpha_1 = \alpha_k$. Hence the proof. ■

A fuzzy graph is a θ-fuzzy graph if its cycle connectivity is its minimum of strong edges, but it need not be a block in general. Figure 8.11 is a fuzzy graph with cycle connectivity, the minimum weight of strong edges, but not a block.

Theorem 8.2.25 *Let X be a minimum cyclic vertex cut of $G = (\sigma, \mu)$ with $|X| = m$. Then,*

(i) there exist at least m strong cycles in G
(ii) there are at least $3m$ pair of vertices with $C^G_{u,v} = CC(G)$
(iii) for $m > 2$, no vertex of G is a cyclic cut vertex
(iv) there are exactly m vertex disjoint strong cycles $C_1, C_2, \cdots, C_m$ such that $s(C_1) = s(C_2) = \cdots = s(C_m) = CC(G)$.

Corollary 8.2.26 *A fuzzy graph has no cyclic cut vertex if and only if any minimum fuzzy vertex cut has more than one element.*

Theorem 8.2.27 *Let G be a connected fuzzy graph with $\kappa_c(G) \geq t$ and M be a minimum cyclic fuzzy vertex cut of G. Let G' be a fuzzy graph obtained by adding a new vertex x to G and joining x to the vertices of M. Let $\mu(z_i x) = \vee \{C^G_{z_i,y} : z_i y$ is an edge in $G\}$. Then $\kappa_c(G') \geq t$.*

Proof Let G be a connected fuzzy graph with $\kappa_c(G') \geq t$ and let $M = \{z_1, z_2, \ldots, z_k\}$ be a minimum cyclic fuzzy vertex cut of G. Let G' be the fuzzy graph obtained by adding a new vertex x to G and joining x to the vertices of M and let

$\mu(z_i x) = \vee \{C^G_{z_i,y} : z_i y$ is an edge in $G\}$. Let M' be a cyclic fuzzy vertex cut of G'.

Claim $S_c(M') \geq \kappa_c(G)$.

Cycle connectivity of a pair of vertices is always less than or equal to its strength of connectedness. Similarly adding a vertex m in aforementioned format never increases the cycle connectivity. Since, $C_{x,u_i} \leq \vee\{C^G_{v,u_i} : \forall v \in \sigma^*\}$.

If $x \in M'$, then $M' \backslash \{x\}$ is a cyclic fuzzy vertex cut of G. Since $\kappa_c(G) \geq t$, strong weight of $M' \backslash \{x\}$ is greater than or equal to $\kappa_c(G)$. So strong weight of M' is greater than $\kappa_c(G)$.

Suppose $x \notin M'$ and $M \subseteq M'$. Then clearly, strong weight of M' is greater than or equal to $\kappa_c(G)$. The case of $M' \subseteq M$ is similar. If $M \backslash M' \neq \emptyset$, then $M \backslash M'$ belongs to a unique connected component of $G' \backslash M'$. Hence, M' is a cyclic fuzzy vertex cut of G. In that case, the strong weight of M is less than or equal to the strong weight of M'. That is, $\kappa_c(G') \geq t$. ■

For any fuzzy cycle C, $\kappa_c(G)$ is the weight of the weakest edge in C and by Theorem 1.2.40 we have the following theorem.

Theorem 8.2.28 *Let $G = (\sigma, \mu)$ be a saturated or β-saturated fuzzy cycle with $|\sigma^*| \geq 3$. Then, $2\kappa_c(G) = \kappa(G)$.*

Theorem 8.2.29 *Let $C = (\tau, \nu)$ be a fuzzy cycle with $|\tau^*| = n \geq 5$ and $\tau(w) = t, t \in (0, 1]$. Then the cyclic vertex connectivity and the cyclic edge connectivity of $C^c = (\gamma, \xi)$ are respectively $\kappa_c(C^c) = t(n-4)$ and $\kappa'_c(C^c) = t\left(\binom{n}{2} - 2(n-1)\right)$.*

Proof Let $C = (\tau, \nu)$ be a fuzzy cycle with $|\tau^*| = n \geq 5$. Let $\tau(w) = t$ with $t \in (0, 1]$. Since C is a fuzzy cycle, it only contains strong edges. Also, $CONN_{C^c}(a, b) = t, \quad \forall a, b \in \tau^*$. Clearly $cd \in \nu^*$ becomes a non strong edge in C^c and between every non adjacent vertices a, b in C, $\xi(ab) = t$. Collection of any 3 non adjacent vertices makes a strong cycle C' with $s(C') = t$. Thus any arbitrary set of 5 vertices have a cycle with strength equals to t.

Thus the removal of a set S of exactly $n-4$ vertices from C results in the inequality $C^{C-S}_{x,y} < C^C_{x,y} \; \forall x, y \in \tau^* \setminus S$. Thus, $\kappa_c(C^c) = t(n-4)$.

As aforesaid all edges $cd \in \nu^*$ are δ-edges in C^c and all edges ab such that $\xi(ab) = t \; \forall ab \notin \nu^*$ are strong edges. Thus there are $\binom{n}{2} - (n-1)$ strong edges in C^c and so every strong cycle in C^c contains only edges from $\xi^* \setminus \nu^*$. Thus, to reduce the cycle connectivity, we need to remove at least $\binom{n}{2} - 2(n-1)$ edges from $\xi^* \setminus \nu^*$, since the remaining $n-1$ edges in $\xi^* \setminus \nu^*$ never contribute to the cycle connectivity of C^c. Thus, the proof follows. ■

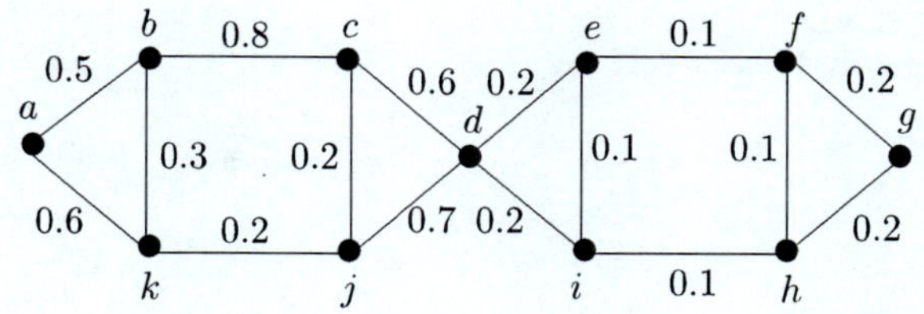

Fig. 8.12 A fuzzy graph with more than one cyclic boost vertex cut

Theorem 8.2.30 *There is a connected fuzzy graph G having* $\kappa_c(G) \geq t,\ \forall t \in R^+$.

Proof First, let $t \in [0, 1)$. A fuzzy cycle $G = (\sigma, \mu)$ can be constructed as follows. Let $\xi = \wedge\{\mu(ab) : ab$ is an edge in $G\}$. If $\xi \geq t$, then, $\kappa_c(G) \geq t$.

Suppose $t \geq 1$. Then $t \in [n, n+1)$ for some $n \in \mathbb{N}$. Let G be a complete fuzzy graph with $n+3$ vertices namely $x_1, x_2, \ldots, x_{n+2}, x_{n+3}$. Let $\sigma(x_i) \geq \frac{t}{(n+2)}$ for $i = n+1, n, \ldots, 2, 1$ in a manner that $\sigma(x_i) = \xi$, a constant for all $i = n+1, n, \cdots, 2, 1$. Then cycle connectivity of G is ξ and any cyclic vertex cut of G contain exactly $n+1$ vertices. Hence any two vertices a and b belong to exactly $n+1$ strong cycles and so $C^G_{a,b} = \xi$. Thus, $\kappa_c(G) = (n+1)\xi \geq (n+1)\frac{t}{(n+2)} \geq t$. ■

It is quite clear from Theorem 1.2.41 and Theorem 1.2.42 that the removal of a set of edges from a fuzzy graph never enhances its vertex or edge connectivity. But it is not true in the case of cycle connectivity. There we can find a set of edges or vertices whose removal increases the cycle connectivity of the fuzzy graph.

Definition 8.2.31 A **cyclic boost vertex cut** $(CBVC)$ of a fuzzy graph $G = (\sigma, \mu)$ is a set of vertices $X \subseteq \sigma^*$ with $CC(G - X) > CC(G)$, provided $CC(G) > 0$, where $CC(G)$ is the cycle connectivity of G.

Example 8.2.32 Consider Fig. 8.12 showing $G = (\sigma, \mu)$ with $\sigma^* = \{a, b, c, d, e, f, g, h, i, j, k\}$, $\sigma(r) = 1$ for all $r \in \sigma^*$ and $\mu(ab) = 0.5$, $\mu(bc) = 0.8$, $\mu(cd) = 0.6$, $\mu(dj) = 0.7$, $\mu(bk) = 0.3$, $\mu(de) = \mu(di) = \mu(kj) = \mu(cj) = \mu(fg) = \mu(gh) = 0.2$ and $\mu(ef) = \mu(fh) = \mu(hi) = \mu(ei) = 0.1$. G has no cyclic boost vertex. But there are pairs $X' = \{a, d\}$ and $X'' = \{d, g\}$, which are cyclic boost vertex cuts in G with $CC(G - X') = 0.2 < 0 = CC(G)$ and $CC(G - X'') = 0.1 < 0 = CC(G)$.

Definition 8.2.33 Let M be a $CBVC$ of G. The **strong weight** of M is defined as $S_c(M) = \sum_{m \in M} \mu(mn)$, where $\mu(mn)$ is the minimum weight of strong edges incident on m. **Cyclic boost vertex connectivity** of a fuzzy graph G, denoted by $\chi_c(G)$, is the minimum strong weight of cyclic boost vertex cuts of G.

In Example 8.2.32, $X_1 = \{a, d\}$ and $X_2 = \{d, g\}$ are $CBVC$ of G with $S_c(X_1) = 0.7$, $S_c(X_2) = 0.4$. Thus the cyclic boost vertex connectivity is 0.4.

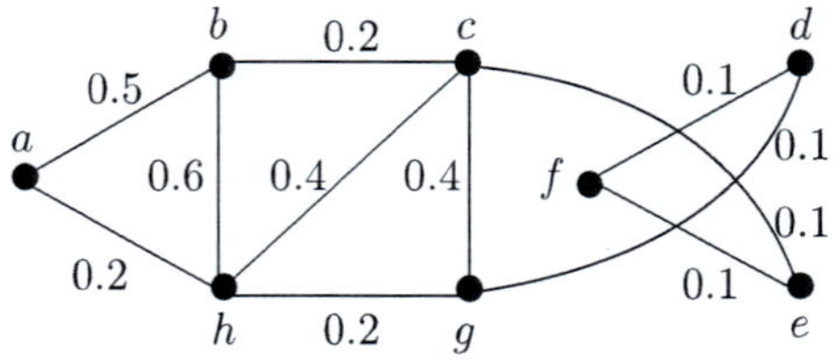

Fig. 8.13 A fuzzy graph G with cyclic boost edge cut

Definition 8.2.34 A **cyclic boost edge cut** $(CBEC)$ of a fuzzy graph $G = (\sigma, \mu)$ is a set of edges $E \subseteq \mu^*$ with $CC(G - E) > CC(G)$, provided $CC(G) > 0$, where $CC(G)$ is the cycle connectivity of G.

Definition 8.2.35 Let E be a cyclic boost edge cut of G. The strong weight of E is defined as $S'_c(E) = \sum_{e_j \in \mu^*} \mu\left(e_j\right)$, where e_j is a strong edge of E. The **cyclic edge connectivity** of G, denoted by $\chi'_c(G)$, is the minimum strong weight of cyclic boost edge cuts of G.

Example 8.2.36 Let $G = (\sigma, \mu)$ be a fuzzy graph with $\sigma^* = \{a, b, c, d, e, f, g, h\}$ and $\sigma(x) = 1$ for all $x \in \sigma^*$. Let $\mu(ab) = 0.5$, $\mu(ah) = \mu(bc) = \mu(hg) = 0.2$, $\mu(cg) = \mu(ch) = 0.4$, $\mu(df) = \mu(dg) = \mu(de) = \mu(ce) = 0.4$ and $\mu(bh) = 0.6$ (Fig. 8.13). $E_1 = \{bh, ch\}$, $E_2 = \{ch, cg\}$ and $E_3 = \{bh, cg\}$ are the only 3 cyclic boost edge cuts of G with $S'_c(E_1) = 1$, $S'_c(E_2) = 0.8$ and $S'_c(E_3) = 1$. Among all cyclic boost edge cuts of G, E_2 has the minimum strong weight and hence $\chi'_c(G) = 0.8$.

Clearly, $1-CVC$ is a cyclic boost vertex of G.

Definition 8.2.37 A fuzzy graph G is said to be **cyclically fairer** if $\chi_c(G) = \chi'_c(G) = 0$.

Theorem 8.2.38 *For a cyclically fairer fuzzy graph,* $\kappa_c(G) \leq \kappa'_c(G)$.

Theorem 8.2.39 *Block fuzzy graphs are cyclically fairer.*

Proof Let G be a block fuzzy graph. Assume that G is not cyclically fairer. Then there exist a set of vertices X in G such that $CC(G \setminus X) > CC(G)$. That is, some non strong cycle C in G becomes a strong cycle in $G \setminus X$ with $s(C) > CC(G)$. Let $e = xy$ be a weakest edge in C. Then $s(C) = \mu(e)$. Since G is a block, by combining Theorems 1.2.59 and 1.2.60 we can say that, x and y are joined by two internally disjoint strongest paths say, P_1 and P_2. Then $C' = P_1 \cup P_2$ is strongest strong cycle with $\mu(e) < s(C')$. Hence, $s(C) = \mu(e) < s(C') < CC(G)$, a contradiction. So, our assumption is wrong. Thus all block fuzzy graphs are cyclically fairer.

8.3 Cycle Cogency of Fuzzy Graphs

In this session, we introduce a new parameter associated with fuzzy graphs, named as cycle cogency.

Definition 8.3.1 Let $G = (\sigma, \mu)$ be a fuzzy graph. For a vertex u of G, there associated a set say $\zeta(u)$ called the **ζ- evaluation** of u and is defined as $\zeta(u) = \{\alpha \mid \alpha \in (0, 1]\}$, where α is the strength of a strong cycle passing through u. G is said to be a **ζ-fuzzy graph** if ζ-evaluation of every vertex in G is either empty or a singleton set. In other words, G is called a ζ-fuzzy graph if for each vertex u, either there is no strong cycle passing through u or all strong cycles passing through u have the same strength.

Definition 8.3.2 Let $G = (\sigma, \mu)$ be a fuzzy graph and let $u \in \sigma^*$. Then, $\zeta_{max}(u) = \vee\{\alpha \mid \alpha \in \zeta(u), u \in \sigma^*\}$ and $\zeta_{min}(u) = \wedge\{\alpha \mid \alpha \in \zeta(u), u \in \sigma^*\}$. Now we denote $\zeta_{max}(u)$ as $Co_G(u)$, and call it as **cycle cogency** of the vertex u, which precisely is the maximum strength of all strong cycles passing through u. Clearly,

$$Co_G(u) = \vee\{C_{u,v}^G : \forall v \in \sigma^*\}.$$

Cycle cogency of a fuzzy graph G is defined by

$$Co(G) = \sum_{u \in \sigma^*} Co_G(u).$$

$Co_G(u)$ can also be written as $Co(u)$ if there is no ambiguity.

Example 8.3.3 Consider the fuzzy graph given in Fig. 8.14 Here, $\sigma^* = \{r_1, r_2, r_3, r_4, r_5, r_6, r_7, r_8, r_9\}$. Let $\mu(r_1r_2) = \mu(r_1r_6) = \mu(r_6r_8) = \mu(r_6r_7) = 0.4$, $\mu(r_2r_3) = \mu(r_3r_4) = 0.6$, $\mu(r_4r_5) = \mu(r_5r_6) = 0.5$, $\mu(r_7r_8) = 0.2$, $\mu(r_6r_9) = 0.9$, $\mu(r_2r_9) = 0.7$, and $\mu(r_4r_9) = 0.8$. We can see that, $Co(G) = 3.8$.

vertex (u)	r_1	r_2	r_3	r_4	r_5	r_6	r_7	r_8	r_9
$Co(u)$	0.4	0.6	0.6	0.6	0.5	0.5	0	0	0.6

Cycle cogency of vertices in Fig. 8.14.

Let G be a fuzzy graph having n vertices with every vertex has cycle cogency 1. Then $Co(G) = n$. Thus, for any fuzzy graph on n vertices, cycle cogency is always less than or equal to n.

A fuzzy graph G and two of its subgraphs H_1 and H_2 are given in Fig. 8.15 Note that $Co(H_1) < Co(G) < Co(H_2)$.

Definition 8.3.4 Strong membership of a vertex x in a fuzzy graph $G = (\sigma, \mu)$ is defined by $\mu_{max}(x) = \vee\{\mu(xu) \mid xu \text{ is a strong edge in } G\}$.

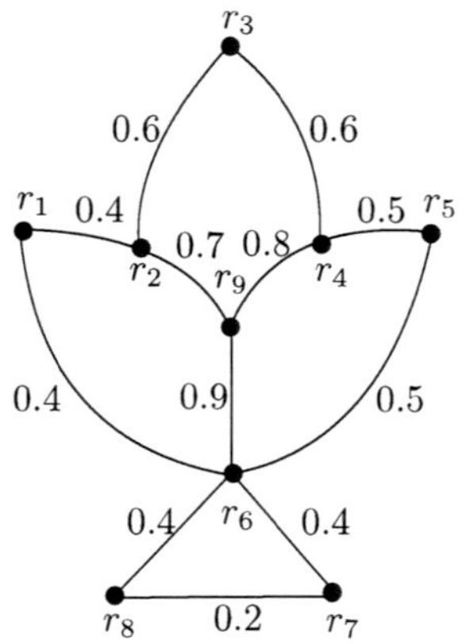

Fig. 8.14 Fuzzy graph with cycle cogency 3.8

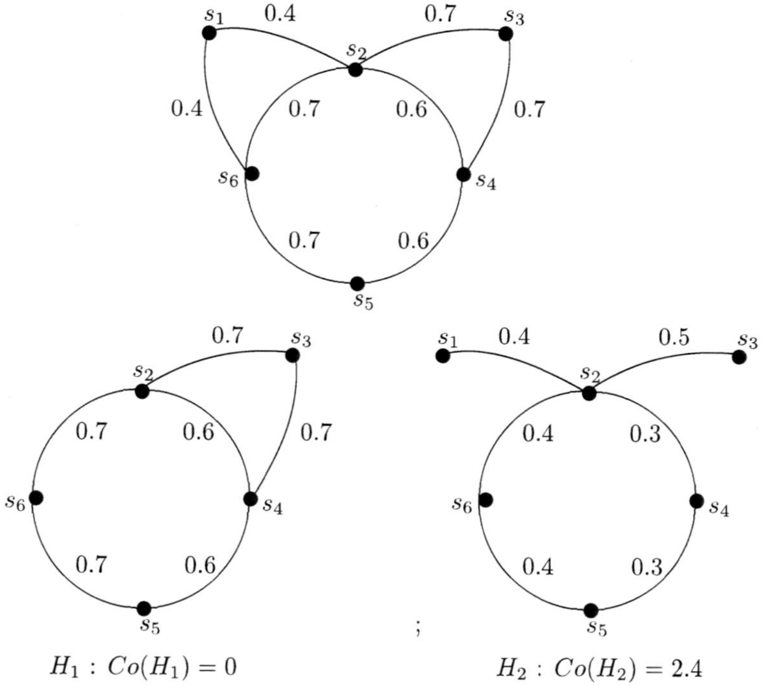

Fig. 8.15 Fuzzy graph G with cycle cogency 1.2 and their subgraphs

Definition 8.3.5 $\beta-$**membership** of a vertex x in a fuzzy graph $G = (\sigma, \mu)$ is defined by $\beta_{max}(x) = \vee\{\mu(xu) | \ xu$ is a β − strong edge in $G\}$.

Theorem 8.3.6 *For a strong fuzzy graph $G = (\sigma, \mu)$ and a partial fuzzy subgraph $H = (\tau, \nu)$ of G, $Co(H) \leq Co(G)$.*

Proof Let $G = (\sigma, \mu)$ be a strong fuzzy graph and $H = (\tau, \eta)$ its partial fuzzy subgraph. Let $x, y \in \tau^*$. By definition of partial fuzzy subgraph, $C^H_{x,y} \leq C^G_{x,y}$ and hence, $Co_H(x) = \vee\{C^H_{x,y} : \forall y \ \in \tau^*\} \leq \vee\{C^G_{x,y} : \forall y \ \in \sigma^*\} = Co_G(x)$. Summing up the cogencies, we get the desired result. ■

Theorem 8.3.7 *Let $G = (\sigma, \mu)$ and $G' = (\sigma', \mu')$ be two fuzzy graphs with $\sigma(u) = \sigma'(t) = 1 \ \forall \ u \in \sigma^*$ and $t \in \sigma'^*$. Let $\mu_{max}(v_i) = \alpha$ and $\mu_{max}(v'_j) = \beta$, where $\alpha \leq \beta$. Then cogency of the vertex $v_i v'_j$ in the Cartesian product $G \times G'$ such that $v_i \in \sigma^*$ and $v'_j \in \sigma'^*$ is $Co(v_i v'_j) = \vee\{Co(v'_j), \alpha\}$, where $\alpha = \mu_{max}(v_i)$ and $\beta = \mu_{max}(v'_j)$, $\alpha \leq \beta$.*

Proof Let G and G' be two fuzzy graphs with $|\sigma^*| = n_1$ and $|\sigma'^*| = n_2$. Let $(\sigma \times \sigma')(u) = 1 \ \forall \ u \in (\sigma \times \sigma')^*$. Clearly, $|(\sigma^* \times \sigma')^*| = n_1 n_2$. Let the vertex set of G be $v_1, v_2, \cdots, v_{n_1}$ and the vertex set of G' be $v'_1, v'_2, \cdots, v'_{n_2}$.

Let $v_p \in \sigma^*$ be an arbitrary vertex in G with cycle cogency w_1. Then there exists a cycle C_1 passing through v_p with $S(C_1) = Co(v_p) = w_1$ and by definition of Cartesian product of two fuzzy graphs, there corresponds exactly n_2 copies of cycles in $G \times G'$ whose strengths equal to w_1. Let C_1 be $C_1 : v_p, v_1, v_2, \ldots, v_p \ p \leq n_1$. Then, $C'_1 : v'_k v_p, v'_k v_1, v'_k v_2, \ldots, v'_k v_p, k \leq n_2$ is an arbitrary cycle in $H = G \times G'$ and by the definition of cross product, $S(C_1) = S(C'_1) = w_1$. Since $v'_k \in \sigma'^*$ is an arbitrary element in G' we can find n_2 such copies of cycles with strength w_1 in H.

Similarly, corresponding to any cycle C'_i with strength w'_i in G', there exist exactly n_1 copies of cycles in H with same strength as that of C'_i. Now we claim that all cycles in H corresponding to a strong cycle in G are strong. Let $C_1 : v_l, v_1, v_2, \cdots, v_l, \ l \leq n_1$ be a strong cycle in G. Make C to be a cycle in H such that, $C : v'_m v_l, v'_m v_1, v'_m v_2, \ldots, v'_m v_l, \ v'_l \in \sigma'^*, m \leq n_2$. Assume that some edges in C are not strong. Let one of them be $e = v'_m v_i, v'_m v_{i+1}$. Then there exists a strongest path P from $v'_m v_i$ to $v'_m v_{i+1}$ with all edges in P having strength greater than $(\mu \times \mu')(v'_m v_i, v'_m v_{i+1})$.

Let P_1 and P_2 be the collection of vertices of G and G' belong to H, respectively. Then, as per property of definition in cross product between two fuzzy graphs, P_1 is a path from v_i to v_{i+1}. But since P is a strongest path, we have $S(P_1) > \mu(v_i v_{i+1})$. This implies $v_i v_{i+1}$ is a $\delta-$edge in G, a contradiction since $v_i v_{i+1}$ is an element of strong cycle C_1. Thus our assumption is wrong and e becomes a strong edge. Thus the cycle C in H corresponding to C_1 in G is strong.

In a similar way we can show each cycle C' in H correspond to cycle C'_1 in G' are also strong. Thus $Co(v_i v'_j) \geq \vee\{Co(v_i), Co(v'_j)\}$.

Next we prove the existence of a strong cycle C passing through $v_i v'_j$ in H with $S(C) = \alpha$. For that, let $e = v_i v_k$ be an edge having end vertex v_i in G with maximum weight α and $e' = v'_j v'_m$ with maximum weight β in G'. Let $\alpha \leq \beta$. Now using the definition of Cartesian product between two fuzzy graphs, we have a cycle $C : v'_j v_i - v'_j v_k - v'_m v_k - v'_m v_i - v'_j v_i$ in H such that $S(C) = \alpha$. As proved before, each edge in C is strong. Thus, $Co(v_i v'_j) \geq \alpha$. Since $\mu_{max}(v_i) = \alpha$, and $Co(v_i) \leq \alpha$, we have $Co(v_i v'_j) \geq \vee\{\alpha, Co(v'_j)\}$. It is generally true that cogency of any vertex is always less than or equal to its μ_{max}. So, $Co(v_i v'_j) \leq \beta$ and if there exists a strong cycle passing through $v_i v'_j$ of strength β, then $Co(v'_j) = \beta$. Thus, $Co(v_i v'_j) = Co(v'_j) = \beta$. Now since no edge incident on $v_i v'_j$ has strength lying between α and β, $Co(v_i v'_j) \leq \alpha$. Thus $Co(v_i v'_j) = \vee\{\alpha, Co(v'_j)\}$. ■

Corollary 8.3.8 *$Co(G) = 0$ if and only if G is a fuzzy tree.*

For a fuzzy cycle $G = (\sigma, \mu)$ with $|\sigma^*| = n$ and $\hat{\beta} = \wedge\{\mu(xy) |\ xy \in \mu^*\}$, $C^G_{x,y} = \hat{\beta}, \forall x, y \in \sigma^*$ and thus, $Co(x) = \hat{\beta} \quad \forall x \in \sigma^*$. Since there are n vertices, $Co(G) = n * \hat{\beta} = CCI(G) - \frac{n(n-3)}{2}$.

Theorem 8.3.9 gives the cycle cogency of a complete fuzzy graph.

Theorem 8.3.9 *In a CFG $G = (\sigma, \mu)$ with $n \geq 3$ vertices, $\sigma(v_i) = w_i$ where $i = 1, 2, \ldots, n$ and $w_1 \leq w_2 \leq \cdots \leq w_n$,*

$$Co(G) = \sum_{i=1}^{n-2} w_i + 2w_{n-2}.$$

Proof By Theorem 1.2.32 all edges in a CFG are strong. Thus all cycles in G are strong. Since edges incident with v_1 have membership value $w_1 \leq w_i \quad \forall i$, all cycles passing through v_1 have strength w_1. Thus $\zeta(v_1) = w_1 = Co(v_1)$. Similarly, any cycle passing through v_2 not containing v_1 has strength w_2. Thus cycles passing through v_2 have strengths w_1 or w_2. Hence, $\zeta(v_2) = \{w_1, w_2\}$. Thus for $1 \leq i \leq n_2$ cycles passing through v_i have strengths from the set $\{w_1, w_2, \ldots, w_i\}$ and hence $\zeta(v_i) = \{w_1, w_2, \ldots, w_i\} \quad \forall i \neq n-1, n$. Hence $Co(v_i) = w_i \quad \forall i \neq n-1, n$. For $v_{n-1}, v_n \in \sigma^*$, cycles passing through these two vertices have strengths $w_1, w_2, \ldots, w_{n-2}$ only. So $\zeta(v_{n-1}) = \zeta(v_n) = \{w_1, w_2, \ldots, w_{n-2}\}$ and $Co(v_{n-1}) = Co(v_n) = w_{n-2}$. Now, $Co(G) = \sum_{1=1}^{n-2} w_i + 2w_{n-2}$. ■

Theorem 8.3.10 *Cycle cogency of a vertex u in a fuzzy graph $G = (\sigma, \mu)$ is always greater than or equal to the maximum weight of $\beta-$strong edges incident at u.*

Proof Consider an arbitrary vertex u. If $e = uw$ is a $\beta-$strong edge, then there exists at least one strongest strong path P between u and w other than the edge uw. Then the cycle $C = P \cup e$ is a strong cycle of G having u as an internal vertex. Since e is $\beta-$strong and C is strong, we have $S(C) = \mu(e)$. Thus, $Co(u) \geq S(C) = \mu(e)$. Since u is an arbitrary vertex, it is true for all vertices. Hence the proof follows. ■

There are times when the cycle cogency of a vertex is strictly greater than the maximum weight of $\beta-$strong edges incident on it. Example 8.3.11 illustrates this.

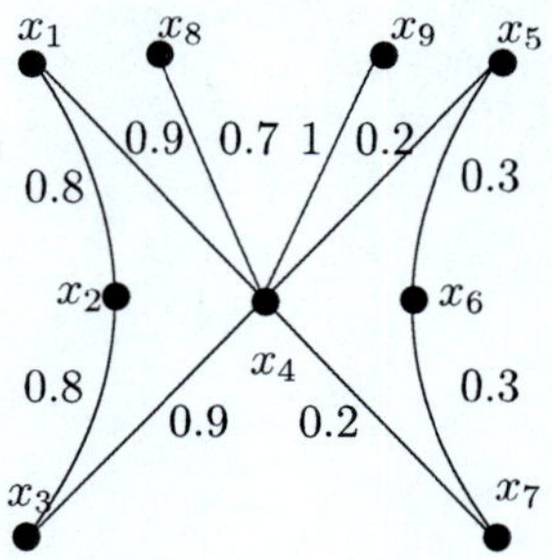

Fig. 8.16 Fuzzy graph G with $Co(x_4) > \beta_{max}(x_4)$

Example 8.3.11 Let $G = (\sigma, \mu)$ be the fuzzy graph given in Fig. 8.16 with $\sigma^* = \{x_1, x_2, x_3, x_4, x_5, x_6, x_7, x_8, x_9\}$. Here,

$$\mu^* = \{x_1x_2, x_2x_3, x_3x_4, x_4x_1, x_4x_8, x_4x_9, x_4x_5, x_5x_6, x_6x_7, x_7x_4\},$$

with $\sigma(x) = 1 \ \forall \ x \in \sigma^*$ and $\mu(x_1x_2) = \mu(x_2x_3) = 0.8$, $\mu(x_3x_4) = \mu(x_4x_1) = 0.9$, $\mu(x_4x_5) = \mu(x_4x_7) = 0.2$, $\mu(x_5x_6) = \mu(x_6x_7) = 0.3$, $\mu(x_4x_8) = 0.7$ andEquality conditioni $\mu(x_4x_9) = 1$. In this fuzzy graph, we can see that for vertex x_4, $Co(x_4) = 0.8 > \beta_{max}(x_4) = 0.2$.

Equality condition in Theorem 8.3.10 trivially holds when G is a block. Thus we have the following theorem.

Theorem 8.3.12 *Cogency of an arbitrary vertex u in a block G is the maximum weight of $\beta-$strong edges incident on u.*

Proof Let u be an arbitrary vertex in a fuzzy block G. We need to prove that the cogency of u is the maximum weight of $\beta-$strong edges incident on u. Let $C : u - v - \cdots - w - u$ be a strong cycle in G such that strength of C equals to the cycle cogency of u. Then at least one of uv or uw is β-strong. If not, the path $v - u - w$ become the unique strongest path, making u a cut vertex. It is not possible since G is a block. So at least one edge uv or uw is $\beta-$strong; let it be uv. Since C is strong and uv is $\beta-$strong, we have $S(C) = \mu(uv) = \beta_{max}(u)$. Hence the proof. ■

Theorem 8.3.12 explicitly shows that, for a vertex u in a block fuzzy graph $Co(u) = \beta_{max}(u)$. Thus, if two vertices have same β_{max}, then they have the same cycle cogency.

Example 8.3.13 shows that the converse of Theorem 8.3.12 is not generally true for separable fuzzy graphs.

Example 8.3.13 Let $G = (\sigma, \mu)$ be the fuzzy graph given in Fig. 8.17 with $\sigma^* = \{w_1, w_2, w_3, w_4, w_5, w_6, w_7\}$ and $\mu^* = \{w_1w_2, w_2w_3, w_3w_4, w_4w_5, w_2w_5, w_2w_6, w_6w_7, w_1w_7\}$. Here, $\mu(w_1w_2) = \mu(w_2w_3) = \mu(w_2w_5) = \mu(w_2w_6) = mu(w_4w_5) = \mu(w_6w_7) = 0.5$, $\mu(w_3w_4) = 0.7$, $\mu(w_1w_7) = 0.6$ and $\sigma(x) = 1 \ \forall x \in \sigma^*$. Then,

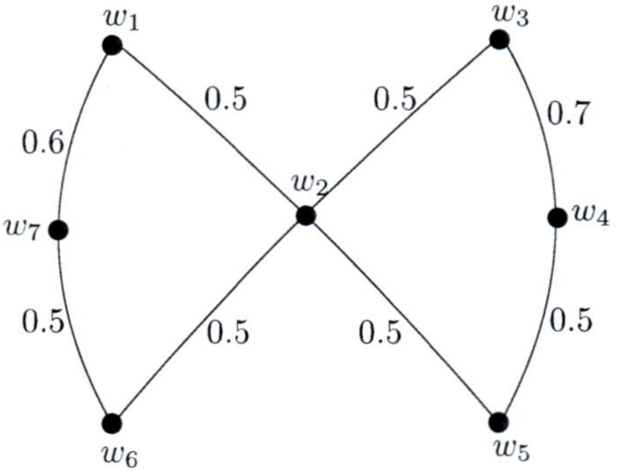

Fig. 8.17 Separable fuzzy graph G with $Co(u) = \beta_{max}(u)$

$CR(G) = CR(G - w_1w_2) = 6$ whereas $CR(G - w_2w_4) = 10.5$ and $CR(G - w_4 w_5) = 6.4$. Thus, w_2w_4, w_5w_6 and w_4w_5 are $CR-$ increasing edges while w_1w_2, $w_2 w_3$ and w_1w_3 are neutral.

We know that if $C^G_{u,v} \neq 0$ for a fuzzy graph G, then there exists a strong cycle C in G passing through both u and v. Thus if G is $\zeta-$evaluated, then $\zeta-$ evaluation for any vertex in G becomes either empty or a singleton set. Thus cogency of u is the strength of C. Thus we have the following theorem.

Theorem 8.3.14 *If $C^G_{u,v} \neq 0$ for some $u, v \in \sigma^*$ of a $\zeta-$fuzzy graph $G = (\sigma, \mu)$, then $C^G_{u,v} = Co(u) = Co(v)$.*

Theorem 8.3.15 *In a θ-fuzzy graph $G = (\sigma, \mu)$ which is a block, cycle cogency of any vertex $u \in \sigma^*$ is the minimum weight of strong edges in G.*

Proof For a θ-fuzzy graph G, $\theta-$evaluation of each pair of vertices is either empty or a singleton set. Furthermore, by Theorem 1.2.60 any two vertices of a block fuzzy graph lie on a common strong cycle. Therefore, $\theta(u, v) \neq \phi\ \forall u, v \in \sigma^*$. Thus, $C^G_{u,v} = \theta(u, v)$. Since G is $\theta-$evaluated and $\theta(u, v) \neq \phi\ \forall u, v \in \sigma^*$ we have, $|\theta(u, v)| = 1$. Now we claim every $\theta-$fuzzy graph is a $\zeta-$fuzzy graph. As we know $\zeta-$evaluation of any vertex u is set of cycle connectivity values between u and other vertices and thus $\zeta(u) \neq \phi$.

To prove our claim, it is enough to prove that $|\zeta(u)| = 1\ \forall u$. In other words, if there is a strong cycle C_1 not containing the vertex y contributing to the cycle connectivity between between u and x, and another cycle C_2 contributing to the cycle connectivity between u and y with $x \notin C_2$, then $S(C_1) = S(C_2)$.

Let $g \in C_1$ be the first common vertex of C_1 and C_2 as one moves along clockwise direction of C_1 starting from x and $h \in C_1$ be the last common vertex between C_1 and C_2. Now if g and h are different from u, then g and h become common vertices of cycles C_1 and C_2. Since $|\theta(g, h)| = 1$, strength of both C_1 and C_2 should be equal. Thus, $C^G_{u,x} = C^G_{u,y}$.

Next case arise when the cycles C_1 and C_2 share no common points. Then, without loss of generality assume $uv \in C_1$ and $uw \in C_2$. By part (iii) of Theorem 1.2.60, uv

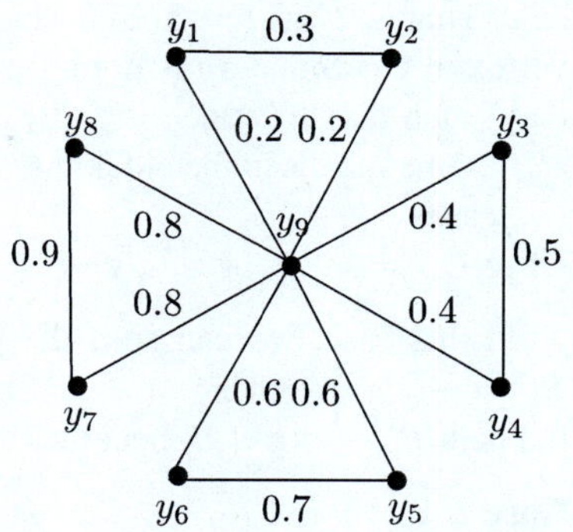

Fig. 8.18 A separable fuzzy graph G

and uw lie on a common strong cycle. Let it be C. Then, $S(C) \leq \mu(uv) \wedge \mu(uw)$. Without loss of generality, assume $\mu(uv) \leq \mu(uw)$. Then, $S(C) = \mu(uv)$. Since the graph is a $\theta-$fuzzy graph, we can arrive at $C^G_{u,v} = S(C) = S(C_1)$. Similarly, $C^G_{u,w} = S(C) = S(C_2)$, which implies, $S(C_1) = S(C_2)$. Hence, $C^G_{u,x} = C^G_{u,y}$. Thus for any arbitrary vertex u, $|\zeta(u)| = 1$.

By Theorem 8.3.12, we have $Co(u)$ is the maximum weight of $\beta-$strong edges incident on u. Let $a_1 \leq a_2 \leq \cdots \leq a_p$ be the strong edges in G and $\beta_{max}(u) = a_i$. Then by above Theorem 8.3.14, $Co(u) = c$, where c is the weight of some $\beta-$strong edge. Thus, $Co(v) = c\ \forall v \in \sigma^*$. If $Co(v) = d \neq c$ for some $v \in \sigma^*$, then $C^G_{v,x} = d\ \forall x \in \sigma^*$. Thus, $C^G_{v,u} = d$, which implies, $\zeta(u) = \{c, d\}$. But we have already proved $|\zeta(u)| = 1$. So, $c = d$. Hence, $Co(u) = Co(v) = c\quad \forall u \in \sigma^*$.

Since a_i's are weights of $\beta-$strong edges, there is at least one edge say $mn \in \sigma^*$ such that $\mu(mn) = a_1$. So, using the cardinality property of $\zeta-$evaluation we have $\zeta(m) = \zeta(n) = a_1$. Thus we have, $Co(u) = a_1\ \forall u \in \sigma^*$. ■

If the cycle connectivity between some pair of vertices in a fuzzy graph is zero, then we can find a cut vertex in the graph. This fact is true for cycle cogency also. Because, any two vertices in a block fuzzy graph lie on a common strong cycle. Thus every vertex in a block should have a non zero cycle cogency. This leads to the following theorem.

Theorem 8.3.16 *If there exists a fuzzy graph with at least one vertex of zero cycle cogency, then the graph is separable.*

Converse of above theorem is not true in general. In Fig. 8.18 we can see that the fuzzy graph G is separable but no vertex in G has cycle cogency zero.

Theorem 8.3.17 *Let $G = (\sigma, \mu)$ be a fuzzy graph. If uv is an $\alpha-$strong edge which is not a fuzzy bond in G, then $Co(u) = Co(v) = C^G_{u,v}$.*

Proof Assume on the contrary that $Co(u) \neq Co(v)$. Without loss of generality, let $Co(u) < Co(v)$. Now let C_1 be a strong cycle of G contributing to the cycle cogency of v. That is, $Co(v) = S(C_1)$. Since $Co(u) < Co(v)$, $u \notin C_1$. Let x be an arbitrary vertex in C_1. Since uv is not a fuzzy bond, there exists a strongest $u - x$ path P_{ux} (say) such that the edge $uv \notin P_{ux}$. Let $y \in C_1$ be the first common vertex of C_1 and P_{ux} as one moves in the clockwise direction along C_1 starting from v. Now there are two cases.

Case 1. $y = v$.

In this case, we can split the path into two sub paths as $P_{ux} = P_1 \cup P_1'$ with $P_1 : u - y = v$ and $P_1' : v - x$. Since uv is a fuzzy bridge, $S(P_1) < \mu(uv)$. Thus, the path $P_{ux}' = uv \cup P_1'$ becomes a path with $S(P_{ux}') > S(P_{ux})$, a contradiction.

Case 2. $y \neq v \in C_1$.

In this case also, partition P_{ux} as $P_{ux} = P_1 \cup P_1'$ as $P_1 : u - y$ and $P_1' : y - x$. Also, split C_1 as $C_1 = P_2 \cup P_2' \cup P_2''$, where $P_2 = v - y$, $P_2' = y - x$ and $P_2'' = x - v$ are paths. First we need to prove that, $S(P_1) = S(P_2)$.

For this we claim, $S(P_1) \geq S(P_2)$. We know, $P_1 \cup P_2$ is a path from u to v. But since uv is $\alpha-$strong, $\mu(uv) > S(P_1) \vee S(P_2)$. If we assume contrary of the claim, then $P_{uy} : uv \cup P_2$ becomes a strong path from u to y with $S(P_1) < S(P_{uy})$. Thus, $u - x$ path $P : P_{uy} \cup P_1'$ has strength $S(P) > S(P_{ux})$, a contradiction. Thus our assumption is wrong and so $S(P_1) \geq S(P_2)$ is always true.

Again we claim that $S(P_1) \not> S(P_2)$. If possible, $S(P_1) > S(P_2)$, then using $\mu(uv) > S(P_1) \vee S(P_2)$ we have $S(vu \cup P_1) > S(P_2)$, which implies that, some edge in P_2 is non strong, a contradiction. Thus, $S(P_1) = S(P_2)$. Now if P_1 is strong, then the cycle $C : vu \cup P_1 \cup P_2' \cup P_2''$ become a strong cycle having $S(C) = S(C_1)$. Now to make C strong, we need to prove that P_1 is strong. (Now for the simplicity take any arbitrary path say $P_6 : m - n$ and o to be a vertex in P_6. Then take a sub path of P_6 from m to o and call it as P_{6o} and a sub path of P_6 from o to n as P_{o6}.) Let P_1 be a non strong path, then there exist at least one non strong edge say $gh \in P_1$. Now rewrite P_1 as $P_1 : P_{1g} \cup gh \cup P_{h1}$. Since uv is not a fuzzy bond and $\mu(gh) \geq S(P_2)$, there exists a strongest $g - h$ path P_3 not containing the edges of the path P_2 and edge uv. Also $v \notin P_3$ else, Case 1 happens.

Thus a path can be constructed as $P_{1modified} : P_{1g} \cup P_3 \cup P_{h1}$. Repeat till all non strong edges are removed from P_1. Let the new path from u to x be $P_{modified}$, such that $v \notin P_{modified}$. Thus, The cycle, $C : vu \cup P_{modified} \cup P_2' \cup P_2''$ becomes a strong cycle having $S(C) = S(C_1)$. Thus, C contributes to the cycle cogency of u, with $Co(u) \geq S(C)$. But if $Co(u) > S(C)$, then the cogency of u becomes lesser than cogency of v, making a contradiction to the first assumption. Thus, $Co(u) = Co(v) = S(C)$. Since both u and v pass through C, by combing the definitions of cycle connectivity and cycle cogency we have, $C_{u,v}^G = S(C)$. ∎

Theorem 8.3.18 *Cycle cogency of isomorphic fuzzy graphs are equal.*

Proof Let G and G' be two isomorphic fuzzy graphs. Then, there exists a bijective function $f : \sigma^* \to \sigma'$, such that for any $x \in \sigma^*$, $\sigma(x) = \sigma'(f(x))$ and for any $xy \in \mu^*$, $\mu(xy) = \mu'(f(x)f(y))$. Let $x, y \in \sigma^*$ and $\zeta(x, y) = \{w_1, w_2, \ldots, w_m\}$

for some integer $m \geq 0$. If we can show that $\zeta-$evaluation of $f(x)$ in G' as $\zeta'(f(x)) = \{w_1, w_2, \ldots, w_m\}$, then the proof is complete.

$w_1 \in \zeta(x)$ implies that there exist at least one strong cycle C passing through x of strength w_1 in G. We need to show that, a strong cycle say C' of strength w_1 in G' exists and passes through $f(x)$. For, consider a strong cycle $C : c_0 = x, c_1, \ldots, c_k, c_{k+1} \ldots, c_p, c_0 = x$ in G of strength w_1 passing through x. If $c_i c_{i+1}$ is a strong edge of strength s_i in G, then $f(c_i) f(c_{i+1})$ will be strong in G' of strength s_i, for $i = 0, 1, \ldots, p-1$. Since G and G' are isomorphic there exists a cycle C' : $f(c_0) = f(x), f(c_1), \ldots, f(c_k), f(c_{k+1}), \ldots, f(c_p), f(c_0) = f(x)$. It is a strong cycle in G' of strength w_1. Hence, $\zeta(x) = \{w_1, w_2, \ldots, w_m\} \subseteq \zeta'(f(x), f(y))$. Because, there might have another strong cycle in G' of strength w' passing through $f(x)$ such that w' does not belong to $\{w_1, w_2, \ldots, w_m\}$. But since G' and G are isomorphic, using the above arguments, one can prove that $\zeta'(f(x)) \subseteq \zeta(x)$.

In other words, there does not exist a strong cycle in G' of strength w' through $f(x)$ whenever w' is not an element of $\zeta(x)$. Thus, $\zeta(x) = \zeta'(f(x))$, which follows that

$$Co_G(x) = \vee \{w_1, w_2, \ldots, w_m\} = Co_{G'}(f(x))$$

Hence, $Co(G) = \sum_{u \in \sigma^*} Co_G(u) = \sum_{f(u) \in \sigma'^*} Co(f(u)) = Co(G')$. ■

Algorithm to Find the Cogency of a Given Fuzzy Graph

In [6] Mathew and Sunitha proposed an algorithm that determines different types of edges in a given fuzzy graph $G = (\sigma, \mu)$ with $|\sigma^*| = n$. Also there are algorithms available to identify cycles in a graph. Now we use the algorithm proposed in [7] to identify cycles in fuzzy graphs.

Construction of the algorithm is as follows.

Step 1: Identify the types of edges in $G = (\sigma, \mu)$ by algorithm in [6].

Step 2: Construct a strong fuzzy subgraph G' of the fuzzy graph of $G = (\sigma, \mu)$ such that G' contains only strong edges of G.

Step 3: Using Gospel's algorithm suggested in [7], identify the cycles of G' as $C_1, C_2, \ldots, C_k$.

Step 4: Let $w_i = \wedge\{\mu(xy) : xy$ is an edge in $C_i\}$ for $i = 1, 2, \ldots, k$.

Step 5: Construct a $k \times n$ matrix with cycles as rows and vertices as columns. If C_r passes through a vertex u, then w_r is the entry corresponding to u in row r and the entry is zero otherwise for $r = 1, 2, \ldots, k$.

Step 6: $Co(u) = \vee \{u_1, u_2, \ldots, u_p\}$, where u_d is the entry in row d corresponding to the column of u for $d = 1, 2, \ldots, k$.

Step 7: Let $Co(G) = \sum_{u \in \sigma^*} Co_G(u)$.

The values obtained in steps 6 and 7 are the required cogency of a vertex and cogency of the given fuzzy graph G.

Illustration for the algorithm

Consider $G = (\sigma, \mu)$ (see Fig. 8.19) where $\sigma^* = \{z_1, z_2, z_3, z_4, z_5, z_6, z_7, z_8\}$ with $\sigma(z) = 1 \forall z \in \sigma^*$ having $\mu(z_1 z_2) = \mu(z_2 z_5) = \mu(z_4 z_7) = 0.7$, $\mu(z_2 z_6) =$

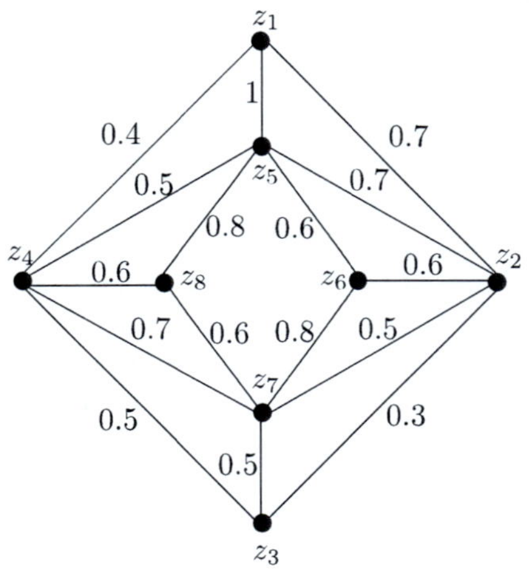

Fig. 8.19 Illustration of the Algorithm

$\mu(z_7z_8) = \mu(z_4z_8) = \mu(z_7z_8) = 0.6$, $\mu(z_2z_7) = \mu(z_3z_5) = \mu(z_3z_4) = \mu(z_4z_5) = 0.5$, $\mu(z_5z_8) = \mu(z_6z_7) = 0.8$ $\mu(z_1z_4) = 0.4$, $\mu(z_2z_3) = 0.3$ and $\mu(z_1z_5) = 1$.

$$\begin{bmatrix} 0.7 & 0.7 & 0 & 0 & 0.7 & 0 & 0 & 0 \\ 0 & 0.6 & 0 & 0 & 0.6 & 0.6 & 0 & 0 \\ 0.6 & 0.6 & 0 & 0 & 0.6 & 0.6 & 0 & 0 \\ 0.6 & 0.6 & 0 & 0 & 0.6 & 0.6 & 0.6 & 0.6 \\ 0.6 & 0.6 & 0 & 0.6 & 0.6 & 0.6 & 0.6 & 0.6 \\ 0.6 & 0.6 & 0.6 & 0.6 & 0.6 & 0.6 & 0.6 & 0.6 \\ 0 & 0 & 0.5 & 0.5 & 0 & 0 & 0.5 & 0.5 \\ 0 & 0 & 0.5 & 0.5 & 0 & 0 & 0.5 & 0 \\ 0 & 0 & 0 & 0.5 & 0 & 0 & 0.5 & 0.5 \end{bmatrix}$$

It can be seen that in G, there are four non strong edges namely, ad, de, bg, bc. Other edges are strong and are part to some cycle C_i. Let $G' = (\sigma', \mu')$ be a strong fuzzy subgraph of G such that G contain all strong edges of G (see Fig. 8.20). Note that $|\sigma'^*| = 8$.

There are nine strong cycles in G. The details are given in Table 8.1 Thus we can construct a 9×8 matrix with cycles as rows and vertices as columns.

But the entries in the columns corresponding to those vertices that do not belong to a common strong cycle will be zero. Thus we can construct a matrix shown with those vertices. From the matrix, it can be said that for a vertex u, $Co(u)$ is the maximum value among all entries in the column corresponding to u. Here, $\zeta(z_1) = \zeta(z_2) = \zeta(z_5) = \{0.6, 0.7\}$, $\zeta(z_3) = \zeta(z_4) = \zeta(z_6) = \zeta(z_7) = \zeta(z_8) = \{0.5, 0.6\}$. Thus, $Co(z_1) = Co(z_2) = Co(z_5) = 0.7$ and $Co(z_3) = Co(z_4) = Co(z_6) = Co(z_7) = Co(z_8) = 0.6$. Thus, $Co(G) = \sum_{z\in\sigma^*} Co_G(z) = 5.1$.

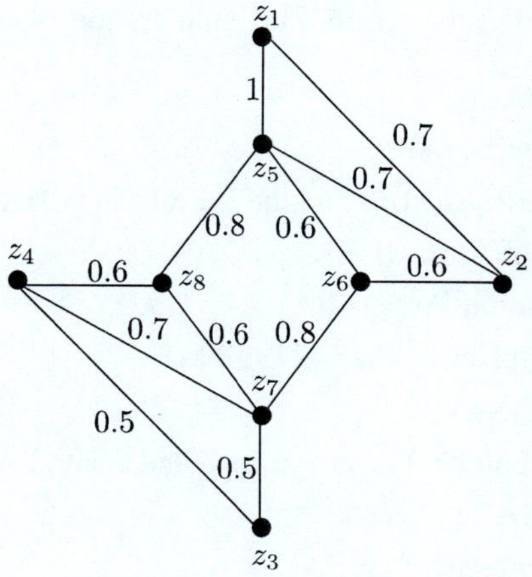

Fig. 8.20 Strong cycles in Fig. 8.19

Table 8.1 Details of strong cycles in Fig. 8.19

Cycle No.	Strong cycle C_i	Strength of cycle C_i
1	z_1, z_2, z_5, z_1	0.7
2	z_2, z_5, z_6, z_2	0.6
3	z_1, z_2, z_6, z_5, z_1	0.6
4	$z_1, z_2, z_6, z_7, z_8, z_5, z_1$	0.6
5	$z_1, z_2, z_6, z_7, z_4, z_8, z_5, z_1$	0.6
6	$z_1, z_2, z_6, z_7, z_3, z_4, z_8, z_5, z_1$	0.5
7	$z_3, z_4, z_8, z_5, z_2, z_6, z_7, z_3$	0.5
8	$z_3, z_4, z_8, z_5, z_6, z_7, z_3$	0.5
9	z_3, z_4, z_8, z_7, z_3	0.5
10	z_3, z_4, z_7, z_3	0.5
11	z_4, z_8, z_7, z_4	0.6

8.4 Application to Human Trafficking

The following is taken from [8]. The International Organization for Migration (IOM) detailed the plight of nearly 3 million migrants stranded worldwide by mid-July. Many more migrants are believed to have been stranded in the subsequent months. This situation posed unprecedented challenges to IOM return and reintegration activities and resulted in a number of adoptions, allowing it to continue to providing return and reintegration support to migrants in need despite health and travel restrictions. The specific changes and innovation practices adopted by IOM offices worldwide in the field of return an reintegration are further discussed in the last chapter of this report, ([8], Introduction).

In [8], IOM considered the return and reintegration of the nine regions given below.

Asia and the Pacific (AP)
Central and North America and the Caribbean (CHAC)
East and Horn of Africa (EHA)
European Economic Area (EEA)
Middle East and North Africa (NMENA)
South America (SA)
South-Eastern Europe, Eastern Europe and Central Asia (SEEECA)
South Africa (SAf)
West and Central Africa (WCA)

IOM provided two tables for each region. One table, presented the number of returns from one region to another for the region as a host region and a second table presenting the number of returns from one region to another for the region as a host region. The numbers are presented in the following table are from [8].

In the following table, the regions heading the rows are returning refugees to the regions heading the columns. The regions heading the rows are the regions of origin. For example, the first column gives the number returnees from the origin regions to AP (Table 8.2).

We next divide every entry in a given column by its column sum. The column sums do not include the elements on the diagonal. The entries in the following table are determined by dividing the entry by the column sum of the previous table. This results in a measure of the returnees for a region as a relation the number of returnees in another region. It allows us to place the situation in a mathematics

Table 8.2 Returns from origin

Returner \ Returnee	AP	CNAC	EHA	EEA	MENA	SA	SEEECA	SAf	WCA
AP	621	16	367	48	27	37	7	29	33
CNAC	5	292		1		59	1	1	
EHA	48	3	768	7		3		4	78
EEA	3678	522	239	476	2153	2056	6908	96	521
MENA	1373	4	2102	6	14	4	6	153	3503
SA	7	3		2		9	6		
SEECA	694	15	11	8	150	10	3349	10	29
SAf	2		258	1			2	323	44
WCA	11	5	28	4	2			17	10936
Col Sum	5818	568	3005	77	2332	2169	6930	310	4208

Table 8.3 Percentage returns from origin

Returner \ Returnee	AP	CNAC	EHA	EEA	MENA	SA	SEEECA	SAf	WCA
AP		0.028	0.122	0.623	0.012	0.017	0.001	0.094	0.008
CNAC	0.001			0.013		0.027	0	0.003	
EHA	0.008	0.005		0.091		0.001		0.013	0.019
EEA	0.632	0.919	0.080		0.923	0.948	0.997	0.310	0.124
MENA	0.236	0.007	0.700	0.078		0.002	0.001	0.494	0.832
SA	0.001	0.005		0.026			0.001		
SEECA	0.119	0.026	0.004	0.104	0.064	0.005		0.032	0.007
SAf	0.000		0.086	0.013			0		0.010
WCA	0.002	0.009	0.009	0.052	0.001			0.055	
Col Sum									

Table 8.4 Returns from host

Returnee \ Returner	AP	CNAC	EHA	EEA	MENA	SA	SEEECA	SAf	WCA
AP	621	5	48	3678	1373	7	694	2	11
CNAC	16	292	3	521	4	3	15		5
EHA	367		768	239	2102		11	258	28
EEA	48	1	7	476	6	2	8	1	4
MENA	27			2153	14		150		2
SA	37	59	3	2056	4	9	10		
SEECA	7	1		6908	6	6	3349	2	
SAf	29	1	4	96	153		10	323	17
WCA	33		78	522	3503		29	44	10936
Col Sum	564	67	143	16173	7151	18	927	307	67

on uncertainty context. The results are given in the following table. For example, $\mu(EEA, AP) = \frac{3678}{5818} = 0.632$ and $\mu(MENA, AP) = \frac{1373}{5818} = 0.236$.

In the following table, the regions heading the columns are returning refugees to the countries heading the rows. The regions heading the columns are host regions. For example, the first column gives the number of returnees from the host region AP to the other regions (Table 8.4).

The columns sums do not include the elements on the diagonal. The entries in the following table are determined by dividing the entry by the column sum of the previous table. For example, $\mu(AP, CNAC) = \frac{16}{564} = 0.028$ and $\mu(AP, EHA) = \frac{367}{564} = 0.651$.

From Tables 8.3 and 8.5 we can define fuzzy directed graphs. Let V denote the set of all regions and let $E = V \times V$. Then $G = (V, E)$ can be considered a directed graph with V the set of vertices and E the set of directed edges, where if $(u, v) \in$

Table 8.5 Percentage returns from host

Returnee \ Returner	AP	CNAC	EHA	EEA	MENA	SA	SEEECA	SAf	WCA
AP		0.075	0.336	0.227	0.192	0.389	0.747	0.007	0.164
CNAC	0.028		0.021	0.032	0.001	0.167	0.016		0.075
EHA	0.651			0.015	0.294		0.012	0.840	0.418
EEA	0.085	0.015	0.049		0.001	0.111	0.009	0.003	0.060
MENA	0.048			0.133			0.162		0.030
SA	0.066	0.880	0.021	0.127	0.001		0.011		
SEECA	0.012	0.015		0.427	0.001	0.333		0.007	
SAf	0.051	0.015	0.028	0.006	0.021		0.011		0.254
WCA	0.059		0.545	0.032	0.490		0.031	0.143	
Col Sum									

$V \times V$, then $(u, v))$ is a directed edge from u to v. Define the fuzzy subsets σ of V and μ of E by $\sigma(v) = 1$ for all $v \in V$ and $\mu(u, v) =$ the entry in Table 8.3 with u heading the row and v heading the column. For example, $\mu(EEA, AP) = 0.632$. Define the fuzzy τ of V and ν of E by for all $v \in V$, $\tau(v) = 1$ and $\nu(u, v) =$ the entry in Table 8.5 with u heading the column and v heading the row. For example, $\mu(AP, CNAC) = 0.028$.

We consider the cycle connectivity of both fuzzy subgraphs. Using Table 8.3 we find that the cycle connectivity of G is 0.236. The cycle involves AP, EEA, and $MENA$. For Table 8.5 we find that the cycle connectivity is 0.164. The cycle involves WCA, AP, and EHA. This result is in keeping with fact that EEA was the main host region with a share of 39% of the total number of migrants assisted to return in 2020. WCA was the main region of origin with a share of 36% of the total number of migrants assisted to return in 2020.

References

1. Mathew, S., Sunitha, M.S.: Cycle connectivity in fuzzy graphs. J. Intell. Fuzzy Syst. **24**, 549–554 (2013)
2. Sunitha, M.S., Vijayakumar, A.: Blocks in fuzzy graphs. J Fuzzy Math **13**(1), 13–23 (2005)
3. Sunitha, M.S., Vijayakumar, A.: Complement of a fuzzy graph, Indian Journal of. Pure Appl. Math. **33**, 1451–1464 (2002)
4. Mathew, S., Mordeson, J.N., Malik, D.S.: Fuzzy Graph Theory, vol. 363. Springer International Publishing (2018)
5. Jicy, N., Mathew, S.: Connectivity analysis of cyclically balanced fuzzy graphs. Fuzzy Inf. Eng. **7**, 245–255 (2015)
6. Mathew, S., Sunitha, M.S.: Types of arcs in a fuzzy graph. Inf. Sci. **179**, 1760–1768 (2009)
7. Read, R.C., Tarjan, R.E.: Bounds on backtrack algorithms for listing cycles, paths and spanning trees. Networks **3**(5), 237–252 (1975)
8. IOM UN Migration, Return and Reintegration Key Highlights (2020)

Chapter 9
Neighborhood Connectivity in Fuzzy Graphs

A new connectivity parameter for fuzzy graphs is discussed in this chapter. It is termed as neighborhood connectivity index. It gives the local connectivity in a very specific manner. Most of the fuzzy graph theoretic structures are studied and their neighborhood connectivity indices are evaluated. Product fuzzy graphs also can be seen towards the end of this chapter. The contents of this chapter are from [1].

9.1 Neighborhood Connectivity Index of Fuzzy Graphs

The potential or capacity of a specific vertex is important in several communication problems. The definition of neighborhood connectivity index given in Definition 9.1.1 is a measure of the potential of a vertex.

Definition 9.1.1 The **neighborhood connectivity index** (NCI), of a fuzzy graph G is defined as $NCI(G) = \sum_{m \in V(G)} d(m)e(m)$, where $d(m)$ is the cardinality of $N(m)$ and $e(m) = \vee\{\mu(mp) : p \in N(m)\}$ with $N(m) = \{p : \mu(mp) > 0, m, p \in \sigma^*\}$. $e(m)$ is termed as the **potential** of the vertex m.

In a fuzzy graph, a vertex with maximum potential is termed as **maximum potential vertex**. We denote potential of a vertex in a fuzzy graph G by $e_G(m)$. Similarly we denote $d_G(m)$, $N_G(m)$ for $d(m)$, $N(m)$ without any conflicts.

Note that $e(m)$ can be defined in terms of connectivity in a different manner. For a vertex m, $e(m) = \vee\{CONN_G(m, p) : p \in V(G)\}$.

For every $x \in \sigma^* \setminus \{m\}$, a strongest $m - x$ path P, (say) contains an edge from $E(m)$, where $E(m) = \{mp : p \in N(m)\}$. If $\vee\{\mu(mp) : p \in N(m)\} = \alpha$, then strength of P is less than or equal to α. In particular if $\mu(mz) = \alpha$ then mz is a strongest path with strength $e(m)$. Therefore both the definitions of $e(m)$ are equivalent.

J. N. Mordeson et al., *Fuzzy Graph Theory*, Studies in Fuzziness and Soft Computing 424, https://doi.org/10.1007/978-3-031-23108-7_9

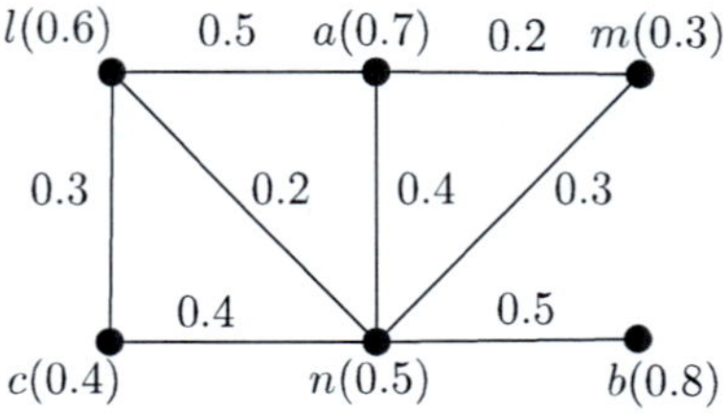

Fig. 9.1 A fuzzy graph G with $NCI(G) = 7.4$

Example 9.1.2 Consider $G = (\sigma, \mu)$ with $\sigma^* = \{l, a, m, b, n, c\}$; $\sigma(l) = 0.6$, $\sigma(a) = 0.7$, $\sigma(m) = 0.3$, $\sigma(b) = 0.8$, $\sigma(n) = 0.5$, $\sigma(c) = 0.4$, and $\mu(la) = 0.5$, $\mu(ln) = 0.2$, $\mu(lc) = 0.3$, $\mu(am) = 0.2$, $\mu(an) = 0.4$, $\mu(mn) = 0.3$, $\mu(bn) = 0.5$, $\mu(nc) = 0.4$.

Now we can find that $d(l) = 3$, $e(l) = \vee\{0.5, 0.3, 0.2\} = 0.5$. Similarly we can find d values and e values for the rest of the vertices also.

Vertex	$d(x)$	$e(x)$	$d(x)e(x)$
l	3	0.5	1.5
a	3	0.5	1.5
m	2	0.3	0.6
b	1	0.5	0.5
n	5	0.5	2.5
c	2	0.4	0.8
$NCI(G)$			7.4

Thus for G in Fig. 9.1, $NCI(G) = 3 \times 0.5 + 3 \times 0.5 + 2 \times 0.3 + 1 \times 0.5 + 5 \times 0.5 + 2 \times 0.4 = 7.4$.

We have an obvious observation, which is given as the next remark.

Remark Neighborhood connectivity index of a fuzzy graph is zero if and only if the cardinality of its edge set is zero.

Proposition 9.1.3 *If $H = (\tau, \nu)$ is a partial fuzzy subgraph of $G = (\sigma, \mu)$, then $NCI(H) \leq NCI(G)$.*

Proof Suppose $H = (\tau, \nu)$ be a partial fuzzy subgraph of $G = (\sigma, \mu)$, with $\sigma^* = \{m_1, m_2, \ldots, m_n\}$. Let m be an arbitrary vertex in τ^*. Then $\nu(mm_i) \leq \mu(mm_i)$ for all other vertices m_i in τ^*. Therefore, $\vee_i\{\nu(mm_i)\} \leq \vee_i\{\mu(mm_i)\}$. Also, $d_H(m) \leq d_G(m)$. Therefore, $NCI(H) = \sum_{m_i} d_H(m_i) \vee_i\{\nu(mm_i)\} \leq \sum_{m_i} d_G(m_i) \vee_i\{\mu(mm_i)\} = NCI(G)$. □

Example 9.1.4 Consider $H = (\tau, \nu)$ in Fig. 9.2. Clearly H is a partial fuzzy subgraph of $G = (\sigma, \mu)$ mentioned in Example 9.1.2. After computing the connectedness between the vertices and cardinality of neighborhood for each vertex, neighborhood connectivity index of H can be calculated as 4.2 which is less than that of G, which is 7.4.

In the following results we establish some bounds for the index.

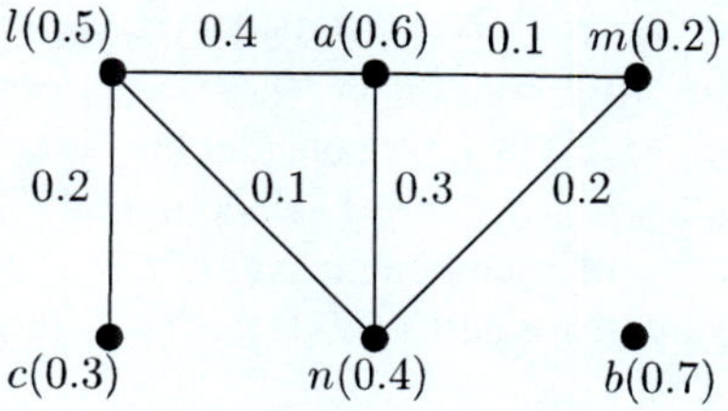

Fig. 9.2 Subgraph H of G with $NCI(H) = 4.2$

Corollary 9.1.5 *For a fuzzy graph $G = (\sigma, \mu)$ with vertex set σ^* and complete fuzzy super graph $G' = (\sigma', \mu')$ spanned by σ^*, we have $0 \leq NCI(G) \leq NCI(G')$.*

Proposition 9.1.6 *For G with $|\sigma^*| = n$, $0 \leq NCI(G) \leq n(n-1)$.*

Proof Consider $G = (\sigma, \mu)$. If $\mu^* = \phi$, then $d(m) = 0$, $e(m) = 0$ for all $m \in \sigma^*$. Thus $NCI(G) = 0$. If $|\mu^*| > 0$, then $0 < d(m) \leq n-1$, $0 < e(m) \leq 1$ for at least one $m \in \sigma^*$. This implies $0 < NCI(G) \leq \sum_{m\in\sigma^*}(n-1) \times 1 = n(n-1)$. The upper bound occurs when the underlying graph is a complete graph and there exist at least one edge incident to each vertex having strength 1. Therefore, $0 \leq NCI(G) \leq n(n-1)$. □

Proposition 9.1.7 *Let $G = (\sigma, \mu)$ be a connected fuzzy graph with n edges. Then $2nt \leq NCI(G) \leq 2ns$ where $t = \wedge\{e(m) : m \in \sigma^*\}$ and $s = \vee\{e(m) : m \in \sigma^*\}$.*

Proof Suppose $G = (\sigma, \mu)$ is a fuzzy graph with n edges. Then $NCI(G) = \sum_{m\in\sigma^*} e(m)d(m) \leq \sum_{m\in\sigma^*} sd(m) = s\sum_{m\in\sigma^*} d(m) = s \times 2n = 2sn$. Similarly, $NCI(G) = \sum_{m\in\sigma^*} e(m)d(m) \geq \sum_{m\in\sigma^*} td(m) = t\sum_{m\in\sigma^*} d(m) = t \times 2n = 2tn$. Therefore, $2nt \leq NCI(G) \leq 2ns$. □

Equality in Proposition 9.1.7 holds only when all the vertices of G possess the same potential. Now we look at the neighborhood connectivity indices of some known structures such as trees, cycles and complete fuzzy graphs.

Corollary 9.1.8 *Consider a fuzzy graph $G = (\sigma, \mu)$ where G^* is a tree. Let $\vee\{d(m) : m \in \sigma^*\} = r$ and let S_i be the set of vertices containing all vertices with degree i, $1 \leq i \leq r$. Then $NCI(G) = \sum_{m\in S_i, i=1}^{r} i \sum_{p\in N(m)} \vee\{\mu(mp)\}$.*

Corollary 9.1.9 *For a cycle $G = (\sigma, \mu)$ having edges $e_1, e_2, \cdots, e_n$ with $\mu(e_i) = t_i$ and $t_{n+1} = t_1$, we have $NCI(G) = 2\sum_{i=1}^{n} \vee\{t_i, t_{i+1}\}$.*

Corollary 9.1.10 *Let $G = (\sigma, \mu)$ be a CFG with $\sigma^* = \{m_1, m_2, \ldots, m_n\}$ such that $t_1 \leq t_2 \leq \cdots \leq t_n$, where $t_i = \sigma(m_i)$, $1 \leq i \leq n$. Then $NCI(G) = (n-1)(t_1 + t_2 + \cdots + t_{n-2} + t_{n-1} + t_{n-1})$.*

Proof Consider the graph G. We know that, for a CFG, $\mu(m_i m_j) > 0$ for all $m_i, m_j \in \sigma^*$. Therefore, $d(m_i) = n-1$ for all m_i, $1 \leq i \leq n$. Now, we can check the potential of vertices. While considering m_1 we see that it is the vertex with minimum membership value. So we can see that $CONN_G(m_1, m_i) = t_1$, $2 \leq i \leq n$. Therefore,

$e(m_1) = t_1$. Next, consider the vertices m_i, $1 < i < n$. Here, $CONN_G(m_s, m_i) \le t_i$ for all $s < i$, $CONN_G(m_r, m_i) = t_i$ for all $r > i$; therefore, $e(m_i) = t_i$, $2 \le i \le n-1$. At last, we consider the vertex m_n. Here we can see that $CONN_G(m_i, m_n) \le t_{n-1}$, $1 \le i \le n-1$, since there is no edge of membership value t_n, but there is an edge of membership value t_{n-1}. Therefore, $e(m_n) = t_{n-1}$. Summing up all those values, we get $NCI(G) = (n-1)(t_1 + t_2 + \cdots + t_{n-2} + t_{n-1} + t_{n-1})$. □

Proposition 9.1.11 *Neighborhood connectivity index of two isomorphic fuzzy graphs are equal.*

Proof Let j be a bijection between the isomorphic fuzzy graphs G_1 and G_2. Since weights of the edges and vertices are preserved by an isomorphism, $N_{G_1}(m) = N_{G_2}(j(m))$, which implies $d_{G_1}(m) = d_{G_2}(j(m))$ for $m \in \sigma_1^*$. Similarly, $CONN_{G_1}(m, p) = CONN_{G_2}(j(m), j(p))$ for $m, p \in \sigma_1^*$. Implying $e_{G_1}(m) = e_{G_2}(j(m))$. Therefore,

$$NCI(G_1) = \sum_{m \in V(G)} d_{G_1}(m) e_{G_2}(m) = \sum_{f(m) \in V(G)} d_{G_2}(j(m)) e_{G_2}(j(m)) = NCI(G_2).$$

i.e., $NCI(G_1) = NCI(G_2)$. □

Theorem 9.1.12 *Consider a fuzzy graph* $G = (\sigma, \mu)$. *If* $0 \le t_1 \le t_2 \le 1$, *then* $NCI(G^{t_2}) \le NCI(G^{t_1})$.

Proof Consider a fuzzy graph $G = (\sigma, \mu)$. In G^{t_2} number of edges with non zero strength incident at a vertex is less than or equal to the number of edges with non zero strength incident at a vertex in G^{t_1}. Therefore, $d_{G^{t_2}}(m) \le d_{G^{t_1}}(m)$. If $\mu_G(mp) \le t_1$, then $\mu_{G^{t_2}}(mp) = \mu_{G^{t_1}}(mp)$. If $t_1 < \mu_G(mp) \le t_2$, then $\mu_{G^{t_2}}(mp) \le \mu_{G^{t_1}}(mp)$. If $\mu_G(mp) > t_2$, then $\mu_{G^{t_2}}(mp) = \mu_{G^{t_1}}(mp)$. Now for $m \in \sigma^*$, $CONN_{G^{t_2}}(m, p) \le CONN_{G^{t_1}}(m, p)$ for all $p \in \sigma^*$. Therefore, $e_{G^{t_2}}(m) \le e_{G^{t_1}}(m)$. Therefore, $NCI(G^{t_2}) = \sum_{m \in V(G)} e_{G^{t_2}}(m) d_{G^{t_2}}(m) \le \sum_{m \in V(G)} e_{G^{t_1}}(m) d_{G^{t_1}}(m) = NCI(G^{t_1})$. □

Next, result focuss on saturated fuzzy cycles.

Theorem 9.1.13 *Consider a saturated fuzzy cycle* G *with* $|V(G^*)| = n$ *for which every* $\alpha-$ *strong edge is of strength* t *and every* $\beta-$ *strong edge is of constant strength, then* $NCI(G) = 2nt$.

Proof Suppose $G = (\sigma, \mu)$ is as in statement of the theorem. Since G^* is a saturated fuzzy cycle, $d(m) = 2$ for any $m \in \sigma^*$. Also from the assumption it follows that t is greater than the constant strength of $\beta-$ strong edges, which implies, $e(m) = t$ for all $m \in \sigma^*$. Therefore, $NCI(G) = \sum_{i=1}^{n} 2t = 2nt$. □

Example 9.1.14 Consider the fuzzy cycle G so that $G^* = C_n$ as given in Fig. 9.3. Clearly it is a saturated fuzzy cycle with $\sigma^* = \{l, a, m, b, n, c, o, d\}$, $\mu(la) = 0.5$, $\mu(am) = 0.3$, $\mu(mb) = 0.5$, $\mu(bn) = 0.3$, $\mu(nc) = 0.5$, $\mu(co) = 0.3$, $\mu(od) = 0.5$, $\mu(dl) = 0.3$. Then neighborhood connectivity index, $NCI(G) = 2 \times 8 \times 0.5 = 8$.

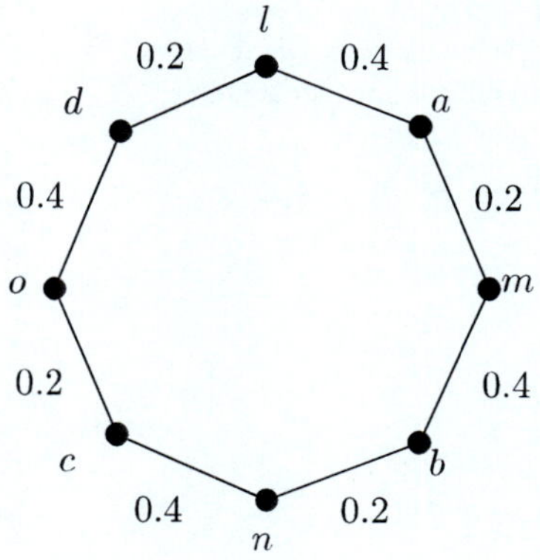

Fig. 9.3 Saturated fuzzy cycle G with $NCI(G) = 8$

Theorem 9.1.15 *There does not exist a connected super fuzzy graph with equal neighborhood connectivity index as that of the parent graph.*

Proof Consider a graph H which has a vertex p in addition to the parent graph G as shown in Fig. 9.4. Let $m \in \sigma_G^*$ then $d_G(m) \leq d_H(m)$, since m may or may not have an edge with p. Now consider p, since $p \notin G, 0 = d_G(p) < d_H(p)$. While considering the potential of the edges. For $m \in \sigma_G^*$, $e_G(m) \leq e_H(m)$, since there may or may not have an edge with strength greater than $e_G(m)$, adjacent to p. While considering p it is obvious that $0 = e_G(p) < e_H(p)$. Therefore, $NCI(G) = \sum_{m \in \sigma_G^*} d_G(m)e_G(m) = \sum_{m \in \sigma_G^*} d_G(m)e_G(m) + d_G(p)e_G(p) < \sum_{m \in \sigma_G^*} d_H(m)e_H(m) + d_H(p)e_H(p) = NCI(H)$. Now we have shown that there does not exist a connected super graph having same neighborhood connectivity index as that of the parent graph when we add a vertex. Next consider a graph H which has an edge e in addition to the parent graph G as shown in Fig. 9.5. There exists at least one vertex m in G to which is e is incident, then $d_G(m) < d_H(m)$. Also we can see that $e_G(m) \leq e_H(m)$. Therefore, $NCI(G) = \sum_{m \in \sigma_G^*} d_G(m)e_G(m) < \sum_{m \in \sigma_G^*} d_H(m)e_H(m) = NCI(H)$. Now we have shown that there does not exist a connected super graph having same neighborhood connectivity index as that of the parent graph when we add an edge. □

Next two theorems propose ways for the construction of fuzzy graphs with a given neighborhood connectivity index value and having some predefined constrains.

Theorem 9.1.16 *For a given $n \in \mathbb{N}, x \in \mathbb{R}$ with $x \leq 2n$, there exists a fuzzy graph $G = (\sigma, \mu)$ of neighborhood connectivity index x with $|\mu^*| = n$.*

Proof Let $|\mu^*| = n$. Construct a fuzzy graph $G = (\sigma, \mu)$ such that $\sigma(m_i) \geq \frac{x}{2n}$ for all $m_i \in \sigma^*$, $\mu(m_i m_j) = \frac{x}{2n}$ for all $m_i m_j \in \mu^*$. Now we can check neighborhood connectivity index of the constructed graph. Here $e(m_i) = \frac{x}{2n}$ for all $m_i \in \sigma^*$. Therefore, $NCI(G) = \sum_{m_i \in V(G)} d(m_i)\frac{x}{2n} = \frac{x}{2n} \sum_{m_i \in V(G)} d(m_i) = \frac{x}{2n} \times 2n = x$. Hence our constructed graph is a fuzzy graph of NCI x with $|\mu^*| = n$. □

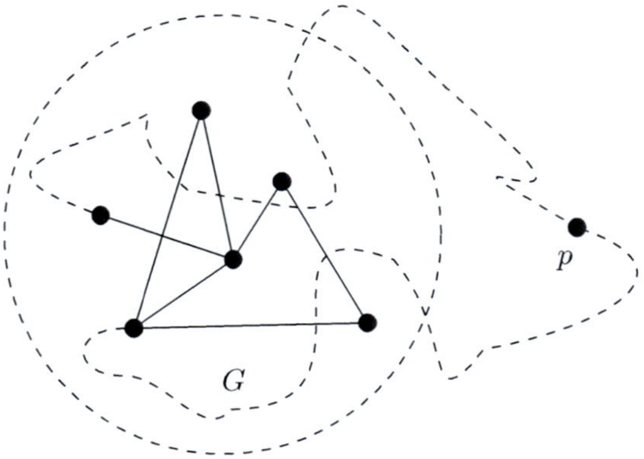

Fig. 9.4 Model of a fuzzy graph which has a vertex in addition to the parent graph

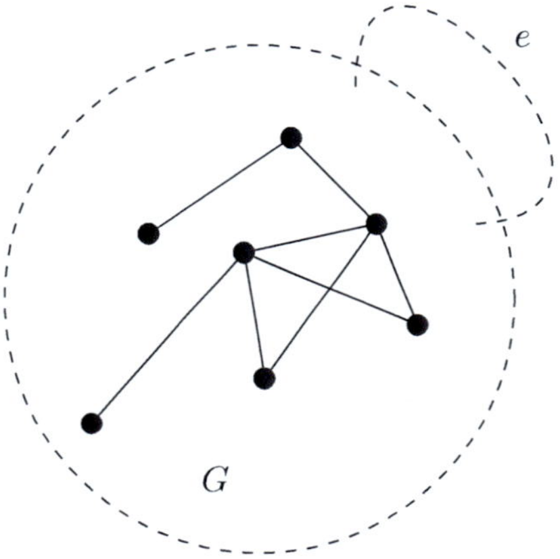

Fig. 9.5 Model of a fuzzy graph which has an edge in addition to the parent graph

Theorem 9.1.17 *For a given $n \in \mathbb{N}$, $x \in \mathbb{R}$ with $x \leq n(n-1)$, there exists a fuzzy graph $G = (\sigma, \mu)$ of neighborhood connectivity index x with $|\sigma^*| = n$.*

Proof We can prove this theorem by a similar construction from Theorem 9.1.16 by taking $|\sigma^*| = n$, $\sigma(m_i) \geq \frac{x}{n(n-1)}$ for all $m_i \in \sigma^*$ and $\mu(m_i m_j) = \frac{x}{n(n-1)}$ for all $m_i m_j \in \mu$. □

Example 9.1.18 Let $|\mu^*| = 4$, $x = 4$. Clearly, $4 \leq 8$. Now we can find a fuzzy graph $G = (\sigma, \mu)$ given in Fig. 9.6 such that $\sigma(l) = 0.8$, $\sigma(a) = 0.6$, $\sigma(m) = 0.5$, $\sigma(b) = 0.6$, $\mu(la) = 0.5$, $\mu(lm) = 0.5$, $\mu(am) = 0.5$, $\mu(ab) = 0.5$ with neighborhood connectivity index, $NCI(G) = 4$.

Proposition 9.1.19 *Consider a fuzzy cycle $G = (\sigma, \mu)$ with $|\sigma^*| = n \geq 4$ and $\sigma(m_i) = t$ for all $m_i \in \sigma^*$. Then $NCI(G^c) - NCI(G) \geq n^2 t - 5nt$, where $G^c = (\sigma^c, \mu^c)$ is the fuzzy complement of the fuzzy graph $G = (\sigma, \mu)$.*

Proof Suppose $G = (\sigma, \mu)$ is a fuzzy cycle. The neighborhood of each vertex in the fuzzy cycle has two vertices. Therefore, $d(m) = 2$. The potential of each vertex

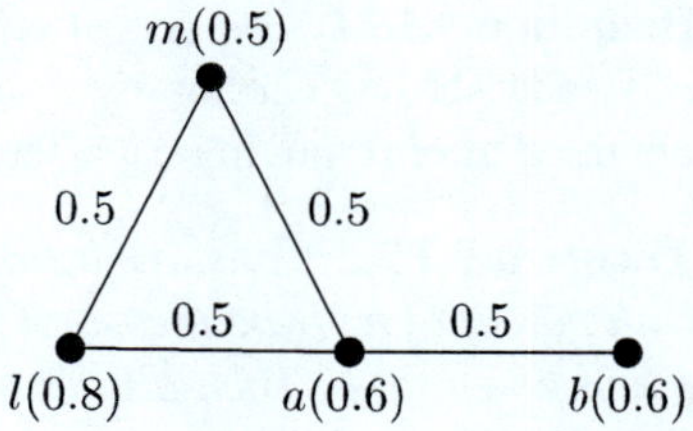

Fig. 9.6 Fuzzy graph with $NCI(G) = 4$, given $m = 4, x = 4$

will always be less than t, since each vertex has strength t. Therefore, $e(m) \leq t$. Therefore,

$$NCI(G) = \sum_{m \in V(G)} d(m)e(m) = 2 \sum_{m \in V(G)} e(m) \leq 2nt \tag{9.1}$$

Now consider the complement $G^c = (\sigma^c, \mu^c)$ of the graph $G = (\sigma, \mu)$. Clearly G^c will have all edges which are not present on the cycle. In addition to that some edges of the cycles can also appear. Therefore, each vertex can have a neighborhood of cardinality greater than $n - 3$. i.e., $d(m) \geq n - 3$ for all $m \in \sigma$. Since all the edges other than those lying on the cycle has strength t, and all others have strength less than t, we can say $e(m) = t$. Therefore,

$$NCI(G^c) = \sum_{m \in V(G)} d(m)e(m) = t \sum_{m \in V(G)} d(m) \geq nt(n-3) = n^2t - 3nt \tag{9.2}$$

From Eqs. (9.1) and (9.2), $NCI(G^c) - NCI(G) \geq n^2t - 3nt - 2nt = n^2t - 5nt$. □

Theorem 9.1.20 *For a fuzzy tree G, which is not a tree, with $F = (\sigma, \nu)$ as its maximum spanning tree, $NCI(F) < NCI(G)$.*

Proof Suppose $G = (\sigma, \mu)$ is a fuzzy tree, which is not a tree. Let $F = (\sigma, \nu)$ be the maximum spanning tree of G.
Claim For each vertex p in G, the edge with maximum strength incident at p will also lie on the maximum spanning tree F of G.

Proof of Claim Suppose not, let p be a vertex in G and pm be the edge with maximum strength incident at p. Suppose pm does not lie on the maximum spanning tree. Then $CONN_F(p, m) < CONN_G(p, m)$, a contradiction. Hence the claim. Now consider an arbitrary vertex m, then $e(m)$ is the maximum of the weight of edges starting from m. Hence by the claim we proved that $e_F(m) = e_G(m)$. Now we will show that $d_F(m) < d_G(m)$. Since our given fuzzy graph is not a tree, the maximum spanning tree of G will be different from G. There will be at least one edge removed from G. Let mp be such an edge. Then clearly, $d_F(m) < d_G(m)$ and $d_F(p) < d_G(p)$. Therefore, $NCI(F) = \sum_{m \in V(G)} d_F(m)e_F(m) = \sum_{m \in V(G)} d_F(m)e_G(m) < \sum_{m \in V(G)} d_G(m)e_G(m) = NCI(G)$. i.e., $NCI(F) < NCI(G)$. □

Definition 9.1.21 Two sets of vertices are called **twinning vertex sets of cardinality** r if each set has cardinality r and neighborhood connectivity index of the graph obtained after removing each set is same.

Theorem 9.1.22 *Consider a fuzzy graph G. Let A be the set of pendant vertices with potential a. B be the set of supporting vertices of vertices from A with degree c and potential b. Then all the sub graphs obtained after removing any one vertex from the set A will have equal neighborhood connectivity index.*

Proof Consider a fuzzy graph G. Let A and B be as defined in the theorem statement. We show that for $u, v \in A$, $NCI(G \setminus u) = NCI(G \setminus v)$. First, we consider vertices which do not belong to B or A. Let f be such a vertex. Then clearly $d_{G\setminus u}(f) = d_{G\setminus v}(f)$ and $e_{G\setminus u}(f) = e_{G\setminus v}(f)$. Let a be the supporting vertex of u and b be the supporting vertex of v. Then $d_{G\setminus u}(a) = d_{G\setminus v}(b)$, since $d_G(a)$ and $d_G(b)$ are equal and removing a pendant vertex reduces it by one and $e_{G\setminus u}(a) = e_{G\setminus v}(b)$, by condition. For those vertices g which belong to A, but they are not u and v and those vertices belong to B but they are not a and b, $d_{G\setminus u}(g) = d_{G\setminus v}(g)$ and $e_{G\setminus u}(g) = e_{G\setminus v}(g)$. Now consider $u, v \in A$. For them we have $d_{G\setminus v}(u) = d_{G\setminus u}(v)$ and $e_{G\setminus v}(u) = e_{G\setminus u}(v)$.
Now, $NCI(G \setminus v) = \sum_{t\notin B, t\notin A} d_{G\setminus v}(t)e_{G\setminus v}(t) + \sum_{t\in A, t\neq u, t\neq v} d_{G\setminus v}(t)e_{G\setminus v}(t)$
$+ \sum_{t\in B, t\neq a, t\neq b} d_{G\setminus v}(t)e_{G\setminus v}(t) + d_{G\setminus v}(u)e_{G\setminus v}(u) + d_{G\setminus v}(b)e_{G\setminus v}(b)$
$= \sum_{t\notin B, t\notin A} d_{G\setminus u}(t)e_{G\setminus u}(t) + \sum_{t\in A, t\neq u, t\neq v} d_{G\setminus u}(t)e_{G\setminus u}(t) + \sum_{t\in B, t\neq a, t\neq b} d_{G\setminus u}(t)$
$e_{G\setminus u}(t)$
$+ d_{G\setminus u}(v)e_{G\setminus u}(v) + d_{G\setminus u}(b)e_{G\setminus u}(b) = NCI(G \setminus u)$. □

The following corollary follows from Theorem 9.1.22.

Corollary 9.1.23 *Consider a fuzzy graph G. Let*

1. *A be the set of pendant vertices with potential a.*
2. *B be the set of supporting vertices of vertices from A with degree c and potential b.*
3. *A_i be the set of vertices of A having same supporting vertex from B.*

Then the neighborhood connectivity index of the sub graph obtained after removing s number of vertices from any A_i will be same.
i.e., such sets are twinning vertex sets of cardinality s.

Example 9.1.24 Consider the fuzzy graph G as in Fig. 9.7 with $\sigma^*=\{l, a, m, b, n, c, o, d, p, e, q, f, r\}$ and $\mu(la) = 0.3$, $\mu(am) = 0.2$, $\mu(ac) = 0.3$, $\mu(mb) = 0.2$, $\mu(md) = 0.3$, $\mu(mo) = 0.4$, $\mu(bn) = 0.4$, $\mu(bd) = 0.1$, $\mu(bq) = 0.2$, $\mu(bf) = 0.1$, $\mu(pq) = 0.3$, $\mu(eq) = 0.3$, $\mu(fr) = 0.4$.

Here $NCI(G) = 8.9$, $NCI(G \setminus l) = 7.9$, $NCI(G \setminus p) = 7.9$, $NCI(G \setminus \{l, c\}) = 7.2$, $NCI(G \setminus \{p, e\}) = 7.2$. It shows that $\{l\}$ and $\{p\}$ are twinning vertex sets of cardinality one and $\{l, c\}$ and $\{p, e\}$ are twinning vertex sets of cardinality two.

Now we compare neighborhood connectivity index with connectivity index and Wiener index.

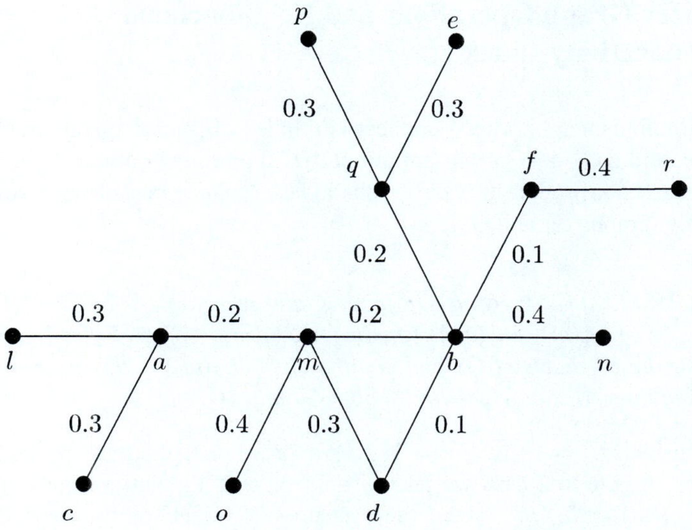

Fig. 9.7 Fuzzy graph having twinning vertex sets

Fig. 9.8 Fuzzy graph with $2CI(G) > NCI(G)$

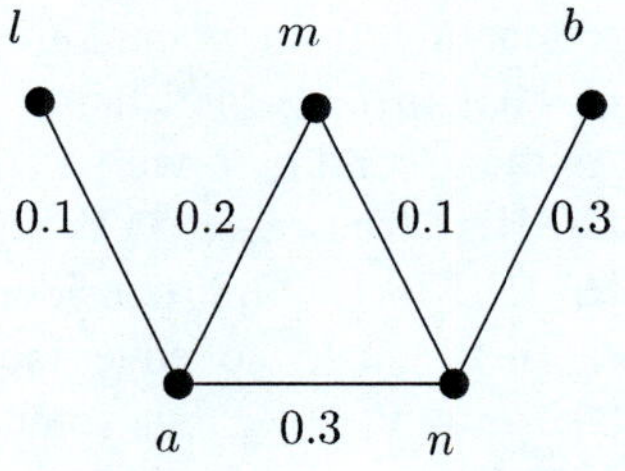

Theorem 9.1.25 *Let $G = (\sigma, \mu)$ be a complete fuzzy graph, $CI(G)$ be the connectivity index of G and $NCI(G)$ the neighborhood connectivity index of G. Then $2CI(G) \leq NCI(G)$.*

Proof Let $G = (\sigma, \mu)$ be a complete fuzzy graph. Then $2CI(G) = 2\sum_{m,p\in\sigma^*} \sigma(m)\sigma(p)CONN_G(m, p) \leq 2\sum_{m,p\in\sigma^*} CONN_G(m, p)$, (since $0 < \sigma(m), \sigma(p) \leq 1$) $\leq \sum_{m,p\in\sigma^*} e(m) + \sum_{m,p\in\sigma^*} e(p)$, (replace one $CONN_G(m, p)$ with $e(m)$ and another with $e(p)$) $= NCI(G)$, since each $e(m)$ repeats $d(m)$ times altogether. □

For a complete fuzzy graph we have $2WI(G) \leq NCI(G)$ since $CI(G) = WI(G)$ by Theorem 1.2.70.

The above result is not always true. Consider the fuzzy graph $G = (\sigma, \mu)$ given in Fig. 9.8 such that $\sigma^* = \{l, a, m, b, n\}$, $\mu(la) = 0.1$, $\mu(am) = 0.2$, $\mu(an) = 0.3$, $\mu(mn) = 0.1$, $\mu(bn) = 0.3$. The neighborhood connectivity index, $NCI(G) = 2.6$ and the connectivity index, $CI(G) = 1.9$. Here $2CI(G) = 3.8 \nleq 2.6 = NCI(G)$. Also note the Wiener index, $WI(G) = 4.2$ which is greater than 2.6.

9.2 Fuzzy Graph Operations and Neighborhood Connectivity Index

This section finds neighborhood connectivity index of product fuzzy graphs. $G_1 \cup G_2$ denote union, $G_1 + G_2$ represent join, $G_1[G_2]$ represent composition, $G_1 \times G_2$ represent Cartesian product of fuzzy graphs and $G_1 \otimes G_2$ represent the tensor product of two fuzzy graphs G_1 and G_2.

Theorem 9.2.1 *Let $G_i = (\sigma_i, \mu_i)$ be fuzzy graphs where $i = 1, 2$. Then $NCI(G_1 \cup G_2) = \sum_m [(\vee\{e_{G_1}(m), e_{G_2}(m)\})\ (d_{G_1}(m) + d_{G_2}(m) - |E_1 \cap E_2(m)|)]$, where E_1 and E_2 are the edge sets of G_1 and G_2 and $|E_1 \cap E_2(m)|$ is the number of edges arising from the vertex m which lies in both G_1 and G_2.*

Proof Consider $G_i = (\sigma_i, \mu_i), i = 1, 2$. We prove this theorem by considering three cases. As the first case we take $m \in V_1$ or $m \in V_2$, but not both. If $m \in V_1$ then $d_{G_1 \cup G_2}(m) = d_{G_1}(m)$ (since there is no new neighbor by construction.)$=$ $d_{G_1}(m) + d_{G_2}(m) - |E_1 \cap E_2(m)|$ (since in this case $d_{G_2}(m) = |E_1 \cap E_2(m)| = 0$). Similar case arises when $m \in V_2$ also. Now consider the potential of the vertex in $G_1 \cup G_2$. For $m \in G_1, e_{G_1 \cup G_2}(m) = e_{G_1}(m)$ (since there is no new edge originating from m and there is no change in weight for the existing edges) $= \vee\{e_{G_1}(m), e_{G_2}(m)\}$ (since in this case $e_{G_2}(m) = 0$). Similarly for $m \in G_2$ also. As the second case we take $m \in V_1 \cap V_2$, but no edge incident at m lies in $E_1 \cap E_2$. Here for $m \in V_1 \cap V_2, d_{G_1 \cup G_2}(m) = d_{G_1}(m) + d_{G_2}(m) = d_{G_1}(m) + d_{G_2}(m) - |E_1 \cap E_2(m)|$. While considering the potential of the vertex, $e_{G_1 \cup G_2}(m) = \vee\{e_{G_1}(m), e_{G_2}(m)\}$, since no edge incident at m lies in $E_1 \cap E_2$. As the third case we take $m \in V_1 \cap V_2$, but some edges incident at m are in $E_1 \cap E_2$. Here for $m \in V_1 \cap V_2, d_{G_1 \cup G_2}(m) = d_{G_1}(m) + d_{G_2}(m) - |E_1 \cap E_2(m)|$. The potential of the vertex m is $e_{G_1 \cup G_2}(m) = \vee\{e_{G_1}(m), e_{G_2}(m)\}$, in this case, since the edges are taking the maximum weight and the maximum will be any of the $e_{G_i}(m), i = 1, 2$. From the three cases, $NCI(G_1 \cup G_2) = \sum_m [(\vee\{e_{G_1}(m), e_{G_2}(m)\})(d_{G_1}(m) + d_{G_2}(m) - |E_1 \cap E_2(m)|)]$. □

Theorem 9.2.2 *Let $G_i = (\sigma_i, \mu_i)$ be the fuzzy graph with $|\sigma_i^*| = n_i$, where $i = 1, 2$. Assuming $V_1 \cap V_2 = \phi$ we have $NCI(G_1 + G_2) = \sum_{m \in G_i, i \neq j} [(d_{G_i}(m) + n_j)(\vee_{p \in G_j}\{\sigma(m) \wedge \sigma(p)\})]$.*

Proof Let $G_i = (\sigma_i, \mu_i)$ be fuzzy graphs with $|\sigma_i^*| = n_i$ where $i = 1, 2$. Suppose $m \in G_1$, then the neighborhood of m has all elements in G_2 in addition to its neighborhood in G_1 itself. Therefore, $d_{G_1 + G_2}(m) = d_{G_1}(m) + n_2$. Similarly if $m \in G_2$, $d_{G_1 + G_2}(m) = d_{G_2}(m) + n_1$. Now we can check the potential of the vertex $m \in G_1$. Since $V_1 \cap V_2 = \phi$ there are two types of edges arising from m. One is those edges whose other end point is in G_1 and other is those edges whose other endpoint is in G_2. Edges of the first case has maximum connectedness $e_{G_1}(m)$. Edges of the second case has connectedness the minimum of the weight of its adjacent vertices. The maximum among them is greater than or

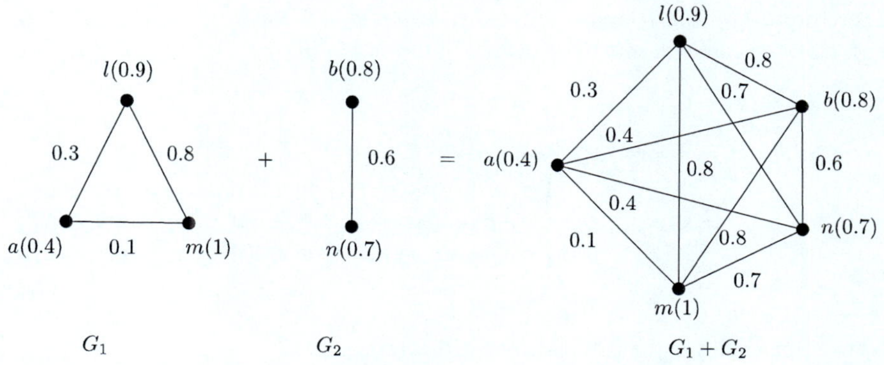

Fig. 9.9 Join of two fuzzy graphs

equal to $e_{G_1}(m)$. Therefore, $e_{G_1+G_2}(m) = \vee_{p\in G_2}\{\sigma(m) \wedge \sigma(p)\}$. The case $m \in G_2$ is similar. Therefore, $e_{G_1+G_2}(m) = \vee_{p\in G_1}\{\sigma(m) \wedge \sigma(p)\}$. Therefore, $NCI(G_1 + G_2) = \sum_{m\in G_1+G_2} d_{G_1+G_2}(m)e_{G_1+G_2}(m) = \sum_{m\in G_1} [(d_{G_1}(m) + n_2)(\vee_{p\in G_2}\{\sigma(m) \wedge \sigma(p)\})] + \sum_{m\in G_2} [(d_{G_2}(m) + n_1)(\vee_{p\in G_1}\{\sigma(m) \wedge \sigma(p)\})] = \sum_{m\in G_i, i\neq j} [(d_{G_i}(m) + n_j)(\vee_{p\in G_j}\{\sigma(m) \wedge \sigma(p)\})]$. □

Example 9.2.3 Consider the fuzzy graphs G_1 and G_2 given in Fig. 9.9 with $\sigma_1^* = \{l, a, m\}$ and $\sigma_2^* = \{b, n\}$ where $\sigma(l) = 0.9,\ \sigma(a) = 0.4,\ \sigma(m) = 1,\ \sigma(b) = 0.8,\ \sigma(n) = 0.7$ and $\mu(la) = 0.3,\ \mu(am) = 0.1,\ \mu(lm) = 0.8,\ \mu(bn) = 0.6$. After finding $G_1 + G_2$ we calculate $NCI(G_1 + G_2) = 4 \times 0.8 + 4 \times 0.4 + 4 \times 0.8 + 4 \times 0.8 + 4 \times 0.7 = 14$. Now using Theorem 9.2.2 we can find this without actually finding $G_1 + G_2$, $NCI(G_1 + G_2) = (2 + 2) \vee \{0.8, 0.7\} + (2 + 2) \vee \{0.4, 0.4\} + (2 + 2) \vee \{0.8, 0.7\} + (1 + 3) \vee \{0.8, 0.4, 0.8\} + (1 + 3) \vee \{0.7, 0.4, 0.7\} = 4 \times 0.8 + 4 \times 0.4 + 4 \times 0.8 + 4 \times 0.8 + 4 \times 0.7 = 14$.

Theorem 9.2.4 *Let $G_i = (\sigma_i, \mu_i)$ be fuzzy graphs with $|\sigma_i^*| = n_i$ where $i = 1, 2$.*

(i) if $\sigma_1 \leq \mu_2$, then $NCI(G_1[G_2]) = \sum_{(m,p)\in V_1\times V_2}[n_2 d_{G_1}(m) + d_{G_2}(p)]\sigma_1(m)$.
(ii) if $\sigma_1 \geq \mu_2, \sigma_2 \geq \mu_1$, then $NCI(G_1[G_2]) = \sum_{(m,p)\in V_1\times V_2}[n_2 d_{G_1}(m) + d_{G_2}(p)][\vee\{e_{G_1}(m), e_{G_2}(p)\}]$.

Proof Let G_1 and G_2 be two fuzzy graphs. Then

$$NCI(G_1[G_2]) = \sum_{(m,p)\in V_1\times V_2} d_{G_1[G_2]}(m, p)e_{G_1[G_2]}(m, p).$$

We can calculate $d_{G_1[G_2]}(m, p)$ and $e_{G_1[G_2]}(m, p)$ separately. The neighborhood of (m, p) consists of three types of vertices. (1) $\{(x, y) : x = m, py \in E_2\}$, the cardinality of this set is $d_{G_2}(p)$. (2) $\{(x, y) : y = p, mx \in E_1\}$, the cardinality of this set is $d_{G_1}(m)$. (3) $\{(x, y) : y \neq p, mx \in E_1\}$, the cardinality of this set is $(n_2 - 1)d_{G_1}(m)$.

Therefore, $d_{G_1[G_2]}(m, p) = d_{G_2}(p) + d_{G_1}(m) + (n_2 - 1)d_{G_1}(m) = n_2 d_{G_1}(m) + d_{G_2}(p)$. Now we can look into the potential of the vertex (m, p).

$$\begin{aligned} e_{G_1[G_2]}(m, p) &= \vee_{(m,p)\in V_1\times V_2}\{\mu_{G_1[G_2]}((m, p)(x, y)) : (x, y) \in V_1 \times V_2\} \\ &= \vee_{(m,p)\in V_1\times V_2} \begin{cases} \sigma_1(m) \wedge \mu_2(py) & : \text{if } \ x = m, py \in E_2 \\ \sigma_2(p) \wedge \mu_1(mx) & : \text{if } \ y = p, mx \in E_1 \\ \sigma_2(p) \wedge \sigma_2(y) \wedge \mu_1(mx) & : \text{if } \ y \neq p, mx \in E_1 \end{cases} \end{aligned} \tag{9.3}$$

Now we can analyze the two parts of the theorem.
Part (i) $\sigma_1 \leq \mu_2$. By Theorem 9.2.1 Eq. (9.3) becomes

$$e_{G_1[G_2]}(m, p) = \vee_{(m,p)\in V_1\times V_2} \begin{cases} \sigma_1(m) & : \text{if } \ x = m, py \in E_2 \\ \mu_1(mx) & : \text{if } \ y = p, mx \in E_1 \\ \mu_1(mx) & : \text{if } \ y \neq p, mx \in E_1 \end{cases} = \sigma_1(m).$$

Therefore, $NCI(G_1[G_2]) = \sum_{(m,p)\in V_1\times V_2}[n_2 d_{G_1}(m) + d_{G_2}(p)]\sigma_1(m)$.
Part (ii) $\sigma_1 \geq \mu_2, \sigma_2 \geq \mu_1$. Here Eq. (9.3) becomes

$$\begin{aligned} e_{G_1[G_2]}(m, p) &= \vee_{(m,p)\in V_1\times V_2} \begin{cases} \mu_2(py) & : \text{if } \ x = m, py \in E_2 \\ \mu_1(mx) & : \text{if } \ y = p, mx \in E_1 \\ \mu_1(mx) & : \text{if } \ y \neq p, mx \in E_1 \end{cases} \\ &= \vee\{e_{G_1}(m), e_{G_2}(p)\}. \end{aligned}$$

Therefore, $NCI(G_1[G_2]) = \sum_{(m,p)\in V_1\times V_2}[n_2 d_{G_1}(m) + d_{G_2}(p)][\vee\{e_{G_1}(m), e_{G_2}(p)\}]$. □

Corollary 9.2.5 *Let $G_i = (\sigma_i, \mu_i)$ be fuzzy graphs with $|\sigma_i^*| = n_i$ where $i = 1, 2$.*

(i) if $\sigma_1 \leq \mu_2$, then $NCI(G_1 \times G_2) = \sum_{(m,p)\in V_1\times V_2}[d_{G_1}(m) + d_{G_2}(p)]\sigma_1(m)$.
(ii) if $\sigma_1 \geq \mu_2, \sigma_2 \geq \mu_1$, then $NCI(G_1 \times G_2) = \sum_{(m,p)\in V_1\times V_2}[d_{G_1}(m) + d_{G_2}(p)][\vee\{e_{G_1}(m), e_{G_2}(p)\}]$.

Proof By construction, the Cartesian product of two fuzzy graphs differ from composition only by the set of edges $\{xy : y \neq p, mx \in E_1\}$. There is no change for $e_{G_1\times G_2}(m, p)$ from $e_{G_1[G_2]}(m, p)$ which can be observed from Eq. (9.3) in Theorem 9.2.4. While considering the neighborhood of the vertex (m, p), third type mentioned in the above proof is missing. Therefore, $d_{G_1\times G_2}(m, p) = d_{G_1}(m) + d_{G_2}(p)$. Hence if $\sigma_1 \leq \mu_2$, then $NCI(G_1 \times G_2) = \sum_{(m,p)\in V_1\times V_2}[d_{G_1}(m) + d_{G_2}(p)]\sigma_1(m)$ and if $\sigma_1 \geq \mu_2, \sigma_2 \geq \mu_1$, then $NCI(G_1 \times G_2) = \sum_{(m,p)\in V_1\times V_2}[d_{G_1}(m) + d_{G_2}(p)]$ $[\vee\{e_{G_1}(m), e_{G_2}(p)\}]$ □

Theorem 9.2.6 *Let $G_i = (\sigma_i, \mu_i)$ be fuzzy graphs with $|\sigma_i^*| = n_i$ where $i = 1, 2$. Then*

$$NCI(G_1 \otimes G_2) = \sum_{(m,p)\in V_1\times V_2} (d(m)d(p))(\wedge\{e(m), e(p)\}).$$

Proof Let $G_i = (\sigma_i, \mu_i)$ be fuzzy graphs with $|\sigma_i^*| = n_i$ where $i = 1, 2$. First, we find $d_{G_1\otimes G_2}(m, p)$ and then $e_{G_1\otimes G_2}(m, p)$. Consider the vertex $(m, p) \in V_1 \times V_2$. In the vertex set of $V_1 \times V_2$ we can find n_2 number of vertices with same first coordinate. Among the n_2 vertices there exists $d(p)$ vertices which has a neighborhood with (m, p). And this case repeats $d(m)$ times. Therefore, $d_{G_1\otimes G_2}(m, p) = d(m)d(p)$.

Now, $e_{G_1\otimes G_2}(m, p) = \vee\{\mu_{G_1\otimes G_2}((m, p)(x, y)); (x, y) \in V_1 \times V_2\} = \vee\{\mu_{G_1}(m x) \wedge \mu_{G_2}(py); mx \in E_1$ and $py \in E_2\} = \wedge\{[\vee\mu_{G_1}(mx), \vee\mu_{G_2}(py)]; mx \in E_1$ and $py \in E_2\} = \wedge\{e(m), e(p)\}$. Therefore,

$$NCI(G_1 \otimes G_2) = \sum_{(m,p)\in V_1\times V_2} (d(m)d(p))(\wedge\{e(m), e(p)\}).$$

□

9.3 Algorithm to Compute NCI

This section discusses an algorithm to find the neighborhood connectivity index of a fuzzy graph and an application related with human trafficking.

Algorithm 9.3.1 Let $G = (\sigma, \mu)$ be a fuzzy graph with n vertices.

1. Construct the matrix $A = [a_{ij}]$ with $a_{ij} = \mu(m_i m_j)$.
2. Find the largest membership value in each row of the matrix. Let it be t_i.
3. Find the number of non-zero entries in each row of the matrix. Let it be s_i.
4. Then $NCI(G) = \sum_{i=1}^{n} t_i \times s_i$.

Illustration of Algorithm: Let $A = (\sigma, \mu)$ be a fuzzy graph in Fig. 9.10 with $\sigma^* = \{l, a, m, b, n, c, o, d\}$ such that $\mu(la) = 0.5$, $\mu(lm) = 0.3$, $\mu(am) = 0.4$, $\mu(mb) = 0.1$, $\mu(bn) = 0.2$, $\mu(nc) = 0.3$, $\mu(no) = 0.6$, $\mu(nd) = 0.6$, $\mu(co) = 0.7$, $\mu(od) = 0.5$.

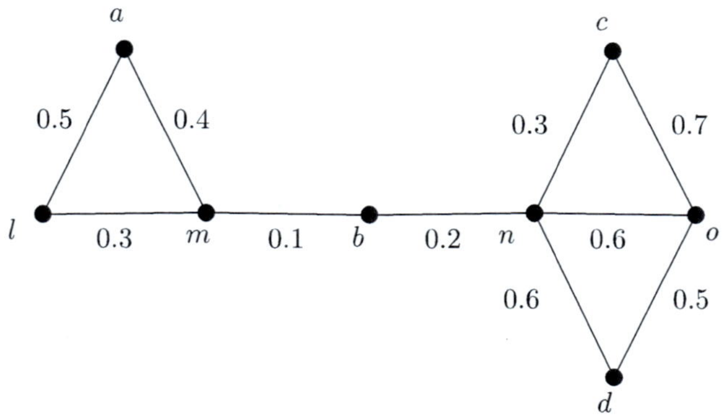

Fig. 9.10 Illustration for algorithm

The matrix representation of the given fuzzy graph is

$$A = \begin{array}{c} \\ l \\ a \\ m \\ b \\ n \\ c \\ o \\ d \end{array} \begin{array}{c} \begin{array}{cccccccc} l & a & m & b & n & c & o & d \end{array} \\ \begin{pmatrix} 0 & 0.5 & 0.3 & 0 & 0 & 0 & 0 & 0 \\ 0.5 & 0 & 0.4 & 0 & 0 & 0 & 0 & 0 \\ 0.3 & 0.4 & 0 & 0.1 & 0 & 0 & 0 & 0 \\ 0 & 0 & 0.1 & 0 & 0.2 & 0 & 0 & 0 \\ 0 & 0 & 0 & 0.2 & 0 & 0.3 & 0.6 & 0.6 \\ 0 & 0 & 0 & 0 & 0.3 & 0 & 0.7 & 0 \\ 0 & 0 & 0 & 0 & 0.6 & 0.7 & 0 & 0.5 \\ 0 & 0 & 0 & 0 & 0.6 & 0 & 0.5 & 0 \end{pmatrix} \end{array}$$

Now neighborhood connectivity index can be calculated by summing the product of highest value of each row and number of non zero entries in each row. Here neighborhood connectivity index, $NCI(G) = 0.5 \times 2 + 0.5 \times 2 + 0.4 \times 3 + 0.2 \times 2 + 0.6 \times 4 + 0.7 \times 2 + 0.7 \times 3 + 0.6 \times 2 = 10.7$.

9.4 Application

Human trafficking has always been widely studied since its impact on human race is huge. In 2017, Mathew and Mordeson discussed about this in [2]. Directed graph technique was used by them to analyze the given data in the Table 9.1. The ✓ in the data represents extremely small flow between regions and thus we neglect that further. The flow within a region is also not considered (Fig. 9.11).

First of all, we construct the directed fuzzy graph S from this data. The vertices represent the regions and the directed edges represent the direction of the transition. Since the adjacency matrix of S is similar to the analyzed data, it is not mentioned again. Now by using the algorithm which we mentioned previously, we calculate

Table 9.1 Flows between different regions

	WC Eur (*l*)	WS Eur (*a*)	C Eur & Bal (*m*)	E Eur & C Asia (*b*)	N Am & C Am & Car (*n*)	S Am (*c*)	E Asia & Pac (*o*)	S Asia (*d*)	S S Afr (*p*)	Mid East (*e*)
WC Eur (*l*)	0.62	0.13								
WS Eur (*a*)		0.16								
C Eur & Bal (*m*)		0.27	0.79		0.05					
E Eur & C Asia (*b*)	0.04	0.04	0.05	0.99						0.06
N Am & C Am & Car (*n*)	0.08	✓	✓		0.59	0.04				
S Am (*c*)		0.07			0.03	0.94				
E Asia & Pac (*o*)	0.07	0.07	✓		0.25	0.01	0.97			0.33
S Asia (*d*)		✓			0.07			0.96		0.18
S S Afr (*p*)		0.16						1.0		0.10
Mid East (*e*)										0.31

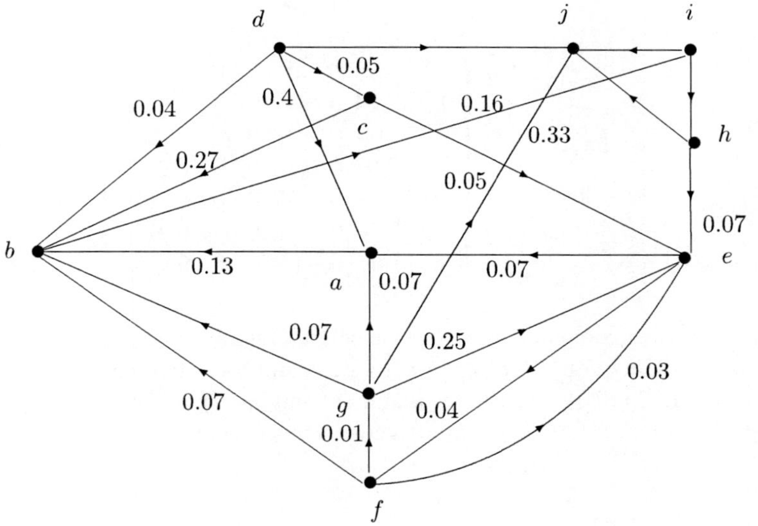

Fig. 9.11 Flow between different regions of the globe

neighborhood connectivity index of S and it is 6.22. After several calculations we came to an assumption that there does not exist a twinning vertex set of cardinality one. Consider the vertex set $\{l, c\}$. Now we construct the adjacency matrix of $S \setminus \{l, c\}$.

$$S \setminus \{l, c\} = \begin{array}{c} a \\ m \\ b \\ n \\ o \\ d \\ p \\ e \end{array} \overset{\begin{array}{cccccccc} a & m & b & n & o & d & p & e \end{array}}{\begin{pmatrix} 0 & 0 & 0 & 0 & 0 & 0 & 0 & 0 \\ 0.27 & 0 & 0 & 0.05 & 0 & 0 & 0 & 0 \\ 0.04 & 0.05 & 0 & 0 & 0 & 0 & 0 & 0.06 \\ 0 & 0 & 0 & 0 & 0 & 0 & 0 & 0 \\ 0.07 & 0 & 0 & 0.25 & 0 & 0 & 0 & 0.33 \\ 0 & 0 & 0 & 0.07 & 0 & 0 & 0 & 0.18 \\ 0.16 & 0 & 0 & 0 & 0 & 1.0 & 0 & 0.10 \\ 0 & 0 & 0 & 0 & 0 & 0 & 0 & 0 \end{pmatrix}}$$

After the computation using algorithm we get the neighborhood connectivity index of $S \setminus \{l, c\}$ as 5.07. Next consider the vertex set $\{m, c\}$ and construct the adjacency matrix of $S \setminus \{m, c\}$.

$$S \setminus \{m, c\} = \begin{array}{c} l \\ a \\ b \\ n \\ o \\ d \\ p \\ e \end{array} \overset{\begin{array}{cccccccc} l & a & b & n & o & d & p & e \end{array}}{\begin{pmatrix} 0 & 0.13 & 0 & 0 & 0 & 0 & 0 & 0 \\ 0 & 0 & 0 & 0 & 0 & 0 & 0 & 0 \\ 0.04 & 0.04 & 0 & 0 & 0 & 0 & 0 & 0.06 \\ 0.08 & 0 & 0 & 0 & 0 & 0 & 0 & 0 \\ 0.07 & 0.07 & 0 & 0.25 & 0 & 0 & 0 & 0.33 \\ 0 & 0 & 0 & 0.07 & 0 & 0 & 0 & 0.18 \\ 0 & 0.16 & 0 & 0 & 0 & 1.0 & 0 & 0.10 \\ 0 & 0 & 0 & 0 & 0 & 0 & 0 & 0 \end{pmatrix}}$$

Here also after the computation we get the neighborhood connectivity index of $S \setminus \{m, c\}$ as 5.07. Therefore, $\{l, c\}$ and $\{m, c\}$ are examples of twinning vertex sets of cardinality two. Similarly, we can see that $\{l, p\}$ and $\{m, d\}$ are also twinning vertex set of cardinality two with neighborhood connectivity index 2.58.

References

1. Jocy, A., Mathew, S., Mordeson, J.N.: Neighborhood connectivity index of a fuzzy graph and its application to human trafficking. Iran. J. Fuzzy Syst. **19**(3), 139–154 (2022)
2. Mathew, S., Mordeson, J.N.: Non-deterministic flow in fuzzy networks and its application in identification of human trafficking chains. New Math. Nat. Comput. **13**, 231–243 (2017)

Chapter 10
Cyclic Connectivity Status and Integrity Index of Fuzzy Graphs

Reachability between vertices is a major topic of study in fuzzy graph theory. Cyclic reachability is also very important in several applications like routing and traffic flow problems. Most of the modern networks like internet, gas and power grids possess cycles of different sizes and capacities. Hence parameters linking the network and their cycles becomes relevant. This chapter focusses on cycles. Two new parameters related with cycles in fuzzy graphs are discussed. Algorithms for their computation are also provided.

10.1 Cyclic Connectivity Status of Fuzzy Graphs

We discuss cyclic connectivity status of fuzzy graphs in this section. It provides a measure of the average stable flow from one node to another cyclically. Certain properties and suitable examples related with this new parameter are also provided.

Definition 10.1.1 Let $\sigma^* = \{w_1, w_2, \ldots, w_m\}$ be an ordered set related with a fuzzy graph $G = (\sigma, \mu)$. For an arbitrary vertex $v \in \sigma^*$, define **Cyclic Connectivity Status** (CCS) as

$$CCS_G(v) = \frac{1}{m-1} \sum_{\substack{r=1 \\ v \neq w_r}}^{m} C^G_{w_r, v}.$$

Cyclic connectivity status of $G = (\sigma, \mu)$ is defined as

$$CCS(G) = \frac{1}{m} \sum_{r=1}^{m} CCS_G(w_r).$$

J. N. Mordeson et al., *Fuzzy Graph Theory*, Studies in Fuzziness and Soft Computing 424, https://doi.org/10.1007/978-3-031-23108-7_10

$CCS_G(v)$ is also written as $CCS(v)$, if no confusion persist regarding the parent fuzzy graph. Vertices like pendent vertices and isolated vertices will have CCS values zero. In a connected graph, each edge has equal μ value and is equal to one. If all the vertices are part of at least one cycle in a connected graph, then CCS of each vertex is equal to 1. Thus inequalities $0 \leq CCS_G(v) \leq 1$ and $0 \leq CCS(G) \leq 1$ are true for any fuzzy graph G.

Example 10.1.2 Consider $G = (\sigma, \mu)$ of Fig. 10.1 with $\sigma^* = \{a, b, c, d, e, f\}$ ordered. $\mu(ad) = \mu(cd) = 0.3$, $\mu(ab) = 0.8$, $\mu(bc) = \mu(ac) = 0.4$, $\mu(ce) = 1$, $\mu(df) = 0.7$. Here, $CCS(G) = 0.14$.

Definition 10.1.3 For $G = (\sigma, \mu)$, define $\theta(w) = \{C^G_{w,z} \mid z \in \sigma^*, z \neq w\}$. If $C^G_{w,z} = 0$, then $\theta(w)$ is defined as an empty set.

First two rows of Table 10.1 are the vertices of Fig. 10.1 and the corresponding CCS of vertices in G respectively. Average value of CCS of vertices present in G is the $CCS(G)$ and is equal to 0.14. Third row shows the CCS of vertices in $\widehat{H}$ where $\widehat{H}$ is the partial fuzzy subgraph of G spanned by $\sigma^* \setminus \{e, f\}$. $CCS(\widehat{H})$ is 0.35. Observe that $\theta(e) = \theta(f) = 0$. Prediction of average cyclic reachability within a network is more accurate for those G in which the number of vertices having their θ values equal to zero is minimum.

We know that strength of a cycle in a fuzzy graph is always a non negative number in [0, 1]. Consider a vertex $w \in \sigma^*$. Let us assume its CCS be zero. If $\frac{1}{m-1} \sum_{\substack{l=1 \\ w \neq w_l}}^{m} CC^G_{w_l, w} = 0$ then, each term of the sum will be equal to zero. Thus, $C^G_{w', w} = 0$ and thus $\theta(w, w') = \emptyset$ for each $w' \in \sigma^* \setminus \{w\}$. On the other hand assume that $\theta(w, w') = \emptyset$ for each $w' \in \sigma^* \setminus \{w\}$. It is straight forward to write no strong

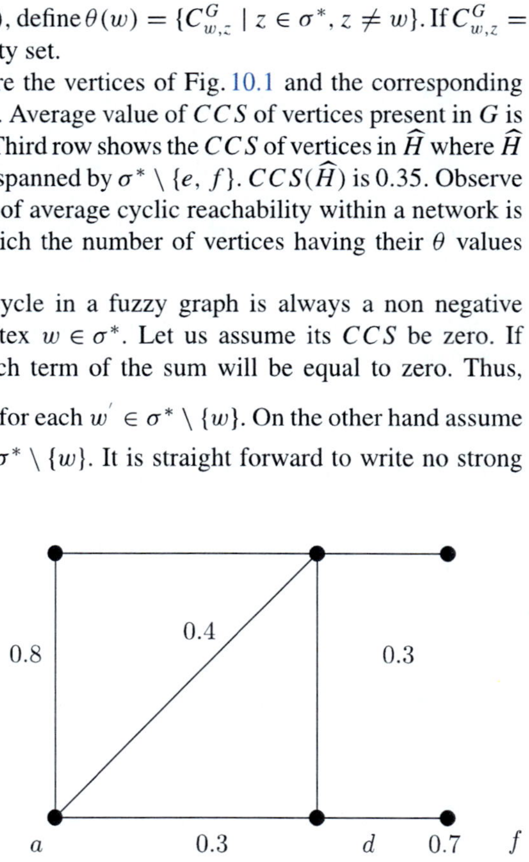

Fig. 10.1 Cyclic connectivity status illustration

Table 10.1 CCS of vertices in Fig. 10.1

Vertex (w)	a	b	c	d	e	f
$CCS_G(w)$	0.22	0.22	0.22	0.18	0.00	0.00
$CCS_{\widehat{H}}(w)$	0.367	0.367	0.367	0.3	–	–

cycle pass through w in G and thus $CCS(w) = 0$. This discussion can be summarized as a proposition, which is a characterization of vertices having zero CCS.

Proposition 10.1.4 *In $G = (\sigma, \mu)$, $CCS(v) = 0$ if and only if $\theta(v, w') = \emptyset$ for $w' \in \sigma^* \setminus \{v\}$.*

Using Proposition 10.1.4, we have a characterization for fuzzy trees.

Corollary 10.1.5 *In a connected fuzzy graph $G = (\sigma, \mu)$, $CCS(w) = 0$ for each $w \in \sigma^*$ if and only if G is a fuzzy tree.*

Next theorem gives a relationship between CCS and CCI of fuzzy graphs with optimum vertex membership values.

Theorem 10.1.6 *In $G = (\sigma, \mu)$, let σ^* has an order $\{w_1, w_2, \ldots, w_m\}$ and vertex membership value is 1 for all vertices in σ^*. Then,*

(i) $CCI(G) = \sum_{k=1}^{m-1} CCS(w_k) - \sum_{l=2}^{m-1}[\sum_{j=1}^{l-1} C^G_{w_l,w_j}]$.
(ii) $CCS(G) = \frac{1}{m}(CCI(G) + \sum_{l=2}^{m}[\sum_{j=1}^{l-1} C^G_{w_l,w_j}])$.

Proof (i) In $G = (\sigma, \mu)$, let $\sigma^* = \{w_1, w_2, \cdots, w_m\}$ and $\sigma(a) = 1$ for all $a \in \sigma^*$. By definition, $CCI(G) = \sum_{p,z\in\sigma*} \sigma(p)\sigma(z)C^G_{p,z} = \sum_{p,z\in\sigma*} C^G_{p,z}$, by the choice of membership values of vertices. Also, $CCI(G) = CCS(w_1) + CCS(w_2) - \{C^G_{w_2,w_1}\} + CCS(w_3) - \{C^G_{w_3,w_1} + C^G_{w_3,w_2}\} + \cdots + CCS(w_{m-1}) - \{C^G_{w_{m-1},w_1} + \cdots + C^G_{w_{m-1},w_{m-2}}\} + CCS(w_m) - \{C^G_{w_m,w_1} + \ldots + C^G_{w_m,w_{m-1}}\}$. Thus $CCI(G) = \sum_{k=1}^{m-1} CCS(w_k) - \{C^G_{w_2,w_1} + C^G_{w_3,w_1} + C^G_{w_3,w_2} + \cdots + C^G_{w_{m-1},w_1} + C^G_{w_{m-1},w_2} + \cdots + C^G_{w_{m-1},w_{m-3}} + C^G_{w_{m-1},w_{m-2}}$. In short, $CCI(G) = \sum_{k=1}^{m-1} CCS(w_k) - \sum_{l=2}^{m-1}[\sum_{j=1}^{l-1} C^G_{w_l,w_j}]$.

(ii) By the definition of $CCS(G)$ and part (i), we can prove this result. As $\sum_{k=1}^{m} CCS(w_k) = mCCS(G)$; $\sum_{k=1}^{m-1} CCS(w_k) = mCCS(G) - CCS(w_m)$. Theorem A can be rephrased as $CCI(G) = mCCS(G) - CCS(w_m) - \sum_{l=2}^{m-1}[\sum_{j=1}^{l-1} C^G_{w_l,w_j}] = mCCS(G) - \sum_{l=2}^{m}[\sum_{j=1}^{l-1} C^G_{w_l,w_j}]$. Thus $CCS(G) = \frac{1}{m}(CCI(G) + \sum_{l=2}^{m}[\sum_{j=1}^{l-1} C^G_{w_l,w_j}])$.

□

Definition 10.1.7 Define $\delta_{CSS}(G) = \wedge\{CSS(v) \mid v \in \sigma^*\}$ as the minimum CCS and $\Delta_{CSS}(G) = \vee\{CSS(v) \mid v \in \sigma^*\}$ as the maximum CCS of $G = (\sigma, \mu)$. Two vertices are said to be of **equi** CCS if they possess equal CCS in G.

Figure 10.1 contains three vertices a, b and c having equi CCS in it.

Proposition 10.1.8 *For a CFG, $G = (\sigma, \mu)$, let $\sigma^* = \{w_1, w_2, \ldots, w_m\}$ has an ordering such that $e_1 \le e_2 \le \ldots \le e_m$ with $\sigma(w_s) = e_s$ for $s = 1, 2, \ldots, m$. Then,*

(i) $CCS(w_1) = e_1$ and $CCS(w_s) = \frac{1}{m-1}[\sum_{k=1}^{s-1} e_k + (m-s)e_s]$ for $2 \le s \le m-2$. More over, $CCS(w_s) = \frac{1}{m-1}$.
(ii) Minimum number of vertices with equi CCS in G is three.
(iii) $CCS(w_1) \le CCS(w_2) \le \ldots \le CCS(w_{m-2}) = CCS(w_{m-1}) = CCS(w_m)$.

(iv) $\delta_{CSS}(G) = e_1$ and $\Delta_{CSS}(G) = CS(w_p)$ for $p \in \{m-2, m-1, m\}$.
(v) Two vertices have equi CCS if they share same σ value.
(vi) $CCS(G) = \frac{1}{m(m-1)}[\sum_{i=1}^{m-2} 2(m-i)e_i + 4e_{m-2}]$.

Proof For a $CFG\ G = (\sigma, \mu)$, choose $\sigma^* = \{w_1, w_2, \ldots, w_m\}$ to be ordered so that $e_1 \le e_2 \le \cdots \le e_m$. Remember that cycles in G are all strong.

(i) Each cycle passing through w_1 has strength e_1, since G is complete and e_1 is the least value among all edge membership values in G. $CCS(w_1) = \frac{1}{m-1}\sum_{s=2}^{m-1} C^G_{w_1,w_s} = \frac{1}{m-1}(m-1)e_1 = e_1$. Now consider $w_s \in \sigma^*, 2 \le s \le m-2$. $C^G_{w_p,w_s} = e_p$, for $p = 1, 2, \cdots, s-1$ and $C^G_{w_p,w_s} = e_s$, for $p = s+1, s+2, \ldots, m$. Thus, $CCS(w_s) = \frac{1}{m-1}[\sum_{p=1}^{s-1} C^G_{w_p,w_s} + \sum_{p=s+1}^{m} C^G_{w_p,w_s} = \frac{1}{m-1}$ $[\sum_{p=1}^{s-1} e_p + (m-s)e_s$. Hence, for $2 \le s \le m-2$, $CS(w_s) = \frac{1}{m-1}[\sum_{k=1}^{s-1} e_k + (m-s)e_s]$.
(ii) It needs a minimum of three vertices to form a cycle. The cycle w_m, w_{m-1}, w_{m-2} has the maximum strength among all cycles present in G. This fact leads to $C^G_{w_m,w_{m-1}} = C^G_{w_m,w_{m-2}} = e_{m-2}$. Thus $CCS(w_m) = CCS(w_{m-1}) = \sum_{p=1}^{m-3} e_p + e_{m-2}$. (i) says that $CCS(w_{m-2})$ also takes this expression. Thus the vertices w_m, w_{m-1} and w_{m-2} have equal CCS in G and hence the minimum number of vertices with equi CCS in G is three.
(iii) We have $e_1 \le e_2 \le \cdots \le e_m$. Now, $(m-1)e_1 = [e_1 + (m-2)e_1] \le [e_1 + (m-2)e_2] = [e_1 + e_2 + (m-3)e_2] \le [e_1 + e_2 + (m-3)e_3] \le \cdots \le [e_1 + e_2 + \cdots + e_{m-3} + 2e_{m-2}]$. This can be written as $(m-1)e_1 \le [e_1 + (m-2)e_2]$ $\le [e_1 + e_2 + (m-3)e_3] \le [e_1 + e_2 + \cdots + e_{m-3} + 2e_{m-2}]$. By the mere definition of CCS of vertices and by (ii), we get $CCS(w_1) \le CCS(w_2) \le \cdots \le CCS(w_{m-2}) = CCS(w_{m-1}) = CCS(w_m)$.
(iv) (iii) impose that $\delta_{CCS}(G) = CCS(w_1)$ and $\Delta_{CCS}(G) = CCS(w_p)$ for $p = m-2, m-1, m$.
(v) Here the vertices are arranged so that their μ values are in increasing order. Let $e_p = e_{p+1}$. When $p = 1$, $CCS(w_2) = \frac{1}{m-1}[e_1 + (m-2)e_1] = e_1 = CCS(w_1)$. When $e_s = e_{s+1}$, $CCS(w_{s+1}) = \frac{1}{m-1}[\sum_{k=1}^{s} e_k + (m-(s+1))e_s] = \frac{1}{m-1}$ $[\sum_{k=1}^{s-1} e_k + (m-s)e_s] = CCS(w_s)$. Hence we proved.
(vi) The cycle containing $w \in \sigma^*$ is the cycle composed of w, w_{m-1}, w_m if $w_{m-1} \ne w \ne w_m$. The strength of this cycle is $\sigma(w)$. w_{m-2}, w_{m-1}, w_m is a cycle in G that has maximum strength and pass through w_{m-1} and w_m. The strength of this cycle is e_{m-2}. This observation along with (i) provides $CCS(G) = \frac{1}{m(m-1)}[\sum_{i=1}^{m-2} 2(m-i)e_i + 4e_{m-2}]$.

□

We see that w_1 has the property that its CCS is equal to the corresponding membership value. Consider the relation $e_s = CCS(w_s)$ for $2 \le i \le m-2$. Then by (ii), $e_s = \frac{1}{m-1}[\sum_{k=1}^{s-1} e_k + (m-s)e_s]$. A rearrangement shows that $e_s = \frac{1}{s-1}\sum_{k=1}^{s-1} e_k$. Thus for any vertex w_i, with $2 \le i \le m-2$ such that its CCS equal to the corresponding membership value happens only if $e_s = \frac{1}{s-1}\sum_{k=1}^{s-1} e_k$. Hence it is quite clear that if the membership value of a vertex $w_i, 2 \leq i \leq m-2$ is equal to the average of membership values of all other vertices preceding it in the ordered set $\{w_1, w_2, \ldots, w_m\}$ then, its CCS and membership values are same.

Theorem 10.1.9 *[1] Let G be a fuzzy graph.*

(i) $CS(v) \geq CCS(v)$ *for each* $v \in \sigma^*$ *and* $CS(G) \geq CCS(G)$.
(ii) *If G is a block containing at least a bridge, then* $CS(G) > CCS(G)$.
(iii) *If G is a block without bridges, then* $CS(G) = CCS(G)$.
(iv) *If G is a fuzzy cycle, then all vertices of G are of equi*CCS.

Proof (i) Let $G = (\sigma, \mu)$ be a fuzzy graph. Let $z, y \in \sigma^*$. If these vertices are joined by a path, then $CONN_G(u, v) > 0$. Since $CONN_G(u, v)$ takes the maximum value among path strengths connecting z, y and hence $CONN_G(u, v) \geq C^G_{z,y}$. Thus, $CS(v) \geq CCS(v)$ for each vertex $v \in \sigma^*$. This in turn gives the relation $CS(G) \geq CCS(G)$.

(ii) Let G be a block containing at least a bridge and $P = \{zz' \in \mu^* \mid zz' \text{ is a bridge in } G\}$. Let $w \in \sigma^*$. $CS(w) = \frac{1}{m-1}(r+s)$ where $r = \sum_{wy \in \mu^*} \mu(wy)$; $wy \in P$ and $s = \sum_{wy \in \mu^*} CONN_G(w, y)$; $wy \in \mu^* \setminus P$. For an edge $zz' \in P$, $\mu(zz') > C^G_{z,z'}$ by the definition of a bridge. Also, in a block G, there exists at least a strongest strong cycle through any two vertices which are not end vertices of a bridge in G [1]. $CONN_G(w, y) = C^G_{w,y}$ if wy does not belong to P. If $r' = \sum_{wy \in \mu^*} C^G_{w,y}$; $wy \in P$ and $s' = \sum_{wy \in \mu^*} C^G_{w,y}$; $wy \in \mu^* \setminus P$, then $r > r'$ and $s = s'$. $\frac{1}{m-1}(r+s) > \frac{1}{m-1}(r'+s')$ and $CS(w) > CCS(w)$ are true for any $w \in \sigma^*$. Thus $CS(G) > CCS(G)$.

(iii) Let G be a block without bridges. This result follows from the proof of (ii). Since G doesn't have bridges, $r = r' = 0$. Moreover, $s = s'$ in a block. Hence $CS(w) = CCS(w)$ for each $w \in \sigma^*$ and hence $CS(G) = CCS(G)$.

(iv) Let G be a fuzzy cycle and l be the strength of a weak edge in it. A fuzzy cycle contains at least two weak edges in it and hence it is a strong cycle with strength equal to l. By this reason, $CONN_G(w, y) = C^G_{w,y}$ and $C^G_{w,y} = l$ for any two vertices $w, y \in G$. Hence $CS(w) = CCS(w) = l$ for each $w \in \sigma^*$. All vertices of G are of equi CCS.

□

10.2 CCS Analysis for Fuzzy Graphs

We discuss different type of vertices in this section. As we have seen CCS of a node is an indicator of average cyclic flow through it. The inactiveness of certain nodes causes depletion in the CCS of some other nodes in the network. Furthermore, it is interesting to observe that inaction of certain nodes results in an enhanced CCS or sometimes a static CCS on other nodes. This section helps to establish the observations as follows.

Relative to a particular vertex $w \in \sigma^*$, let us partition the vertex set σ^* of $G = (\sigma, \mu)$ as CCS elevating vertex set (EV), CCS depleting vertex set (DV) and CCS static vertex set (SV). This kind of partition sets is an essential basic information with regard to nodes of higher priority (like technically assisting work stations, significant routers) in a network. This help to improve fault tolerance of the respective network to a great extent.

Definition 10.2.1 $B \subset \sigma^*$ is said to be a CCS **elevating vertex set** (EV), of a vertex $v \in \sigma^*$ if $CCS_G(v) < CCS_{G-b}(v)$ for each $b \in B$. If $CCS_G(v) = CCS_{G-b}(v)$ for each $b \in B$, then B is termed as the CCS **static vertex set** (SV) of v. If $CCS_G(v) > CCS_{G-b}(v)$ for each $b \in B$, then B is called CCS **depleting vertex set** (DV) of v.

Related to an arbitrary $z \in \sigma^*$, a partition of $\sigma^* \setminus \{z\}$ is possible based on the impact of inaction of vertices on z in terms of CCS. For an illustration, consider a fuzzified network model $G = (\sigma, \mu)$ of a man made real world network (Fig. 10.2). Here, $\sigma^* = \{w_1, w_2, w_3, w_4, w_5, w_6\}$ has an ordering. The edge membership values are as follows. $\mu(w_1w_2) = 0.6 = \mu(w_4w_2)$, $\mu(w_1w_3) = \mu(w_5w_3) = \mu(w_5w_6) = \mu(w_6w_3) = 0.9$, $\mu(w_3w_4) = \mu(w_1w_4) = \mu(w_6w_4) = 0.8$.

Figure 10.3 shows all vertex deleted fuzzy subgraphs of Fig. 10.2. Cyclic connectivity status of vertices of σ^* is displayed in Table 10.2. The same table contains $CCS_{G-w}(x)$; $w, x \in \sigma^*, w \neq x$. $CCS_G(w_2) = 0.6$, $CCS_G(w_1) = CCS_G(w_4) = 0.76$ and $CCS_G(w) = 0.8$ for all $w \in \{w_3, w_5, w_6\}$. $CCS_G(w_2)$ stays the same even at the removal of w_5 from G. Thus w_5 belongs to $SV(w_2)$. Removal of any vertex from $\{w_1, w_3, w_4, w_6\}$ diminish $CCS(w_2)$ and hence $DV(w_2) = \{w_1, w_3, w_4, w_6\}$. In other words, inaction of each vertex in the set $\{w_1, w_3, w_4, w_6\}$ reduces $CCS(w_2)$. Also no vertex in σ^* on its removal enhances the $CCS(w_2)$. Thus$EV(w_2) = \emptyset$.

Table 10.3 shows the cyclic connectivity status analysis of vertices of G. Note that union of $SV(w)$, $EV(w)$ and $DV(w)$ together with $\{w\}$ compose σ^* for each $w \in \sigma^*$.

Consider a θ-fuzzy graph $G = (\sigma, \mu)$. Let $| \sigma^* |= m$ and $v \in \sigma^*$. Remember that in a $\theta-$fuzzy graph, if there is a strong cycle passing through two distinct vertices, then all other strong cycles through them will have unique strength. Let us look at an interesting property of vertices of θ-fuzzy graphs if θ value and CCS value are known. Three possibilities and their indications are as follows.

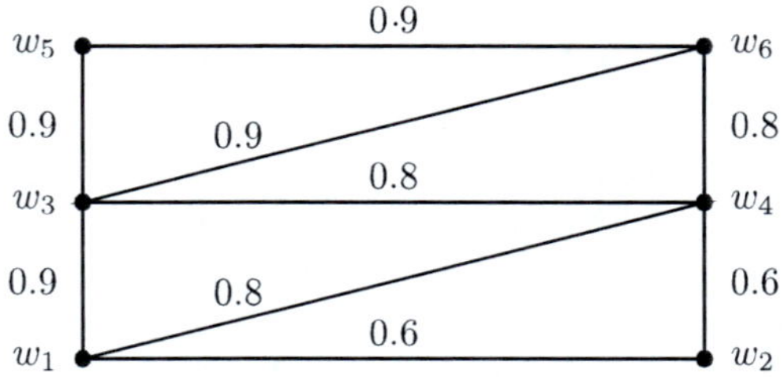

Fig. 10.2 CCS analysis of vertices

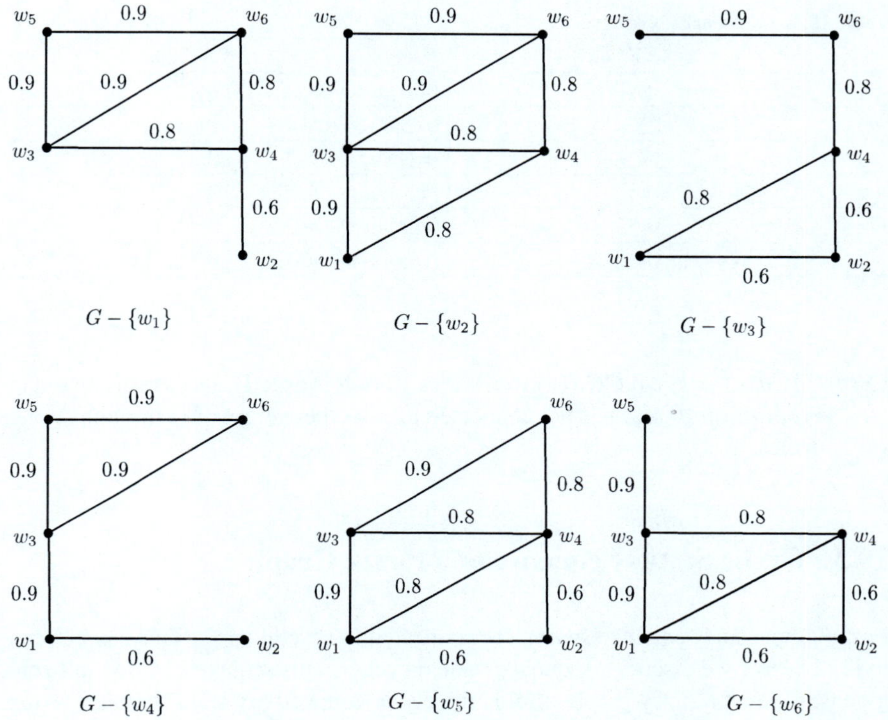

Fig. 10.3 Vertex deleted fuzzy subgraphs of Fig. 10.2

Table 10.2 CCS of vertices—variations

w	w_1	w_2	w_3	w_4	w_5	w_6
$CCS_G(w)$	0.76	0.6	0.8	0.76	0.8	0.8
$CCS_{G-w_1}(w)$	–	0	0.65	0.6	0.65	0.65
$CCS_{G-w_2}(w)$	0.8	–	0.85	0.8	0.85	0.85
$CCS_{G-w_3}(w)$	0.3	0.3	–	0.3	0	0
$CCS_{G-w_4}(w)$	0	0	0.45	–	0.45	0.45
$CCS_{G-w_5}(w)$	0.75	0.6	0.75	0.75	–	0.75
$CCS_{G-w_6}(w)$	0.55	0.45	0.55	0.55	0	–

Case (i). $\theta(v) = k$ with $CCS(v) = k$ for $k \neq 0$: w is connected to each vertex in σ^* by at least a strong cycle of strength k.

Case (ii). $\theta(v) = k$ with $CCS(v) = k^{'}$ for $k, k^{'} \neq 0$ and $k^{'} < k$: v is connected to p vertices in σ^* by at least a strong cycle of strength k, where $p = \frac{k^{'}(m-1)}{k}$.

Table 10.3 CCS analysis of vertices

w	$EV(w)$	$DV(w)$	$SV(w)$
w_1	$\{w_2\}$	$\{w_3, w_4, w_5, w_6\}$	$\emptyset$
w_2	$\emptyset$	$\{w_1, w_3, w_4, w_6\}$	$\{w_5\}$
w_3	$\{w_2\}$	$\{w_1, w_4, w_5, w_6\}$	$\emptyset$
w_4	$\{w_2\}$	$\{w_1, w_3, w_5, w_6\}$	$\emptyset$
w_5	$\{w_2\}$	$\{w_1, w_3, w_4, w_6\}$	$\emptyset$
w_6	$\{w_2\}$	$\{w_1, w_3, w_4, w_5\}$	$\emptyset$

Case (iii). $\theta(v) = k$ and $CCS(v) = k$ with $k = 0$. It is exactly the case discussed in Proposition 10.1.3. v is not connected to any other vertex of G through strong cycles.

10.3 Cyclic Status Sequence of a Fuzzy Graph

Fuzzy graphs and sequence spaces are connected in several ways. This can be seen in [2, 3]. Now we discuss a sequence named cyclic status sequence (CSS) in fuzzy graphs. CSS of a fuzzy graph unfolds the cyclic connectivity status of each vertex of G at a glance. To make it more convenient, it is desirable to have an ordering for σ^* and the same order is kept for the entries of the sequence also.

Definition 10.3.1 For $G = (\sigma, \mu)$, let $\sigma^* = \{w_1, w_2, \ldots, w_m\}$ be ordered and $CCS(w_l) = p_l$ for $l = 1, 2, \ldots, m$. Define **cyclic status sequence** (CSS) of G as $CSS(G) = (p_1, p_2, \ldots, p_m)$.

It is easier to find $CCS(G)$ from its CSS. Average value of all the entries of $CSS(G)$ is the $CCS(G)$. If each vertex belongs to at least a strong cycle of G then, each entry of $CSS(G)$ is non zero. Correspondingly, any fuzzy tree has a zero sequence as its CSS. In reverse, if we think of a fuzzy graph G with $CSS(G)$ as a zero sequence, it can be seen that its underlying structure will be that of a fuzzy cycle. If $CSS(G) = (0, 0, \ldots, 0)$ then, $CCS(w) = 0$ for each $w \in \sigma^*$. The statement 'each vertex of G is not part of a strong cycle' infers that G is a fuzzy tree [4].

Consider Fig. 10.2 of Sect. 10.2. Column 2 of Table 10.2 can be used to write $CSS(G)$ as $(0.6, 0.6, 0.72, 0.72, 0.72, 0.72)$. $CCS(G) = \frac{1}{6}[2(0.6) + 4(0.72)] = 0.68$. This value tells us that the average cyclic reachability between any two vertices of G is 0.68.

Theorem 10.3.2 *Let $\sigma^* = \{w_1, w_2, \ldots, w_m\}$ is an ordered set associated with a strong fuzzy graph $G = (\sigma, \mu)$. Also let $ba \in \mu^*$. Assume that $CSS(G)$ and $CSS(G - ba)$ are sequences $(p_1, p_2, \ldots, p_m)$ and $(r_1, r_2, \ldots, r_m)$ respectively. $r_l \neq p_l$ for some l; $l = 1, 2, \ldots, m$ if and only if ba is an $L-$cyclic bridge of G.*

Proof Associated with the strong fuzzy graph $G = (\sigma, \mu)$, let $\sigma^*=\{w_1, w_2, \ldots, w_m\}$ has an order. As G is strong, for c and e from σ^*, $\theta(c, e) \mid_{G-ba} \leq \theta(c, e) \mid_G$. This leads to the relation $C^{G-ba}_{c,e} \leq C^G_{c,e}$, for any edge $ba \in \mu^*$. Assume that ba is an $L-$cyclic bridge. By definition, removal of ba reduces cycle connectivity between some pair of vertices of G. Let w_l and w_q be such a pair. Hence $C^{G-ba}_{w_l,w_q} < C^G_{w_l,w_q}$. Thus for $w = w_l$ and w_q, $CCS_{G-ba}(w) = \frac{1}{m-1}\sum^m_{\substack{k=1\\ w\neq w_k}} C(w_k, w)^{G-ba} > \frac{1}{m-1}\sum^m_{\substack{k=1\\ w\neq w_k}} C(w_k, w)^G >= CCS_G(w)$. Hence $r_l \neq p_l$ and $r_q \neq p_q$.

Conversely assume that $r_l \neq p_l$ for some $l = 1, 2, \cdots, m$. Then, two cases may arise.

Case 1. $CCS_G(w_l) < CCS_{G-ba}(w_l)$.

Remember that the values $C^{G-ba}_{w_l,w}$ and $C^G_{w_l,w}$ are either positive or equal to zero. Since $CCS_G(w_l) < CCS_{G-ba}(w_l)$, existence of (at the minimum) a pair of vertices say z and y in σ^* so that $k = C^G_{z,y} < C^{G-ba}_{z,y} = k'$. This statement has the following meaning. Removal of an edge ba from G increases cycle connectivity between z and y in G. Assume that D is the cycle passing through z and y of strength of strength k'. As k is the cycle connectivity of z and y in the strong fuzzy graph G, the value k' must be equal to k. This is a contradiction to the statement $k = C^G_{z,y} < C^{G-ba}_{z,y} = k'$. We can conclude that there is no possibility for $CCS_G(w_l) < CCS_{G-ba}(w_l)$.

Case 2. $CCS_G(w_l) > CCS_{G-ba}(w_l)$.

By the assumption,

$$\frac{1}{m-1}\sum^m_{\substack{k=1\\ w\neq w_k}} C^{G-ba}_{w_k,w} < \frac{1}{m-1}\sum^m_{\substack{k=1\\ w\neq w_k}} C^G_{w_k,w}.$$

Since each term on both sides of this inequality is non negative, there should be at least a pair of vertices in σ^* such that their cycle connectivity get reduced when ba is deleted from G. Thus ba function as an $L-$cyclic bridge. □

10.4 Algorithms

Interconnection networks may have extensive number of nodes and links. CCS of vertices and that of a fuzzy graph are significant parameters linked with a fuzzified normalized network. Decisions to improve the cyclic reachability lead to strengthened connectivity and stable flow in a network. With the help of CCS such decisions can be taken in precision. Following algorithms will be of great help in the computation of CCS and cyclic connectivity analysis of nodes in a network.

Algorithm 10.4.1 An algorithm for the computation of CCS in a fuzzy graph.

This algorithm helps to find out the values of CCS of a fuzzy graph and that of its vertices. In $G = (\sigma, \mu)$, let σ^* has an order say $\{w_1, w_2, \ldots, w_m\}$. Let $w \in \sigma^*$. Following steps can be used to find CCS of G and that of its vertices.

Step 1. Locate strong edges of G with an algorithm provided in [4].
Step 2. Use the data obtained from step 1 to construct $\widehat{G}$ by deleting all non-strong edges from G.
Step 3. With the help of Algorithm A (till step 6) of [5], find out $C^{\widehat{G}}_{z,x}$ for $z, x \in \sigma^*$.
Step 4. Construct an $m \times m$ matrix with vertices of G (in the order maintained in σ^*) as rows and columns. $C^{\widehat{G}}_{w_a,w_b}$ is the entry corresponding to column b, row a; $a \neq b$. Let the diagonal entries of this matrix be zero.
Step 5. Define $CCS_G(w_a) = \frac{1}{m-1}\sum_{r=1}^{m} C^{\widehat{G}}_{w_a,w_r}$ which is the sum of all entries of row a.
Step 6. Define $CCS(G) = \frac{1}{m}\sum_{r=1}^{m} CCS_G(w_r)$.
At the end of step 5 we get CCS of a desired vertex of G and step 6 provides $CCS(G)$.

Algorithm 10.4.2 An algorithm to identify SV, DV and EV of a desired vertex.
In $G = (\sigma, \mu)$, let $v \in \sigma^*$. Consider a vertex $z \in \sigma^*$, $z \neq v$.

Step 1: Find $CS_G(v)$ and $CS_{G-z}(v)$ using Algorithm A.
Step 2: Let $\tilde{\zeta} = \frac{CS_G(v)}{CS_{G-z}(v)}$
Step 3: (a) If $\tilde{\zeta} = 1$, $z \in SV(v)$ (b) If $\tilde{\zeta} > 1$, $z \in DV(v)$ (c) If $\tilde{\zeta} < 1$, $z \in EV(v)$.

Accordingly we can identify $SV(w)$, $DV(w)$ and $EV(w)$.

Illustration of Algorithms

Consider Fig. 10.2. It is a fuzzy graph having an ordered vertex set $\sigma^* = \{w_1, w_2, w_3, w_4, w_5, w_6\}$. Here, each edge of G is strong. Hence $\widehat{G}$ of step 2 in Algorithm A is G itself. A 6×6 matrix containing cycle connectivity of pairs of vertices of $\sigma^* \times \sigma^*$ (as mentioned in step 4 of Algorithm A) is made.

$$
\begin{array}{c} \\ w_1 \\ w_2 \\ w_3 \\ w_4 \\ w_5 \\ w_6 \end{array}
\begin{array}{c} \begin{array}{cccccc} w_1 & w_2 & w_3 & w_4 & w_5 & w_6 \end{array} \\
\begin{bmatrix} 0 & 0.6 & 0.8 & 0.8 & 0.8 & 0.8 \\ 0.6 & 0 & 0 & 0.6 & 0.6 & 0.6 \\ 0.8 & 0.6 & 0 & 0.8 & 0.9 & 0.9 \\ 0.8 & 0.6 & 0.8 & 0 & 0.8 & 0.8 \\ 0.8 & 0.6 & 0.9 & 0.8 & 0 & 0.9 \\ 0.8 & 0.6 & 0.9 & 0.8 & 0.9 & 0 \end{bmatrix} \end{array}
$$

Since our fuzzy graph is undirected, the matrix obtained is symmetric. Using Step 5 of Algorithm A, row (or column) sum of corresponding vertices will provide the required CCS and these values are shown in column 1 of Table 10.2. Average of all the entries of column 1—Table 10.2 gives $CCS(G)$ and it is 0.753.

Consider $w = w_2 \in \sigma^*$. Let $z = w_5$. Refer row 2 of Table 10.2 for the following. We use Algorithm B to identify the effect of removal of z on w. $\tilde{m} = \frac{CS_G(w)}{CS_{G-z}(w)} = 1$. Thus w_5 belongs to SV of $w = w_2$. Now let $z = w_3$. $\tilde{m} = \frac{0.6}{0.3} = 2 > 1$ and hence w_3 belongs to DV of w_2.

10.5 Integrity Index of Fuzzy Graphs

Connectivity and related problems in graph theory are important due to its enormous applications in network science. This section discuss a connectivity parameter in fuzzy graphs named integrity index, that can serve as an indicator of the average flow along cycles through each vertex. As the number of cycles through a pair of vertices increases, the connectivity between them gets stronger.

Man made networks have a great influence in our daily lives. Studies related to connectivity parameters in fuzzy graphs are highly beneficial for the improvement in fault tolerance of networks. Integrity index of a vertex in a fuzzy graph is a connectivity parameter of the average flow through that vertex along strongest cycles.

Definition 10.5.1 Let $\sigma^* = \{a_1, a_2, \ldots, a_m\}$ be the vertex set of a fuzzy graph $G = (\sigma, \mu)$ in some order. **Integrity index** of a vertex $a \in \sigma^*$ is defined by

$$I_G(a) = \frac{1}{m-1} \sum_{\substack{l=1 \\ a \neq a_l}}^{m} CONN^G_{a_l,a}$$

where $CONN^G_{a_l,a}$ refers the strength of a strongest cycle passing through a_l and a in G.

Also integrity index of $G = (\sigma, \mu)$ is defined as

$$I(G) = \frac{1}{m} \sum_{l=1}^{m} I_G(a_l).$$

In short, we may write $I(a)$ to denote $I_G(a)$. In case of graphs, $I(G) = 0$ if G is a tree and $I(G) = 1$ if G is complete. Thus, we can write the relation $0 \leq I(G) \leq 1$ for a fuzzy graph. Consider a fuzzy graph $G = (\sigma, \mu)$. In order to have a nonzero value for $CONN^G_{a',a}$, we need a minimum of two disjoint strongest paths in G between a' and a. Thus for any partial fuzzy subgraph H of G, $CONN^H_{a',a} \leq CONN^G_{a',a}$. This leads to the relation $I_H(a) \leq I_G(a)$ for each vertex a in H and $I(H) \leq I(G)$.

Proposition 10.5.2 *$I(H) \leq I(G)$ for any partial fuzzy subgraph H of a fuzzy graph G.*

Note that for a fuzzy tree $G = (\sigma, \mu)$, $I(G)$ need not be zero. Cycles can also be present in the corresponding crisp structure G^*. It needs a $\delta-$ edge along with the strong edges of MST to constitute a cycle in G. Thus each such cycle has a strength equal to the membership value of one of the $\delta-$ edges of G. Hence we are able to write the following relation. For any $a \in \sigma^*$, $I(a) \leq q$ and $q = \vee\{\mu(aa') \mid aa'$ is a $\delta-$ edge in $G\}$ and that $I(G) \leq q$. Proposition 10.5.3 proposes a bound for integrity index of vertices and Corollary 10.5.4 gives an upper bound for $I(G)$.

Proposition 10.5.3 *In a fuzzy graph $G = (\sigma, \mu)$, $0 \leq I(a) \leq \sigma(a)$ for every $a \in \sigma^*$.*

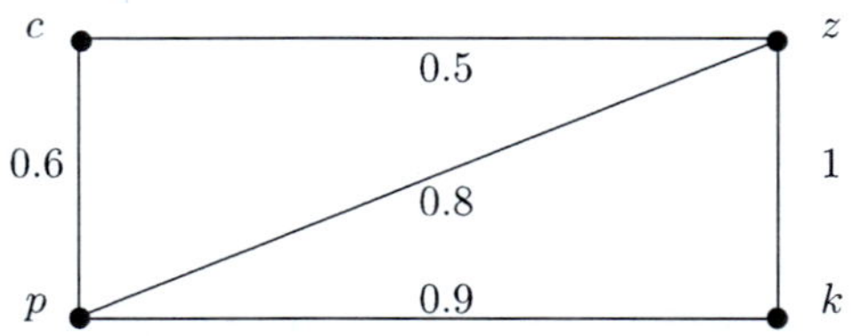

Fig. 10.4 Integrity index illustration

Table 10.4 Integrity index of vertices in Fig. 10.4

Vertex (a)	p	c	z	k
$I(a)$	0.7	0.5	0.7	0.7

Proof Consider $a \in \sigma^*$ in $G = (\sigma, \mu)$. Then we have $\sigma(a) > 0$. If there are no cycles passing through a, then $CONN^G_{a',a} = 0$ for each $a' \in \sigma^* \setminus \{a\}$. Thus $0 = I(a) \leq \sigma(a)$ holds. Consider the case when at least one cycle pass through a. Recall that $\sigma(a)$ is an upper bound for the membership value of any edge incident with a. Strength of a cycle will not be greater than the μ value of its weakest edge. Thus each cycle through a has a strength that never exceeds $\sigma(a)$. This leads to the relation that $\sum_{\substack{l=1 \\ a \neq a_l}}^{m} CONN^G_{a_l,a} \leq (m-1)\sigma(a)$. Hence $I(a) = \frac{1}{m-1}\sum_{\substack{l=1 \\ a \neq a_l}}^{m} CONN^G_{a_l,a} \leq \sigma(a)$. □

Corollary 10.5.4 is straight forward from Proposition 10.5.2.

Corollary 10.5.4 *In a fuzzy graph $G = (\sigma, \mu)$ with $|\sigma^*| = m$, $I(G) \leq \frac{1}{m}\sum_{a \in \sigma^*} \sigma(a)$.*
Following is an example for integrity index of vertices of a fuzzy graph. We calculate $I(G)$ of Fig. 10.4 by taking the average value of integrity index of its vertices (Table 10.4).

Example 10.5.5 Consider $G = (\sigma, \mu)$ in Fig. 10.4 with an ordered set of vertices $\sigma^* = \{p, c, z, k\}$; $\mu(cz) = 0.5$, $\mu(zk) = 1$, $\mu(kp) = 0.9$, $\mu(pc) = 0.6$, $\mu(pz) = 0.8$. Thus, $I(G) = \frac{1}{4}[I(p) + I(c) + I(z) + I(k)] = 0.65$.

Integrity index of each vertex in a fuzzy cycle has the same value as we can see in Proposition 10.5.6 (Table 10.4).

Proposition 10.5.6 *If $G = (\sigma, \mu)$ is a fuzzy cycle of strength s, then $I(a)$ will have a unique value for each $a \in \sigma^*$ and $I(a) = I(G)$ in G.*

Proof Let $a \in \sigma^*$ and $|\sigma^*| = m$. Consider an element a' of σ^* such that $a \neq a'$. Since structure of G is that of a fuzzy cycle, there are exactly two internally disjoint paths connecting a and a' in G. Union of the said paths is G itself. Thus $CONN^G_{a',a} = s$. By definition, $I_G(a) = \frac{1}{m-1}\sum_{\substack{l=1 \\ a \neq a_l}}^{m} CONN^G_{a_l,a} = \frac{1}{m-1}(m-1)s = s$. Since $I(a) = s$ for each $a \in \sigma^*$, $I(G) = s$. □

Relationship of integrity index with other connectivity parameters in fuzzy graphs is also important. Connectivity status (CS) of a fuzzy graph is discussed in [6]. Proposition 10.5.7 gives a relationship between CS and integrity index.

Proposition 10.5.7 *The relation $I(G) \leq CS(G)$ holds in any fuzzy graph $G = (\sigma, \mu)$.*

Proof Consider a fuzzy graph $G = (\sigma, \mu)$ with $| \sigma^* | = m$. We prove this result in two steps. Step (i) will prove existence of the same relationship for vertices. Second step proves the relationship between $I(G)$ and $CS(G)$.

Step (i). Let $b \in \sigma^*$. A vertex connected to b contributes a quantity ranging in (0, 1] to $CS(w)$. Whereas a vertex connected to b contributes to $I(b)$ only if the number of internally disjoint strongest paths between them should be at least two. Thus $CONN^G_{b,a} \leq CONN_G(b, a)$ for each $a \in \sigma^* \setminus \{b\}$. This in turn gives us

$$\sum_{\substack{l=1 \\ b \neq a_l}}^{m} CONN^G_{a_l,b} \leq \sum_{\substack{l=1 \\ b \neq a_l}}^{m} CONN_G(a_l, b).$$

Average of both quantities obey the same inequality. Hence $I(b) \leq CS(b)$.

Step (ii). Average of integrity index of vertices of σ^* gives the integrity index of G. Similarly, average value of connectivity status of vertices gives $CS(G)$. Hence we are able to conclude the relation $I(G) \leq CS(G)$.

□

Related to strong edges, there are sequence representations in fuzzy graphs. In [3], the authors discuss about α- sequence $\alpha_s(G)$, β- sequence $\beta_s(G)$ and strong sequence in fuzzy graphs. Theorem 10.5.8 discusses the equality of cyclic connectivity status and integrity index of a block fuzzy graph.

Theorem 10.5.8 *If $G = (\sigma, \mu)$ is a block such that each entry in $\alpha_s(G)$ is zero, then $I(G) = CS(G)$.*

Proof Let $\sigma^* = \{a_1, a_2, \ldots, a_m\}$. Assume that each entry in $\alpha_s(G)$ is zero. Consequently no vertex in σ^* is incident with any $\alpha-$strong edge in G. Let a and a' be any two elements of σ^*. Since G does not contain $\alpha-$strong edges, aa' is not a bridge. Hence these two vertices will be connected by at least two different strongest paths P_1 and Q_1 such that intersection of internal vertex sets of P_1 and Q_1 is empty. Thus $CONN^G_{a,a'} = CONN_G(a, a')$ for any $a, a' \in \sigma^*$. This leads to

$$\frac{1}{m-1} \sum_{\substack{l=1 \\ b \neq a_l}}^{m} CONN^G_{a_l,b} = \frac{1}{m-1} \sum_{\substack{l=1 \\ b \neq a_l}}^{m} CONN_G(a_l, b)$$

for any vertex $b \in \sigma^*$. Thus $I(b) = CS(b)$ for each $b \in \sigma^*$. By the logic used as in Step (ii) of Proposition 10.5.7, we can conclude that $I(G) = CS(G)$. □

Theorem 10.5.9 shows some general properties related to a connected fuzzy graph satisfying some conditions.

Theorem 10.5.9 *In a connected fuzzy graph $G = (\sigma, \mu)$, let $|\sigma^*| = m$. Let $a \in \sigma^*$ be such that a has degree two in G^* and one of the edges incident at a is not strong. If k is the strength of the mentioned non strong edge, then*

(*i*) $I(a)$ will never take the value zero.
(*ii*) $\frac{2k}{m-1} \leq I(a) \leq k$.

Proof Consider a connected fuzzy graph $G = (\sigma, \mu)$. Let $a \in \sigma^*$ be such that a has degree two in G^*. It is clear that a is incident with only two edges. Among them, let ap is not a strong edge and is of strength k.

(*i*) If possible assume that integrity index of a is zero. Then $CONN^G_{a',a} = 0$ for each $a' \in \sigma^* \setminus \{a\}$. It is because of the fact that $CONN^G_{a',a}$ is always non negative. This indicates that there is no cycle of non zero strength passing through a. Consequently both the edges incident at a should be $\alpha-$strong. If ap is $\beta-$strong, then a will be part of a cycle and it is not possible. In brief, ap is $\alpha-$strong and it is against our assumption that ap is not strong. Thus $I(w)$ can not be zero.
(*ii*) Proof of (*i*) says that a is always part of a cycle (of positive strength). Since a is incident with exactly two edges, it is true that these two edges will always be part of each cycle passing through a. Note that any cycle is formed by at least three vertices. More over, for each vertex $a' \neq a$ in the cycle, $CONN^G_{a',a} = k$. In the case when exactly two vertices are connected to the vertex a cyclically, we have $I(a) = \frac{1}{m-1}(k + k) = \frac{2k}{m-1}$. If there is a cycle passing through all the vertices of G then, $I(a) = \frac{1}{m-1}(m-1)k = k$. Hence, $\frac{2k}{m-1} \leq I(a) \leq k$.

□

Theorem 10.5.10 shows certain properties satisfied by blocks.

Theorem 10.5.10 *Let $G = (\sigma, \mu)$ be a block and let $a \in \sigma^*$. Then the following statements are true.*

(i) Bridges are not incident with end vertices of G.
(ii) $I(a) \neq 0$.
(iii) Bounds for $I(a)$ is determined by the bridges incident with a.

Proof Let $G = (\sigma, \mu)$ be a block and $a \in \sigma^*$. Let $B = \{ab \in \sigma^* \mid ab \text{ is a bridge in } G\}$.

(i) Let qs be a bridge in G. If possible qs be such that q is an end vertex of G. Since G is connected, s will act as a cutvertex in G. Which is not possible in a block. Hence bridges are not incident with end vertices of G.

(ii) If possible assume that $I(a) = 0$. Then $CONN^G_{a,q} = 0$ for each $q \in \sigma^* \setminus \{a\}$. Thus there is no cycle passing through a in G. Also $CONN_G(a, q) \neq 0$ as G is connected. Let q be a vertex in σ^* such that aq does not belong to μ^*. Moreover, aq is not an element of B. Since G is a block and aq is not a bridge, there are at least two strongest paths connecting a and q in G. This contradicts $CONN^G_{a,q} = 0$.
Now the only possibility for $q \in \sigma^*$ satisfying conditions $CONN^G_{a,q} = 0$ and $CONN_G(a, q) \neq 0$ is that $aq \in \mu^*$ with $aq \in B$. In other words, a is a neighbor of each vertex in σ^*. Let $p, p' \in \sigma^*$. Then ap and ap' are in μ^*. If $pp' \in \mu^*$, then a, p, p', a will form a cycle through a and that contradicts $CONN^G_{a,q} = 0$ for each $q \in \sigma^* \setminus \{a\}$. Now let pp' is not an edge in G, then there exists at least a path connecting p and p' as G is connected. This path connecting p and p' along with pap' forms a cycle through a. This also contradicts $CONN^G_{a,q} = 0$ for each $q \in \sigma^* \setminus \{a\}$. Hence for each $a \in \sigma^*$, $I(a)$ is a non zero value.

(iii) Let $B = \{aq \in \mu^* \mid aq$ is a bridge in $G\}$. We can partition $\sigma^* \setminus \{a\}$ in to two sets namely R_1 and R_2 so that for each $b \in R_1$, $ab \in B$ and for each $c \in R_2$, ac does not belong to B. Now, $I(a) = \sum_{b \in R_1} CONN^G_{a,b} + \sum_{c \in R_2} CONN^G_{a,c}$. Let $R = \sum_{c \in R_2} CONN^G_{a,c}$. If M and m are the maximum and minimum values of the set $\{CONN^G_{a,q} \mid aq \in B\}$ respectively, then

$$\frac{1}{m-1}(km + R) \leq I(a) \leq \frac{1}{m-1}(kM + R).$$

Hence the bounds for $I(a)$ is determined by the bridges incident with a. □

It is the peculiarity of a complete fuzzy graph that there are cycles connecting any pair of vertices in it. Since each edge in a CFG is strong, Proposition 2.8 of [6] is true in the case of integrity index also.

Theorem 10.5.11 *In a CFG $G = (\sigma, \mu)$, assume that the vertex set σ^* is arranged so that their σ values are in ascending order. If $\sigma^* = \{a_1, a_2, \ldots, a_m\}$ then*

(i) Minimum number of vertices with equal integrity index in G is three.
(ii) $I(a_1) \leq I(a_2) \leq \ldots \leq I(a_{m-2}) = I(a_{m-1}) = I(a_m)$.

10.6 Integrity Analysis of Vertices in a Fuzzy Graph

Real life networks usually face the problem of unexpected reduction of flow in the whole network or in a part. Sometimes the disconnection can also happen. Certain nodes are highly critical so as to keep a stable flow in the whole network. If the impact of idleness of other nodes over the critical node is predictable, the efficiency of the network can be assured. Related to a vertex, we partition the vertex set in to three subsets as follows.

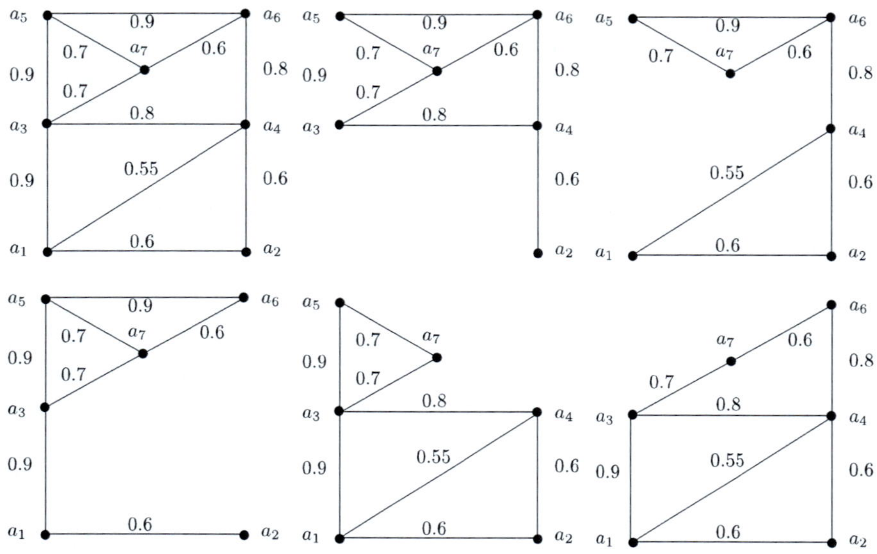

Fig. 10.5 $G = (\sigma, \mu)$ and its vertex deleted fuzzy subgraphs

Definition 10.6.1 In a fuzzy graph $G = (\sigma, \mu)$, let $a \in \sigma^*$. A vertex $b \in \sigma^*$ is called **integrity elevating vertex** IEV of a if $I_{G-b}(a) > I_G(a)$. If $I_{G-b}(a) < I_G(a)$, then b is termed as **integrity depleting vertex** IDV of a. If $I_{G-b}(a) = I_G(a)$, then b is called **integrity neutral vertex** INV of a.

Figure 10.5 consists of six fuzzy graphs. The fuzzy graph at the upper left most corner is the parent fuzzy graph. Rest of them are vertex deleted partial fuzzy subgraphs of the parent fuzzy graph. To illustrate the integrity analysis of vertices, let us consider a fuzzy graph $G = (\sigma, \mu)$ in Fig. 10.5. Vertices in σ^* are $\{a_1, a_2, \ldots, a_7\}$. Row 2 of Table 10.5 shows the integrity index of vertices of G. Integrity index of each vertex in vertex deleted subgraphs are shown in other rows of Table 10.5. It can be seen that $I_G(a_3) = 0.7167$. Removal of a_2 and a_7 increases its integrity index to 0.73 and 0.72 respectively. Hence a_2 and a_7 are $IEV(a)$. Removal of any of the vertices a_1, a_4, a_5 and a_6 reduces integrity index of a_3. IEV, IDV and INV of all the vertices of σ^* is showed in Table 10.6.

Algorithm 10.6.2 Algorithm to find integrity index of vertices in fuzzy graphs.

Given a fuzzy graph $G = (\sigma, \mu)$. Let $|\sigma^*| = m$ and $a \in \sigma^*$. To compute $I(a)$, do the following. Let $\sigma^* = \{a, b_1, b_2, \ldots, b_{m-1}\}$ be in order. We can make use of algorithms as in [7–9] to locate cycles between two vertices (for Step 1).

Let $i = 1$ and $K = 0$.

Step 1: With the help of a cycle detection algorithm, identify cycles through a and b_i.

Table 10.5 Variation of connectivity status of vertices

a	a_1	a_2	a_3	a_4	a_5	a_6	a_7
$I_G(a)$	0.6	0.6	0.7167	0.7167	0.7167	0.7167	0.6667
$I_{G-a_1}(a)$	–	0	0.62	0.62	0.62	0.62	0.56
$I_{G-a_2}(a)$	0.55	–	0.73	0.73	0.73	0.73	0.67
$I_{G-a_3}(a)$	0.22	0.22	–	0.22	0.24	0.24	0.24
$I_{G-a_4}(a)$	0	0	0.4	–	0.4	0.56	0.4
$I_{G-a_5}(a)$	0.6	0.6	0.6	0.6	–	0.6	0.6
$I_{G-a_6}(a)$	0.36	0.36	0.64	0.36	0.28	–	0.28
$I_{G-a_7}(a)$	0.6	0.6	0.72	0.72	0.72	0.72	–

Table 10.6 Cyclic status analysis

a	a_1	a_2	a_3	a_4	a_5	a_6	a_7
$IEV(a)$	$\emptyset$	$\emptyset$	a_2, a_7	a_2, a_7	a_2, a_7	a_2, a_7	a_2
$IDV(a)$	a_2, a_3, a_4, a_6	a_1, a_3, a_4, a_6	a_1, a_4, a_5, a_6	a_1, a_3, a_5, a_6	a_1, a_3, a_4, a_6	a_1, a_3, a_4, a_5	a_1, a_3, a_4, a_5, a_6
$INV(a)$	a_5, a_7	a_5, a_7	$\emptyset$	$\emptyset$	$\emptyset$	$\emptyset$	$\emptyset$

Step 2: Let k_i be the maximum value among strengths of all cycles through a and b_i.
Step 3: Do $K = K + k_i$.
Step 4: Do i=i+1.
Step 5: If $i = m$, go to Step 6.
else go to Step 1.
Step 6: $I(a) = \frac{1}{m-1} K$.

Algorithm 10.6.3 Algorithm to find integrity index of a fuzzy graph.
We can compute integrity index of a fuzzy graph if we have integrity index of its vertices. This is the logic behind the following algorithm. Given a fuzzy graph $G = (\sigma, \mu)$ with $|\sigma^*| = m$.

Step 1: Using Algorithm 5.1, compute integrity index of its vertices.
Step 2: Compute $I(G) = \frac{1}{m-1} \sum_{a \in \sigma^*} I(a)$.
Step 2 will give the required integrity index.

Illustration of Algorithm
Consider the fuzzy graph $G = (\sigma, \mu)$ in Fig. 10.6 with six vertices; $\sigma^* = \{r, s, t, e, f, g\}$. The membership values of edges are as given in the figure. With the help of cycle detection algorithm, we find out the values of $\{k_i\}$ for $i = 1, 2, \cdots, 5$ for each vertex. As per the assumption in our algorithm, we have to order σ^* so that the vertex whose integrity index has to be calculated should be in the first position. All other vertices will be kept in the order $\{r, s, t, e, f, g\}$. For example, if we are

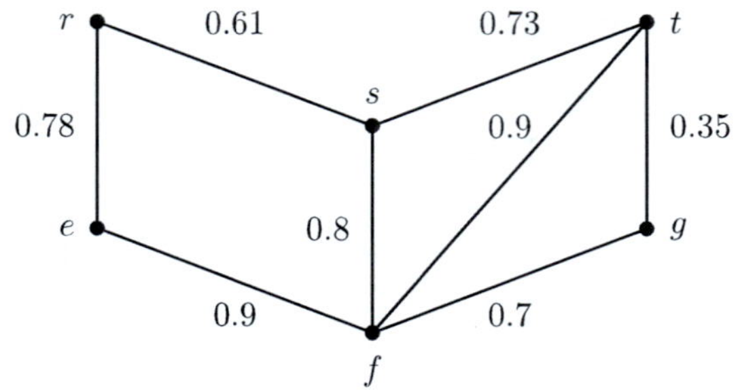

Fig. 10.6 Illustration of algorithm

Table 10.7 Illustration of algorithm

	r	s	t	e	f	g
k_1	0.61	0.61	0.61	0.61	0.61	0.35
k_2	0.61	0.73	0.73	0.61	0.73	0.35
k_3	0.61	0.61	0.61	0.61	0.73	0.35
k_4	0.61	0.73	0.73	0.61	0.61	0.35
k_5	0.35	0.35	0.35	0.35	0.35	0.35
K	2.79	3.03	3.03	2.79	3.03	2.1

Table 10.8 Integrity index of vertices

Vertex (a)	r	s	t	e	f	g
$I(a)$	0.558	0.606	0.606	0.558	0.606	0.35

calculating $I(e)$, we will order σ^* as $\{e, r, s, t, f, g\}$. For $I(s)$, the order of σ^* should be $\{s, r, t, e, f, g\}$ and so on. Table 10.7 gives the values of K and $\{k_i\}$ for $i = 1, 2, \ldots, 5$ corresponding to each vertex in G. Table 10.8 shows integrity index of vertices of Fig. 10.6.

10.7 Applications to Human Trafficking and Internet

Parameters provide numerical information related to a system. Suitable choice of parameters help to get accurate predictions about a system. Here we present two natural situations where cyclic connectivity status can be applied.

10.7.1 Application to Human Trafficking

We consider the main routes of trafficking from South America to the United States. The routes and the amount of flow can be found in [10, 11]. In [11], details of the reported trafficking in persons were provided. Information was provided with respect to the reported human trafficking in terms of origin, transit, and destination countries, according to a certain index. The directed graph under consideration is as follows:

Columbia → Costa Rica → El Salvador → Guatemala → Mexico → United States.

Columbia → Ecuador → Honduras → Guatemala → Mexico → United States.

Nicaragua → Costa Rica → El Salvador → Guatemala → Mexico → United States.

Dominican Republic → Panama → Costa Rica → El Salvador → Guatemala → Mexico → United States.

Let $G = (V, E)$ be a graph. Suppose $|V| = m$. Let (σ, μ) be a fuzzy subgraph of G. Assume the elements of V are ordered so that $\sigma(w_1) \leq \sigma(w_2) \leq \cdots \leq \sigma(w_m)$. We assume that G is complete and that for all $uv \in E, \mu(uv) > 0$. Since G is complete, we have for all $u, v \in V$ that $uv \in E$. In fact, assume that $\mu(uv) = \sigma(u) \wedge \sigma(v)$. With the help of Proposition 2.8, we are able to write, $CCS(G) = \frac{1}{m(m-1)}[\sum_{i=1}^{m-2} 2(m-i)e_i + 4e_{m-2}]$.

In our application, we consider the government response of a country to human trafficking. We assume that the government response of any two countries will have a joint influence on human trafficking even if there is no edge connecting the two countries. Hence we examine the complete fuzzy graph determined by the σ values of the countries, where the σ value represents the country's response. The σ values are taken from [10] and then normalized. They are given in Table 10.9.

Table 10.9 Integrity index of vertices in the illustration

Country	Govt. response (σ)
Col.	0.53
Cost.	0.55
El Sal.	0.43
Guat.	0.56
Mex.	0.62
U. S.	0.88
Ecu.	0.51
Hon.	0.39
Nic.	0.59
Dom. Rep.	0.63
Pan.	0.48

Thus,

$$\begin{aligned} CCS(G) &= \frac{1}{11}\frac{1}{10}[2(10)(0.39) + 2(9)(0.43) + 2(8)(0.48) \\ &\quad + 2(7)(0.51) + 2(6)(0.53) + 2(5)(0.55) + 2(4)(0.56) \\ &\quad + 2(3)(0.59) + 2(2)(0.62) + 4(0.62)] \\ &= \frac{1}{110}[54.96] = 0.50. \end{aligned}$$

We see that there is room for improvement for the government response by countries.

10.7.2 Application to Internet

We consider a network in internet with 8 routers. Weight of each link connecting any two routers denotes the maximum assignable bandwidth between them. Fuzzification process of a weighted network can be executed by normalizing the weights. Figure 10.7 shows the fuzzified network G.

We named the routers using alphabets r, t, s, u, w, v, a, k. Consider a set of routers among them say $S = \{t, r, u, v, w\}$ by assuming flow between them is crucial in the functionality of the network. The matrix shown has entries that correspond to cycle connectivity between nodes of G. If we consider routers in the order r, t, s, u, w, v, a, k, then $\{0.63, 0.45, 0, 0.45, 0.45, 0.428, 0.18, 0.18\}$ is the CSS of G.

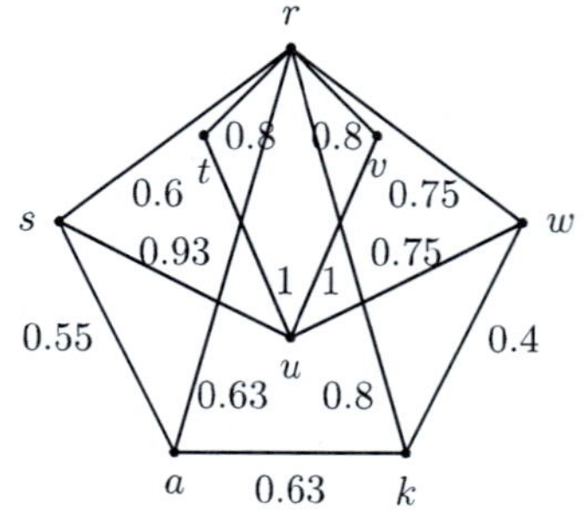

Fig. 10.7 The fuzzified network in Sect. 10.7.2

$$
\begin{array}{c} \\ r \\ s \\ t \\ u \\ v \\ w \\ a \\ k \end{array}
\begin{array}{c}
\begin{array}{cccccccc} r & s & t & u & v & w & a & k \end{array} \\
\begin{bmatrix}
- & 0 & 0.8 & 0.8 & 0.8 & 0.75 & 0.63 & 0.63 \\
0 & - & 0 & 0 & 0 & 0 & 0 & 0 \\
0.8 & 0 & - & 0.8 & 0.8 & 0.75 & 0 & 0 \\
0.8 & 0 & 0.8 & - & 0.8 & 0.75 & 0 & 0 \\
0.8 & 0 & 0.8 & 0.8 & - & 0.75 & 0 & 0 \\
0.75 & 0 & 0.75 & 0.75 & 0.75 & - & 0 & 0 \\
0.63 & 0 & 0 & 0 & 0 & 0 & - & 0.63 \\
0.63 & 0 & 0 & 0 & 0 & 0 & 0.63 & -
\end{bmatrix}
\end{array}
$$

If we are interested in the flow within the vertices that belong to S irrespective of the presence of other routers, then let us look at the spanning fuzzy subgraph $\widehat{H}$ of G spanned by S. Corresponding matrix of cycle connectivity is considered.

$$
\begin{array}{c} \\ r \\ t \\ u \\ v \\ w \end{array}
\begin{array}{c}
\begin{array}{ccccc} r & t & u & v & w \end{array} \\
\begin{bmatrix}
- & 0.8 & 0.8 & 0.8 & 0.75 \\
0.8 & - & 0.8 & 0.8 & 0.75 \\
0.8 & 0.8 & - & 0.8 & 0.75 \\
0.8 & 0.8 & 0.8 & - & 0.75 \\
0.75 & 0.75 & 0.75 & 0.75 & -
\end{bmatrix}
\end{array}
$$

The CSS of $\widehat{H}$ whose order is as that in S is $\{0.7875, 0.7875, 0.7875, 0.7875, 0.75\}$. Vertex t is reachable cyclically to all other vertices of S through strong cycles of strength 0.7875 on average. Similar comments can be made on other vertices also. $CCS(\widehat{H}) = 0.78$ indicates that all vertices of S are cyclically reachable to one another with strong cycles of strength 0.78 on average. Even if the failure of routers outside S occur for instance, the graph $\widehat{H}$ remains stable with a reachability 0.78 on average in G.

Presence of vertices in G such that they are not part of any strong cycle is the reason for the reduced values of CCS of vertices of S in G when compared to their values of CCS in $\widehat{H}$. $\{0.45, 0.63, 0.45, 0.45, 0.45, 0.428\}$ is the CSS related with S whose order is as that in S. Each router in S is cyclically reachable to other routers with an average connectivity greater than 0.428.

References

1. Mathew, S., Mordeson, J.N., Malik, D.S.: Fuzzy Graph Theory, p. 363. Springer International Publishing (2018)
2. Mathew, S., Sunitha, M.S.: Node connectivity and arc connectivity in fuzzy graphs. Inf. Sci. **180**, 519–531 (2010)
3. Mathew, J.K., Mathew, S.: Some special sequences in fuzzy graphs. Fuzzy Inf. Eng. **8**, 31–40 (2016)
4. Mathew, S., Sunitha, M.S.: Types of arcs in a fuzzy graph. Inf. Sci. **179**, 1760–1768 (2009)
5. Binu, M., Mathew, S., Mordeson, J.N.: Cyclic connectivity index of fuzzy graphs. IEEE Trans. Fuzzy Syst. **29**(6), 1340–1349 (2021)

6. Binu, M., Mathew, S., Mordeson, J.N.: Cyclic connectivity status of fuzzy graphs. IEEE Trans. Fuzzy Syst. (2022) (In press)
7. Alon, N., Yuster, R., Zwick, U.: Finding and counting given length cycles. Algorithmica **17**(3), 209–223 (1997)
8. Read, R.C., Tarjan, R.E.: Bounds on backtrack algorithms for listing cycles, paths and spanning trees. Networks **3**(5), 237–252 (1975)
9. Tong, Z., Zheng, D.: An algorithm for finding the connectedness matrix of a fuzzy graph. Congressus Numerantium **120**, 189–192 (1996)
10. Trafficking in Persons: Global Patterns. Appendices-United Nations Office on Drug and Crime, Citation Index (2006)
11. Trafficking in Persons: Global Patterns. The United Nations Office for Drugs and Crime, Trafficking in Persons Citation Index

Index

J. N. Mordeson et al., *Fuzzy Graph Theory*, Studies in Fuzziness and Soft Computing 424, https://doi.org/10.1007/978-3-031-23108-7